AF471834

Theory of Reflection

Dear MyCopy Customer,

This Springer book is a monochrome print version of the eBook to which your library gives you access via SpringerLink. It is available to you at a subsidized price since your library subscribes to at least one Springer eBook subject collection.

Please note that MyCopy books are only offered to library patrons with access to at least one Springer eBook subject collection. MyCopy books are strictly for individual use only.

You may cite this book by referencing the bibliographic data and/or the DOI (Digital Object Identifier) found in the front matter. This book is an exact but monochrome copy of the print version of the eBook on SpringerLink.

Developments in Electromagnetic Theory and Applications

J. Heading, *Managing Editor*

The University College of Wales, Department of Mathematics, Aberystwyth, U.K.

1. Caldwell J and Bradley R, eds: Industrial Electromagnetics Modelling. 1983. ISBN 90-247-2889-4
2. Booker HG: Cold Plasma Waves. 1984. ISBN 90-247-2977-7
3. Lekner J: Theory of Reflection. 1987. ISBN 90-247-3418-5

Theory of Reflection

of Electromagnetic and Particle Waves

by

John Lekner

Department of Physics
Victoria University of Wellington
New Zealand

SPRINGER-SCIENCE+BUSINESS MEDIA, B.V. 

Library of Congress Cataloging in Publication Data

Lekner, John.
Theory of reflection.

(Developments in electromagnetic theory and applications ; 3)
Includes indexes.
1. Reflection (Optics) 2. Electromagnetic waves--Transmission. I. Title. II. Title: Particle waves. III. Series.
QC425.L55 1987 530.1'24 86-24511

DOI 10.1007/978-94-015-7748-9

Copyright

© 1987 by Springer Science+Business Media Dordrecht
Originally published by Martinus Nijhoff Publishers, Dordrecht in 1987
MyCopy version of the original edition 1987

All rights reserved. No part of this publication may be reproduced, stored in a retrieval system, or transmitted in any form or by any means, mechanical, photocopying, recording, or otherwise, without the prior written permission of the publishers, Springer-Science+Business Media, B.V.
www.springer.com/mycopy

For my parents

Contents

Preface

This book is written for scientists and engineers whose work involves wave reflection or transmission. Most of the book is written in the language of electromagnetic theory, but, as the title suggests, many of the results can be applied to particle waves, specifically to those satisfying the Schrödinger equation. The mathematical connection between electromagnetic s (or TE) waves and quantum particle waves is established in Chapter 1. The main results for s waves are translated into quantum mechanical language in the Appendix. There is also a close analogy between acoustic waves and electromagnetic p (or TM) waves, as shown in Section 1-4. Thus the book, though primarily intended for those working in optics, microwaves and radio, will be of use to physicists, chemists and electrical engineers studying reflection and transmission of particles at potential barriers. The techniques developed here can also be used by those working in acoustics, oceanography and seismology.

Chapter 1 is recommended for all readers: it introduces reflection phenomena, defines the notation, and previews (in Section 1-6) the contents of the rest of the book. This preview will not be duplicated here. We note only that applied topics do appear: two examples are the important phenomenon of attenuated total reflection in Chapter 8, and the reflectivity of multilayer dielectric mirrors in Chapter 12. The subject matter is restricted to linear classical electrodynamics in non-magnetic media, and the corresponding particle analogues. Phenomena in non-linear and quantum optics are not covered. Even with these restrictions the book has grown larger than originally planned.

My interest in the theory of reflection was stimulated by David Beaglehole's studies of interfaces by polarization modulation ellipsometry, and work in this field was made more enjoyable by the many discussions we have had.

The editor of this series, John Heading, has kindly made many comments and suggestions as the book was being written. This generosity with his time and expertise is greatly appreciated, especially in view of his widespread commitments.

Almost all the book was written while I was a Visiting Fellow at the Department of Applied Mathematics of the Australian National University. The warm hospitality of Barry Ninham and his colleagues combined with the beauty of the Australian bush, coast and wildlife to make the year here a delight. My family only wished that more time could be given to exploring Australia, and less to the book!

I was fortunate to have overlapped with Colin Pask for part of the stay here. Through incisive comments he influenced the form and content of the book, especially the chapters on Riccati-type equations, inversion problems, and pulse and beam reflection.

Finally, special thanks are due to Kayleen Scott and Diana Wallace, who produced a professional word-processed text while coping with all the usual demands on their time.

JOHN LEKNER
Canberra, June 1986

1

Introducing reflection

Electromagnetic, acoustic and particle waves all scatter, diffract and interfere. Reflection is the result of the constructive interference of many scattered or diffracted waves originating from scatterers in a stratified medium. This fundamental many-body approach is hard to apply (two illustrations are given in Section 1–5). Usually one replaces the collection of scatterers by an effective medium whose properties are represented, as far as wave propagation is concerned, by a function of position and frequency (or energy), such as the dielectric function ε in the electromagnetic case, or the effective potential V in the quantum particle case. Electromagnetic and particle waves then satisfy the same kind of linear partial differential equation, with ε and V playing similar roles.

In a medium with planar stratification the functions ε and V depend on only one spatial variable, and the partial differential equations then separate. Snell's Law is a direct consequence of this separability of the spatial dependence. The differential equations, and the elementary reflection properties which follow from them, are derived for electromagnetic, particle, and acoustic waves in the first four Sections. The many-body, constructive interference, aspect of reflection is outlined in Section 1-5. Finally, Section 1-6 previews some of the main results in Chapters 1 to 13.

1-1 The electromagnetic *s* wave

The reflection of a plane electromagnetic wave at a planar interface between two media is completely characterized when solutions for two mutually perpendicular polarizations are known. The polarizations conventionally chosen are: one with its electric vector perpendicular to the plane of incidence (labelled *s*, from the German *senkrecht*, perpendicular), and the other with its electric vector parallel to the plane of incidence (labelled *p*).

We consider monochromatic waves, of angular frequency ω. The reflection of a general electromagnetic wave (a pulse, for example) can be analyzed as that of a superposition of monochromatic waves. For a given ω the time dependence of all fields is carried in the factor $e^{-i\omega t}$. (This is the convention in quantum and solid state physics, and much of optics. In radio and electrical engineering the factor $e^{i\omega t}$ is often used. With the convention used here the dielectric function has positive imaginary part in the case of absorption.) We will consider only *non-magnetic* media in this book. The electrodynamic properties of a medium are then contained in the dielectric function $\varepsilon(\mathbf{r}, \omega)$ which is the ratio of the permitivity of the medium at position **r** and angular frequency ω to that of the vacuum. The wave equations follow from Maxwell's two curl equations relating the electric field **E** and the magnetic field **B**:

$$\nabla \times \mathbf{E} = i\omega \mathbf{B} \qquad \text{or} \qquad \nabla \times \mathbf{E} = i\frac{\omega}{c}\mathbf{B}, \tag{1}$$

$$\nabla \times \mathbf{B} = -i\varepsilon\frac{\omega}{c^2}\mathbf{E} \qquad \text{or} \qquad \nabla \times \mathbf{B} = -i\varepsilon\frac{\omega}{c}\mathbf{E}. \tag{2}$$

(The equations on the left are in SI units, those on the right in Gaussian units; the difference lies in the positioning of the speed of light c. In reflection studies, theory and experiment deal in dimensionless ratios, and the choice of units is irrelevant. Even the formal distinction disappears from equation (5) onward.)

For a planar interface lying in the xy plane, and an electromagnetic wave propagating in the x and z directions, the s wave has $\mathbf{E} = (0, E_y, 0)$ and (1) gives

$$-\frac{\partial E_y}{\partial z} = i\frac{\omega}{c}B_x, \qquad \frac{\partial E_y}{\partial x} = i\frac{\omega}{c}B_z, \tag{3}$$

and $B_y = 0$. The other curl equation gives

$$\frac{\partial B_x}{\partial z} - \frac{\partial B_z}{\partial x} = -i\varepsilon\frac{\omega}{c}E_y. \tag{4}$$

On eliminating B_x and B_z from (3) and (4), we obtain a second order partial differential equation for E_y,

$$\frac{\partial^2 E_y}{\partial x^2} + \frac{\partial^2 E_y}{\partial z^2} + \varepsilon\frac{\omega^2}{c^2}E_y = 0. \tag{5}$$

For planar stratifications the dielectric function depends on one spatial variable, z. The partial differential equation is then separable, with

$$E_y(x, z, t) = e^{i(Kx - \omega t)}E(z), \tag{6}$$

where $E(z)$ satisfies the ordinary differential equation

$$\frac{d^2E}{dz^2} + q^2E = 0, \qquad q^2 = \varepsilon\frac{\omega^2}{c^2} - K^2 = k^2 - K^2. \tag{7}$$

The meanings of k, K and q are evident from (5), (6) and (7): $k = \varepsilon^{1/2}\omega/c$ is the local value of the wavevector, $K = k_x$ is the component of the wavevector along the interface, and $q = k_z$ is the component of the wavevector normal to the interface. For a plane wave incident from medium 1 as shown in Figure 1-1,

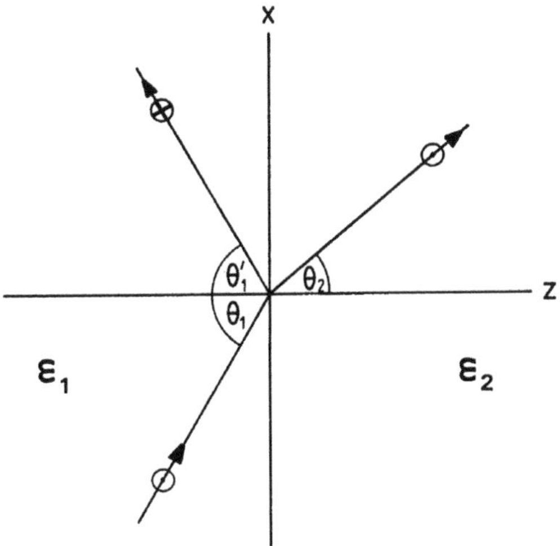

Figure 1-1. Reflection of the electromagnetic *s* wave at a planar interface between media characterized by dielectric constants ε_1 and ε_2. The figure is drawn for the air–water interface at optical frequencies, with $\varepsilon_1 \simeq 1$, $\varepsilon_2 \simeq (4/3)^2$.

the existence of the separation-of-variables constant $K(=k_{1x} = k'_{1x})$ implies

$$\varepsilon_1^{1/2} \sin \theta_1 = \varepsilon_1^{1/2} \sin \theta'_1 = \varepsilon_2^{1/2} \sin \theta_2, \tag{8}$$

where θ_1, θ'_1 and θ_2 are the angles of incidence, reflection, and transmission (or refraction).

Thus the fact that ε is a function of one spatial coordinate only, and the consequent separation of variables, implies the laws of reflection and refraction: the angle of reflection is equal to the angle of incidence, and the angles of incidence and refraction are related by Snell's Law. The refractive indices n_1 and n_2 of the two media, defined as coefficients in Snell's Law $n_1 \sin \theta_1 = n_2 \sin \theta_2$, are $\sqrt{\varepsilon_1}$ and $\sqrt{\varepsilon_2}$. Note that the laws of reflection-refraction do not depend on the transition between the two media being sharp: they are valid for an arbitrary variation of $\varepsilon(z)$ between the asymptotic values ε_1 and ε_2.

As ε attains its limiting values ε_1 and ε_2, $q = (\varepsilon\omega^2/c^2 - K^2)^{1/2}$ takes the limiting values

$$q_1 = \varepsilon_1^{1/2} \frac{\omega}{c} \cos \theta_1, \qquad q_2 = \varepsilon_2^{1/2} \frac{\omega}{c} \cos \theta_2. \tag{9}$$

(For $\theta_1 > \theta_c = \arcsin (n_2/n_1)$ there is total reflection, q_2 is imaginary, and θ_2 is complex. This is discussed along with the particle case in Section 1-3.) Snell's Law and the relationships between the wavevector components are incorporated together in Figure 1–2.

We now define the reflection and transmission amplitudes r_s and t_s in terms of the limiting forms of the solution of (7):

$$e^{iq_1z} + r_s\, e^{-iq_1z} \leftarrow E(z) \rightarrow t_s\, e^{iq_2z}. \tag{10}$$

The reflection amplitude is thus defined as the ratio of the coefficient of e^{-iq_1z} to that of e^{iq_1z}, the transmission amplitude as the coefficient of e^{iq_2z} when the incident wave

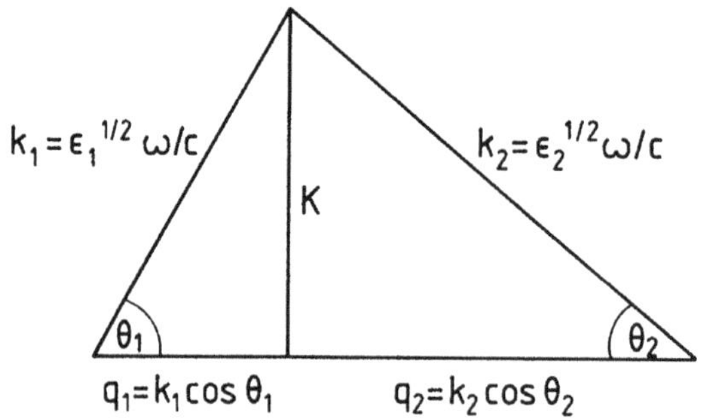

Figure 1-2. Graphical representation of $k_i^2 = q_i^2 + K^2$ and $K = k_1 \sin \theta_1 = k_2 \sin \theta_2$. The figure is drawn for the air–water interface, as in Figure 1-1.

e^{iq_1z} has unit amplitude. Theory aims to obtain general properties of the reflection and transmission amplitudes, and to develop methods for calculating these for a given dielectric function profile. The calculation is simple for the important step profile

$$\varepsilon_0(z) = \begin{cases} \varepsilon_1 & (z < 0) \\ \varepsilon_2 & (z > 0). \end{cases} \tag{11}$$

For this profile we obtain r_s and t_s from the continuity of E and $\mathrm{d}E/\mathrm{d}z$ at $z = 0$. (If, for example, $\mathrm{d}E/\mathrm{d}z$ were discontinuous, $\mathrm{d}^2E/\mathrm{d}z^2$ would have a delta function part, and (7) would not be satisfied.) For the step profile, E is given by the left and right sides of (10) for $z < 0$ and $z > 0$, respectively. The continuity of E and $\mathrm{d}E/\mathrm{d}z$ at the origin gives

$$1 + r_{s0} = t_{s0}, \qquad iq_1(1 - r_{s0}) = iq_2 t_{s0}. \tag{12}$$

Thus

$$r_{s0} = \frac{q_1 - q_2}{q_1 + q_2}, \qquad t_{s0} = \frac{2q_1}{q_1 + q_2}. \tag{13}$$

On using (8) and (9), the expressions (13) may be put into the Fresnel forms (Fresnel, 1823)

$$r_{s0} = \frac{\sin(\theta_2 - \theta_1)}{\sin(\theta_2 + \theta_1)}, \qquad t_{s0} = \frac{2 \sin\theta_2 \cos\theta_1}{\sin(\theta_2 + \theta_1)}. \tag{14}$$

The phases of the reflected and transmitted waves are specified only when the phase of the incident wave *and* the location of the interface are specified. The above equations are for the discontinuity in $\varepsilon(z)$ located at $z = 0$. In general, for the step located at z_1,

$$r_{s0} = e^{2iq_1z_1} \frac{q_1 - q_2}{q_1 + q_2}, \qquad t_{s0} = e^{i(q_1 - q_2)z_1} \frac{2q_1}{q_1 + q_2}. \tag{15}$$

A special situation arises at grazing incidence ($\theta_1 \to \pi/2$, $q_1 \to 0$), when the incident and reflected waves are propagating in the same direction. Then the phase of the reflected wave *is* well-defined without specification of the interface location, and

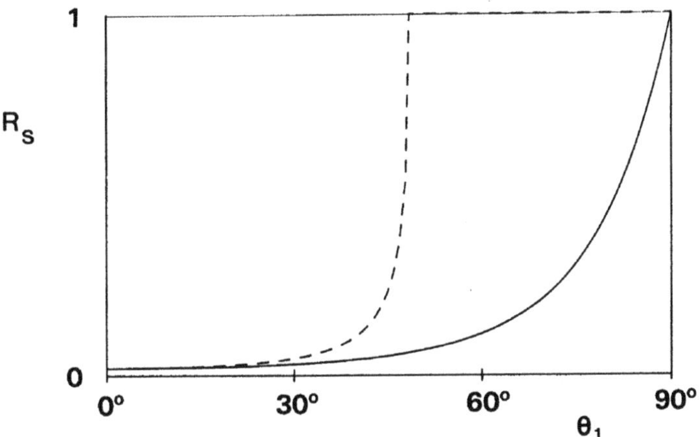

Figure 1-3. Step profile reflectivity for the s wave. The parameters are for the air–water interface at optical frequencies, as in Figures 1-1 and 1-2. The full curve is for light incident from air; the dashed curve for light incident from water shows total internal reflection for $\theta_1 > \theta_c \simeq \arcsin(3/4) \simeq 48.6°$.

$r_{s0} \to -1$ (even in the case of the total internal reflection, when q_2 is imaginary). The fact that $r_s \to -1$ at grazing incidence is a general property of all interfaces, as will be shown in Section 2-3.

The classical electromagnetic fields **E** and **B** are real quantities, and the complex notation is used for mathematical convenience. (Complex fields are intrinsic in the quantum theory of particles, however.) The physical reflected s wave is, for unit amplitude of the incident wave,

$$\operatorname{Re}\{r_s \exp i(Kx - q_1 z - \omega t)\} = \operatorname{Re}(r_s)\cos(Kx - q_1 z - \omega t) - \operatorname{Im}(r_s)\sin(Kx - q_1 z - \omega t).$$

The intensity is proportional to the time average of the square of this, namely

$$\tfrac{1}{2}[\operatorname{Re}(r_s)]^2 + \tfrac{1}{2}[\operatorname{Im}(r_s)]^2 = \tfrac{1}{2}|r_s|^2.$$

The incident intensity is proportional to the time average of $\cos^2(Kx + q_1 z - \omega t)$, which is 1/2. Thus, $R_s = |r_s|^2$ is the ratio of the reflected intensity to the incident intensity. This quantity is called the reflectivity, or reflectance. Figure 1-3 shows R_s for a sharp transition between air and water, with light incident from air, and from water.

1-2 The electromagnetic *p* wave

We again take the incident and reflected waves propagating in the zx plane, and the stratifications lying in xy planes. For the p wave, $\mathbf{B} = (0, B_y, 0)$; the Maxwell equation (1) gives

$$\frac{\partial E_x}{\partial z} - \frac{\partial E_z}{\partial x} = i\frac{\omega}{c} B_y, \tag{16}$$

while (2) implies $E_y = 0$ and

$$\frac{\partial B_y}{\partial z} = i\varepsilon \frac{\omega}{c} E_x, \qquad \frac{\partial B_y}{\partial x} = -i\varepsilon \frac{\omega}{c} E_z. \tag{17}$$

Elimination of E_x and E_z gives

$$\frac{\partial}{\partial x}\left(\frac{1}{\varepsilon}\frac{\partial B_y}{\partial x}\right) + \frac{\partial}{\partial z}\left(\frac{1}{\varepsilon}\frac{\partial B_y}{\partial z}\right) + \frac{\omega^2}{c^2} B_y = 0. \tag{18}$$

When ε is a function of one spatial coordinate z, the laws of reflection and refraction again follow from the separability of (18). We set

$$B_y(x, z, t) = \mathrm{e}^{i(Kx-\omega t)} B(z), \tag{19}$$

where K has the same meaning as for the s wave; then $B(z)$ satisfies the ordinary differential equation

$$\frac{\mathrm{d}}{\mathrm{d}z}\left(\frac{1}{\varepsilon}\frac{\mathrm{d}B}{\mathrm{d}z}\right) + \left(\frac{\omega^2}{c^2} - \frac{K^2}{\varepsilon}\right) B = 0. \tag{20}$$

When ε is constant (outside the interfacial region), the p wave equation has the same form as the s wave equation, with the same wavevector component q perpendicular to the interface. But within the interface there is an additional term proportional to the product of $\mathrm{d}\varepsilon/\mathrm{d}z$ and $\mathrm{d}B/\mathrm{d}z$. This term may be removed (and (20) converted to the form of the s wave equation (7)) in two ways. The first involves defining a new dependent variable

$$b = \left(\frac{\varepsilon_1}{\varepsilon}\right)^{1/2} B. \tag{21}$$

(The factor $\varepsilon_1^{1/2}$ makes identical the limiting forms of b and B in medium 1.) The equation satisfied by b is

$$\frac{\mathrm{d}^2 b}{\mathrm{d}z^2} + q_b^2 b = 0, \qquad q_b^2 = q^2 - \varepsilon^{1/2}\frac{\mathrm{d}^2\varepsilon^{-1/2}}{\mathrm{d}z^2} = q^2 + \frac{1}{2\varepsilon}\frac{\mathrm{d}^2\varepsilon}{\mathrm{d}z^2} - \frac{3}{4}\left(\frac{1}{\varepsilon}\frac{\mathrm{d}\varepsilon}{\mathrm{d}z}\right)^2. \tag{22}$$

This form of the p polarization equation is useful for special profiles, in particular the exponential profile, which has $\log\varepsilon$ linear in z, and the Rayleigh profile, which has $\varepsilon^{-1/2}$ linear in z. These are discussed in Chapter 2. It is also useful at short wavelengths, in the derivation of a perturbation theory for the p wave (Chapter 6).

The second transformation which removes the $(\mathrm{d}\varepsilon/\mathrm{d}z)\,(\mathrm{d}B/\mathrm{d}z)$ term is a dilation of the z variable in proportion to the local value of $\varepsilon(z)$: we define a new independent variable Z by

$$\mathrm{d}Z = \varepsilon\, \mathrm{d}z. \tag{23}$$

Then, as may be seen on division of (20) by ε, the p wave equation reads

$$\frac{\mathrm{d}^2 B}{\mathrm{d}Z^2} + Q^2 B = 0, \qquad Q^2 = \frac{1}{\varepsilon}\frac{\omega^2}{c^2} - \frac{K^2}{\varepsilon^2}. \tag{24}$$

This equation, in terms of the dilated z variable, and a reduced normal component of the wavevector, $Q = q/\varepsilon$, will be useful in many applications thoughout this book.

The p wave reflection and transmission amplitudes are defined in terms of the limiting forms of $B(z)$:

$$e^{iq_1z} - r_p\, e^{-iq_1z} \leftarrow B(z) \rightarrow \left(\frac{\varepsilon_2}{\varepsilon_1}\right)^{1/2} t_p\, e^{iq_2z} \tag{25}$$

The reason for the factors -1 and $(\varepsilon_2/\varepsilon_1)^{1/2}$ multiplying r_p and t_p is that we wish r_s and r_p and t_s and t_p to refer to the same quantity, here chosen to be the electric field. (This is not the only convention in use: some authors have the opposite sign on r_p.) The electric field components for the p wave are found from (2), (19) and (25) to have the limiting forms

$$\varepsilon_1^{-1/2} \cos\theta_1\, e^{i(Kx-\omega t)}\, (e^{iq_1z} + r_p\, e^{-iq_1z}) \leftarrow E_x \rightarrow \varepsilon_1^{-1/2} \cos\theta_2 t_p\, e^{i(Kx+q_2z-\omega t)}, \tag{26}$$

$$-\varepsilon_1^{-1/2} \sin\theta_1\, e^{i(Kx-\omega t)}\, (e^{iq_1z} - r_p\, e^{-iq_1z}) \leftarrow E_z \rightarrow -\varepsilon_1^{-1/2} \sin\theta_2 t_p\, e^{i(Kx+q_2z-\omega t)}. \tag{27}$$

The x-component of the electric field (tangential to the interface) thus has the reflection amplitude r_p, while the z-component (normal to the interface) has reflection amplitude $-r_p$.

At normal incidence there is no physical difference between the s and p polarizations: both have electric and magnetic fields tangential to the interface. For our geometry, E_z is zero at normal incidence, and (1) implies $\partial E_x/\partial z = i(\omega/c)B_y$. Thus B, the solution of (20) and (25), must be proportional to dE/dz, where E is the solution of (7) and (10). On substituting dE/dz for B in (20) (with K set equal to zero) the left side becomes

$$\frac{d}{dz}\left\{\frac{1}{\varepsilon}\left(\frac{d^2E}{dz^2} + \varepsilon\frac{\omega^2}{c^2}E\right)\right\},$$

and this is zero, by (7). Thus (20) is satisfied by dE/dz at normal incidence. The proportionality of B and dE/dz at normal incidence, when applied to the limiting forms (10) and (25), gives the equality of r_p with r_s and of t_p with t_s. (Proportionality of B and dE/dz could be replaced by equality of B and $(c/i\omega)\, dE/dz$, but then (25) would have to be modified by the factor $\varepsilon_1^{1/2}$.)

At a discontinuity in the dielectric function, B and $dB/\varepsilon dz = dB/dZ$ are continuous (from (20) or (24)). For the step profile $\varepsilon_0(z)$ defined by (11), B is equal to

$$B_0(z) = \begin{cases} e^{iq_1z} - r_{p0}\, e^{-iq_1z} & (z < 0) \\ \left(\dfrac{\varepsilon_2}{\varepsilon_1}\right)^{1/2} t_{p0}\, e^{iq_2z} & (z > 0). \end{cases} \tag{28}$$

The continuity of B and $dB/\varepsilon\, dz$ at the origin gives

$$1 - r_{p0} = \left(\frac{\varepsilon_2}{\varepsilon_1}\right)^{1/2} t_{p0}, \tag{29}$$

$$iQ_1(1 + r_{p0}) = iQ_2\left(\frac{\varepsilon_2}{\varepsilon_1}\right)^{1/2} t_{p0}, \tag{30}$$

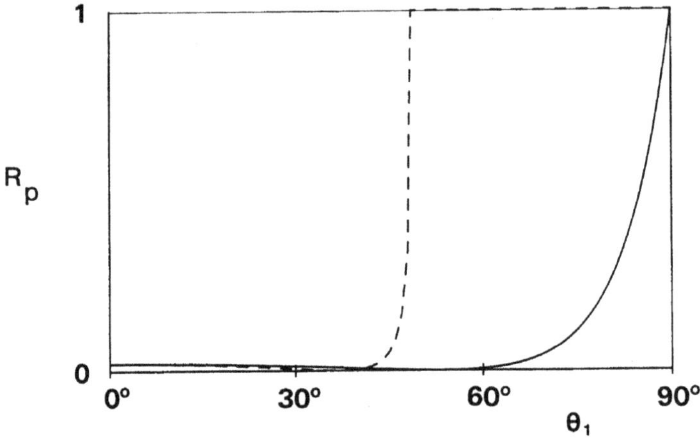

Figure 1-4. Step profile reflectivity for the p wave, for the air–water interface. The full curve is for light incident from air, the dashed curve for incidence from water. Note the zeros at the Brewster angles, arctan (4/3) $\simeq$ 53.1° and arctan (3/4) $\simeq$ 36.9°, respectively.

where $Q_1 = q_1/\varepsilon_1$ and $Q_2 = q_2/\varepsilon_2$. Thus (compare (13))

$$-r_{p0} = \frac{Q_1 - Q_2}{Q_1 + Q_2}, \qquad \left(\frac{\varepsilon_2}{\varepsilon_1}\right)^{1/2} t_{p0} = \frac{2Q_1}{Q_1 + Q_2}. \tag{31}$$

On using (8) and (9) we obtain the Fresnel forms

$$r_{p0} = \frac{\tan(\theta_2 - \theta_1)}{\tan(\theta_2 + \theta_1)}, \qquad t_{p0} = \frac{2\sin\theta_2\cos\theta_1}{\sin(\theta_2 + \theta_1)\cos(\theta_2 - \theta_1)}. \tag{32}$$

The reflectivity of the p polarization off a discontinuity in the dielectric function is shown in Figure 1-4.

From (31) we see that the p wave shows zero reflection when $Q_1 = Q_2$, that is at the Brewster angle

$$\theta_B = \arctan\left(\frac{\varepsilon_2}{\varepsilon_1}\right)^{1/2}. \tag{33}$$

It is apparent from (24) that this angle has special significance not only for a sharp transition between two media, but for diffuse profiles as well. This is because the wave equation in the dilated variable Z links two media with effective wavevector components Q_1 and Q_2, which are equal at this angle. The difference between the s and p effective wavevector components q and Q, and the reason for small p reflectivity at the Brewster angle, are illustrated in Figure 1-5. There we show q^2 versus z and Q^2 versus Z for the hyperbolic tangent profile

$$\varepsilon(z) = \tfrac{1}{2}(\varepsilon_1 + \varepsilon_2) - \tfrac{1}{2}(\varepsilon_1 - \varepsilon_2)\tanh z/2a, \tag{34}$$

for which

$$Z = \tfrac{1}{2}(\varepsilon_1 + \varepsilon_2)z - (\varepsilon_1 - \varepsilon_2)a \log\cosh(z/2a). \tag{35}$$

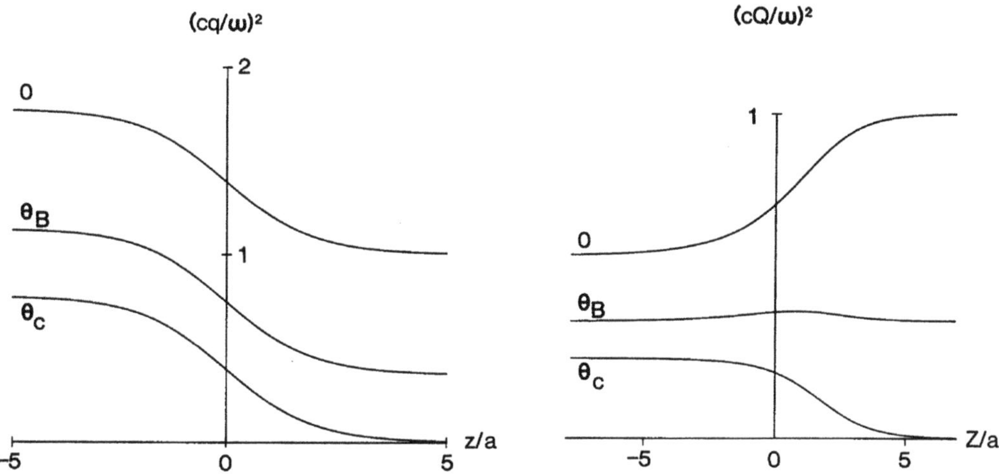

Figure 1-5. Squares of the normal wavevector component q and of the effective normal component Q for the s and p waves. The figure shows $q^2(z)$ and $Q^2(Z)$ for the hyperbolic tangent dielectric function profile, at three angles of incidence. The upper curve (in each case) is for normal incidence, the middle curve is at the Brewster angle $\theta_B = \arctan(\varepsilon_2/\varepsilon_1)^{1/2}$, and the lower curve is at the critical angle for total internal reflection, $\theta_c = \arcsin(\varepsilon_2/\varepsilon_1)^{1/2}$. The dielectric constants $\varepsilon_1 = (4/3)^2$ and $\varepsilon_2 = 1$ approximate the water–air interface. Water is on the left in both diagrams.

At the Brewster angle θ_B,

$$Q_1^2 = Q_2^2 = \frac{(\omega/c)^2}{\varepsilon_1 + \varepsilon_2} = Q_B^2, \tag{36}$$

$$K^2 = \varepsilon_1 \varepsilon_2 Q_B^2 = K_B^2. \tag{37}$$

From (24), a general profile $\varepsilon(z)$ has Q^2 at the Brewster angle given by

$$Q^2(\theta_B, z) = \frac{\omega^2}{c^2}\left\{\varepsilon(z) - \frac{\varepsilon_1\varepsilon_2}{\varepsilon_1 + \varepsilon_2}\right\}\Big/\varepsilon^2(z). \tag{38}$$

Thus the bump in Q^2 at the Brewster angle (see Figure 1-5) has the analytic form

$$Q^2(\theta_B, z) - Q_B^2 = \frac{\omega^2}{c^2}\frac{(\varepsilon_1 - \varepsilon)(\varepsilon - \varepsilon_2)}{\varepsilon^2(\varepsilon_1 + \varepsilon_2)}. \tag{39}$$

The p wave equation in the Z,Q notation has reflection at θ_B due to the small variation in the effective wavevector component Q as given by (39). For the step profile, ε is either ε_1 or ε_2, and there is no variation in Q and thus no reflection.

A common explanation for the small reflection of the p polarization at θ_B is in terms of the angular dependence of the dipole radiation from each atom or molecule which produces the transmitted and reflected waves. The far-field radiation pattern of a dipole has zero amplitude along the line of oscillation of the dipole (see Section 1-5, equation (78)); this is the reason for the polarization of light from the sky. We see from (32) that r_{p0} is zero when $\theta_1 + \theta_2 = \pi/2$, that is when the refracted and reflected waves are at a right angle (see Figure 1-6). The argument goes that at this angle of incidence there is no radiation from the accelerated electrons in the material to produce a p-polarized signal in the direction of specular

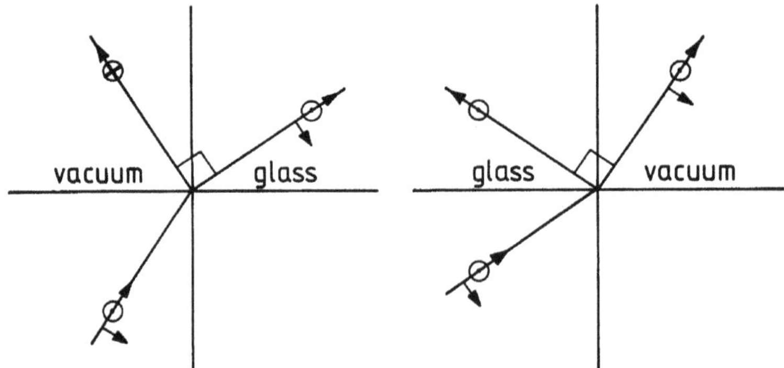

Figure 1-6. Illustrating complete transmission of the p wave at the Brewster angle.

reflection (left-hand part of Figure 1-6.). But transparency also exists in the reverse case of material to vacuum (right-hand side of the figure). In this case the explanation in terms of electrons radiating along the transmitted beam to produce (or fail to produce) the reflected beam does not apply. Futher, a similar case of transparency at the interface between two unlike media occurs with acoustic waves (as will be discussed in Section 1-4), and in that case the radiation from each scatterer does not have a dipole character.

1-3 Particle waves

In non-relativistic quantum mechanics, the motion of a particle of mass m and energy $\mathscr{E}$ in a potential V is determined by Schrödinger's equation for the probability amplitude Ψ,

$$-\frac{\hbar^2}{2m}\nabla^2\Psi + V\Psi = \mathscr{E}\Psi. \tag{40}$$

Here $\hbar$ is Planck's constant divided by 2π. We will consider reflection at a planar stratified boundary region between two uniform media characterized by potentials V_1 and V_2. Examples of the particles and interfaces to which this description applies are: electrons at a junction between two metals (with possibly an oxide layer in between); neutrons reflecting off a solid or liquid surface; and helium atoms reflecting at a liquid helium surface. In each of these examples the potential V in the single-particle equation (40) is an effective potential, representing the net effect of all the interactions between the particle and the scatterers in the medium through which it moves. An example of how this effective potential is determined is given in Section 1-5.

We again consider plane waves propagating in the zx plane, incident on a planar interface, with stratification in the z direction. For this geometry, V depends on one spatial variable z, and Ψ is independent of y. The z, x variable dependence in (40) is then separable, with

$$\Psi(z, x) = e^{iKx}\,\psi(z) \tag{41}$$

(it is usual to suppress the time dependence $e^{-i\mathscr{E}t/\hbar}$). Substitution of (41) into (40) gives an ordinary differential equation for ψ:

$$\frac{d^2\psi}{dz^2} + q^2\psi = 0, \qquad q^2(z) = \frac{2m}{\hbar^2}[\mathscr{E} - V(z)] - K^2. \tag{42}$$

From (41), K is the x-component of the wavevector in either medium, and is an invariant of the motion. The laws of reflection follows from this invariance, which is due to the absence of transverse components of the force, $\partial V/\partial x = 0 = \partial V/\partial y$. If the angles of incidence, reflection and refraction are θ_1, θ_1' and θ_2, the laws of reflection and refraction follow from the invariance of $K = k_{1x} = k'_{1x} = k_{2x}$:

$$k_1 \sin\theta_1 = k_1 \sin\theta_1' = k_2 \sin\theta_2, \tag{43}$$

where

$$k_i^2 = K^2 + q_i^2 = \frac{2m}{\hbar^2}[\mathscr{E} - V_i]. \tag{44}$$

As before, q is the component of the wavevector normal to the interface, with limiting values

$$k_1 \cos\theta_1 = q_1 \leftarrow q(z) \rightarrow q_2 = k_2 \cos\theta_2. \tag{45}$$

These relations are summarized in Figure 1-7.

On comparison of (7) and (42) we see that there is a one-to-one correspondence between the reflection problems for the electromagnetic s wave and particle waves obeying Schrödinger's equation, with the replacement

$$\varepsilon(z)\frac{\omega^2}{c^2} \leftrightarrow \frac{2m}{\hbar^2}[\mathscr{E} - V(z)]. \tag{46}$$

The reflection amplitude r and the transmission amplitude t are defined in terms of the limiting forms of the solution of (42):

$$e^{iq_1z} + r\,e^{-iq_1z} \leftarrow \psi(z) \rightarrow t\,e^{iq_2z}. \tag{47}$$

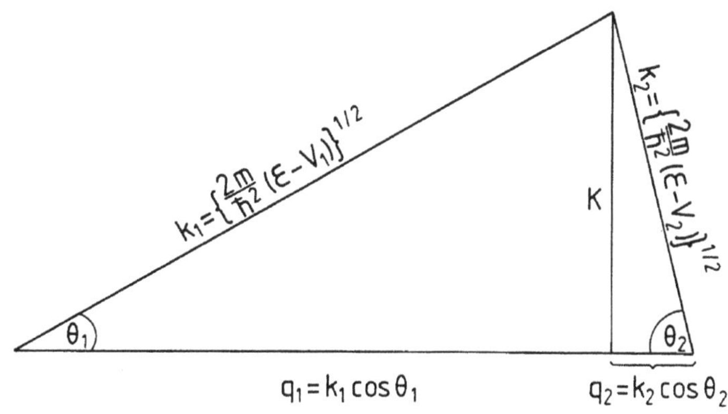

Figure 1-7. Graphical representation of $k_i^2 = K^2 + q_i^2$ and $K = k_1 \sin\theta_1 = k_2 \sin\theta_2$. The figure is drawn for electrons at 10 eV above the Fermi level in bulk aluminium, at the aluminium–vacuum interface. ($\mathscr{E}_F - V_1 \simeq 11.7$ eV, so $\mathscr{E} - V_1 \simeq 21.7$ eV; $V_2 - \mathscr{E}_F \simeq 4.2$ eV, so $\mathscr{E} - V_2 \simeq 5.8$ eV; the ratio of the refractive indices is $\{(\mathscr{E} - V_2)/(\mathscr{E} - V_1)\}^{1/2} \simeq 0.517$.)

For example, for the potential step

$$V_0(z) = \begin{cases} V_1 & (z < 0) \\ V_2 & (z > 0), \end{cases} \tag{48}$$

continuity of y and dy/dz at $z = 0$ gives the Fresnel-type equations

$$r_0 = \frac{q_1 - q_2}{q_1 + q_2}, \qquad t_0 = \frac{2q_1}{q_1 + q_2}. \tag{49}$$

Note that, as in the case of electromagnetic waves, the boundary conditions follow from the differential equations; they are not an additional assumption of the theory.

A refractive index can be defined for particles. From (43) and (44) we see that the refractive index is proportional to $(\mathscr{E} - V)^{1/2}$, that is to the square root of the kinetic energy, or to the local value of the wavevector k. The proportionality to $(\mathscr{E} - V)^{1/2}$ is also a classical result: the equations for the conservation of energy and transverse momentum for a particle incident at angle θ_1 onto a planar stratification between media 1 and 2 read

$$\tfrac{1}{2}mv_1^2 + V_1 = \mathscr{E} = \tfrac{1}{2}mv_2^2 + V_2. \tag{50}$$

$$mv_1 \sin\theta_1 = mv_2 \sin\theta_2. \tag{51}$$

Equation (51) shows that the refractive indices are proportional to v_i, which from (50) are equal to $[2(\mathscr{E} - V_i)/m]^{1/2}$. However, partial reflection does not exist for classical particles: there is either total reflection (when $V > \mathscr{E} - \tfrac{1}{2}m(v_1 \sin\theta_1)^2$ anywhere), or no reflection (when $V < \mathscr{E} - \tfrac{1}{2}(v_1 \sin\theta_1)^2$ everywhere).

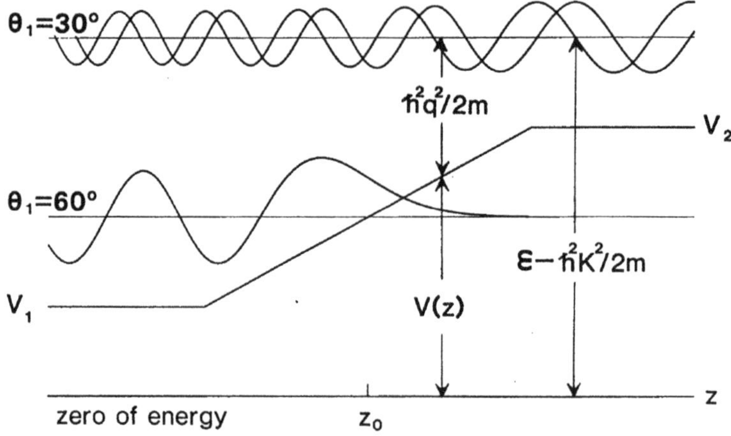

Figure 1-8. Probability amplitudes, at two angles of incidence, for particle waves incident from the left onto a linear ramp potential. The energy and potential values are such that $\theta_c = 45°$. The upper two waves are the real and imaginary parts of the probability amplitude ψ for incidence at 30°. The net flux to the right of the barrier, $q_2|t|^2$, is the same as the net flux on the left, $q_1(1 - |r|^2)$, despite the visible increase in the real and imaginary parts of the probability amplitude to the right. The lower curve is minus the real part of the probability amplitude for a totally reflected wave, incident at 60°. Only the real part is shown, since the real and imaginary parts of ψ are proportional to each other in total reflection: $\mathrm{Im}\,\psi/\mathrm{Re}\,\psi = \tan\delta/2$ when $r = e^{i\delta}$. The classical turning point z_0 (where $q^2 = 0$) is halfway up the ramp.

In contrast, total reflection occurs in the wave theory only if $V_2 > \mathscr{E} - \hbar^2 K^2/2m$ (q_2 is then imaginary, leading to exponential decay of the probability amplitude in medium 2); regions of imaginary q (negative $q^2 = (2m/\hbar^2)(\mathscr{E} - V) - K^2$) where $V > \mathscr{E} - \hbar^2 K^2/2m$ do not lead to *total* reflection when q_2 is real. Electromagnetic waves are likewise totally reflected when q_2 is imaginary, that is when $\varepsilon_2 \omega^2/c^2 < K^2$, or $\sin^2\theta_1 > \varepsilon_2/\varepsilon_1$. Thus the critical angle for total reflection is given by

$$\theta_c = \arcsin\left(\frac{\varepsilon_2}{\varepsilon_1}\right)^{1/2}, \qquad \theta_c = \arcsin\left(\frac{\mathscr{E} - V_2}{\mathscr{E} - V_1}\right)^{1/2}, \tag{52}$$

in the electromagnetic and particle wave cases. Partial and total reflection of particle and electromagnetic s waves is compared in Figures 1-8 and 1-9.

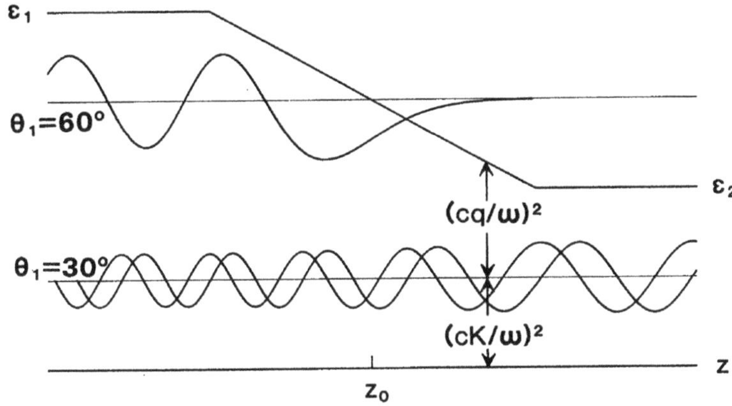

Figure 1-9. The electromagnetic s wave at two angles of incidence onto a linear dielectric function profile. The radiation is incident from the left. The dielectric constants are $\varepsilon_1 = 2$, $\varepsilon_2 = 1$, so that $\theta_c = 45°$. The lower two waves are the real and imaginary parts of $E(z)$, at 30° angle of incidence. The upper curve is the real part of $E(z)$ for a totally reflected wave incident at 60°. The figure is drawn for $(\omega/c)\Delta z = 6\pi$; the profile thickness Δz is three times the wavelength in the second medium, at normal incidence. The wavefunctions are the same as in Figure 1-8.

1-4 Acoustic waves

There is an interesting close correspondence between the reflection of sound and the reflection of the electromagnetic p wave. This will be demonstrated in the simplest case of fluid, non-viscous media.

Sound waves propagate changes in density and pressure which are usually very small compared to the mean values. The equations of motion, continuity, and state can then be linearized by setting

$$\text{density} = \varrho + \varrho_a, \qquad \text{pressure} = p + p_a, \tag{53}$$

where ϱ and p are the mean local values of the density and pressure, and ϱ_a and p_a are the small excess time-dependent values due to the presence of acoustic waves. On dropping second order terms in ϱ_a, p_a and the velocity of a fluid element, and neglecting the force due to gravity apart from its effect on stratification according to density, one obtains the equation (Bergmann, 1946)

$$\nabla^2 p_a - \frac{1}{v^2}\frac{\partial^2 p_a}{\partial t^2} - \frac{1}{\varrho}\nabla\varrho \cdot \nabla p_a = 0. \tag{54}$$

Here $v^2 = (\partial p/\partial \varrho)_s$ is the adiabatic derivative of the pressure with respect to density, and gives the square of the phase velocity in a uniform medium.

Consider now the reflection of sound at a (sharp or diffuse) interface characterized by a density profile $\varrho(z)$ and an adiabatic pressure derivative $(\partial p/\partial \varrho)_s = v^2(z)$. For a plane monochromatic wave propagating in the zx plane, we have

$$p_a(z, x, t) = P(z)\, e^{i(Kx - \omega t)}. \tag{55}$$

K is again the component of the wavevector along the interface, and is a constant of the motion:

$$K = \frac{\omega}{v_1}\sin\theta_1 = \frac{\omega}{v_2}\sin\theta_2, \tag{56}$$

where v_1, v_2 are the limiting values of $\{(\partial p/\partial \varrho)_s\}^{1/2}$ in the two media, and θ_1, θ_2 are the angles of incidence and refraction. The differential equation for P is

$$\varrho\frac{d}{dz}\left(\frac{1}{\varrho}\frac{dP}{dz}\right) + q^2 P = 0, \tag{57}$$

with

$$q^2(z) = \frac{\omega^2}{v^2(z)} - K^2; \tag{58}$$

q is again the normal component of the wavevector, with limiting values $q_1 = (\omega/v_1)\cos\theta_1$, $q_2 = (\omega/v_2)\cos\theta_2$.

The term $(d\varrho/dz)(dP/dz)$ in (57) may be removed by introducing a new dependent variable $P/\sqrt{\varrho}$, as Bergmann notes. This is analogous to the transformation to $B/\sqrt{\varepsilon}$ discussed in Section 1-2. A more fruitful approach is analogous to the transformation to a dilated z variable in the p wave case: (57) has the same form as the electromagnetic p wave equation

$$\varepsilon\frac{d}{dz}\left(\frac{1}{\varepsilon}\frac{dB}{dz}\right) + q^2 B = 0.$$

In terms of a new independent variable Z, defined by $dZ = \varrho dz$, (57) becomes

$$\frac{d^2 P}{dZ^2} + Q^2 P = 0, \tag{59}$$

where $Q = q/\varrho$. (As defined here, Z and Q no longer have the dimensions of length and (length)$^{-1}$; this can be altered by respectively dividing Z and multiplying Q by some density, for example $(\varrho_1 + \varrho_2)/2$.)

It is clear from the form of (59), and the discussion of reflection at the Brewster angle given in Section 1-2, that weak reflection of acoustic waves (zero reflection, in the case of a sharp transition between the two media) is expected whenever $Q_1 = Q_2$. This holds when

$$\frac{\cos\theta_1}{\varrho_1 v_1} = \frac{\cos\theta_2}{\varrho_2 v_2}. \tag{60}$$

This result was first given (for a sharp interface) by George Green (1838). On eliminating θ_2 from (60) and Snell's Law (56), one finds that weak reflection occurs at an angle of incidence (which we will call Green's angle) $\theta_1 = \theta_G$, given by

$$\tan^2\theta_G = \frac{(\varrho_2 v_2)^2 - (\varrho_1 v_1)^2}{\varrho_1^2(v_1^2 - v_2^2)}. \tag{61}$$

In contrast to the electromagnetic p wave case, weak reflection of acoustic waves does not happen at a certain angle at a boundary between *any* two media: the quantities $\varrho_1 v_1 - \varrho_2 v_2$ and $v_1 - v_2$ must have opposite signs.

At Green's angle θ_G (where $Q_1 = Q_2$), K^2 is equal to

$$K_G^2 = \frac{\omega^2}{\varrho_1^2 - \varrho_2^2}\left\{\left(\frac{\varrho_1}{v_2}\right)^2 - \left(\frac{\varrho_2}{v_1}\right)^2\right\}, \tag{62}$$

and the common value of Q_1 and Q_2 is given by

$$Q_G^2 = \frac{\omega^2}{\varrho_1^2 - \varrho_2^2}\left\{\frac{1}{v_1^2} - \frac{1}{v_2^2}\right\}. \tag{63}$$

According to (59), the acoustic wave in the Z variable then reflects from the bump in Q^2, given by

$$Q^2 - Q_G^2 = \frac{\omega^2}{\varrho^2(\varrho_1^2 - \varrho_2^2)}\left\{\frac{\varrho_1^2 - \varrho_2^2}{v^2} - \left(\frac{\varrho_1}{v_2}\right)^2 + \left(\frac{\varrho_2}{v_1}\right)^2 - \varrho^2\left(\frac{1}{v_1^2} - \frac{1}{v_2^2}\right)\right\}. \tag{64}$$

One can define an acoustic reflection amplitude r and a transmission amplitude t in terms of the pressure:

$$e^{iq_1 z} + r\, e^{iq_1 z} \leftarrow P(z) \rightarrow t\, e^{iq_2 z}. \tag{65}$$

For a sharp transition between media 1 and 2, P and $dP/\varrho dz = dP/dZ$ are continuous at the boundary. (This is evident from (57); note that, as in the electromagnetic and particle wave cases, the differential equation dictates the boundary conditions, which are not an additional input to the theory.) Thus, for a sharp boundary located at the origin,

$$r = \frac{Q_1 - Q_2}{Q_1 + Q_2}, \qquad t = \frac{2Q_1}{Q_1 + Q_2}. \tag{66}$$

These may be rewritten as

$$r = \frac{\varrho_2\tan\theta_2 - \varrho_1\tan\theta_1}{\varrho_2\tan\theta_2 + \varrho_1\tan\theta_1}, \qquad t = \frac{2\varrho_2\tan\theta_2}{\varrho_2\tan\theta_2 + \varrho_1\tan\theta_1}. \tag{67}$$

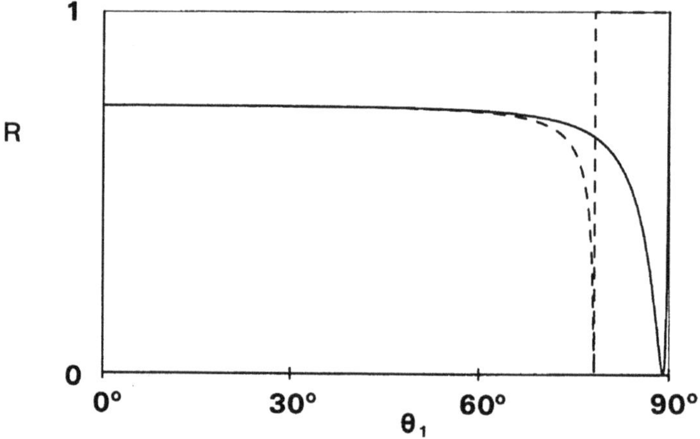

Figure 1-10. Reflectivity of acoustic waves at a mercury-water interface, according to (66). Dashed line: sound incident from the slower medium (mercury). Full line: sound incident from the faster medium (water). The curves are drawn for $\varrho_{Hg}/\varrho_{H_2O} = 13.57$, $v_{Hg}/v_{H_2O} = 0.9789$; the critical and Green's angles are $\theta_c \simeq 78.21°$, $\theta_G \simeq 78.18°$, 89.12°.

Total reflection occurs for angles of incidence greater than

$$\theta_c = \arcsin (v_1/v_2). \tag{68}$$

(This result follows on setting $\theta_2 = \pi/2$ in (56); it holds for any interface, no matter how diffuse, provided absorption can be neglected.) The critical angle θ_c will be close to Green's angle θ_G, if the latter exists, when $v_1 \simeq v_2$. The reflectivity of a step profile then rapidly goes from zero at θ_G to unity at θ_c and beyond, as illustrated for the mercury–water interface in Figure 1–10.

When $\varrho_1 v_1 \simeq \varrho_2 v_2$ the reflectivity at normal incidence is small, and θ_G (if it exists) will also be small. This is the case for carbon tetrachloride and water, illustrated in Figure 1-11.

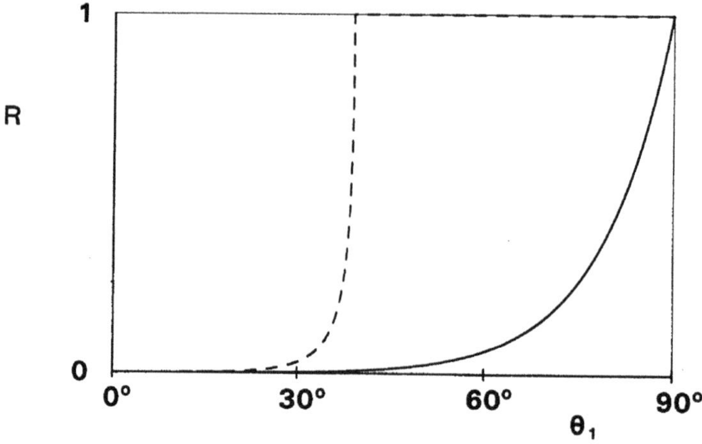

Figure 1-11. Reflectivity of acoustic waves at a water-carbon tetrachloride interface, obtained from (66). Dashed line: sound incident from the slower medium (CCl_4). Full line: sound incident from the faster medium (water). The curves are drawn for $\varrho_{CCl_4}/\varrho_{H_2O} = 1.595$, $v_{CCl_4}/v_{H_2O} = 0.6274$; the critical and Green's angles are $\theta_c \simeq 38.86°$, $\theta_G \simeq 1.73°$, 2.76°.

1-5 Scattering and reflection

Most of the results in this book come from analysis of the differential equations for waves in material media, the media being characterized by a dielectric function, or an effective potential, or the density and speed of sound, in the case of electromagnetic, particle or acoustic waves. This approach hides the many-body complexity of the real physics: specular reflection, for example, is the result of the constructive interference of many scattered or re-radiated waves. A discussion of reflection from this point of view will be given here; it leads to values for the functions characterizing the media, such as ε and V, in terms of the properties of the particles comprising the system. Such approaches go back to Lorentz (1909), Darwin (1924) and Hartree (1928) in the electromagnetic case. We will begin with an adaptation of Fermi's (1950) argument for the effective potential of a collection of neutron scatterers, since this is simpler.

Consider the reflection of a beam of neutrons by a thin slab of material. The neutrons interact with the nuclei in the slab. For slow neutrons this interaction is characterized by a length b, the scattering length for neutrons off a bound scatterer. An incident plane wave e^{ikz} causes each scatterer to radiate a spherical wave $-b\, e^{ikr}/r$. The reflected wave is found by summing up the scattered waves from all parts of the slab. The geometry is illustrated in Figure 1-12.

If n is the number density of the scatterers, $(2\pi\varrho d\varrho\Delta z)n$ is the number of scatterers within an annulus between ϱ and $\varrho + d\varrho$. The reflected wave at P is thus

$$\psi_r = \int_0^\infty d\varrho\; 2\pi\varrho\Delta z\, n\left(-\frac{b\, e^{ikr}}{r}\right). \tag{69}$$

For fixed z we have $\varrho d\varrho = r dr$, so that

$$\psi_r = -2\pi n\Delta z\, b \int_{-z}^{\infty} dr\, e^{ikr}. \tag{70}$$

The integral over r is not defined as it stands, because we have used e^{ikz} as the incident wave, namely a plane wave extending to infinity in the x and y directions.

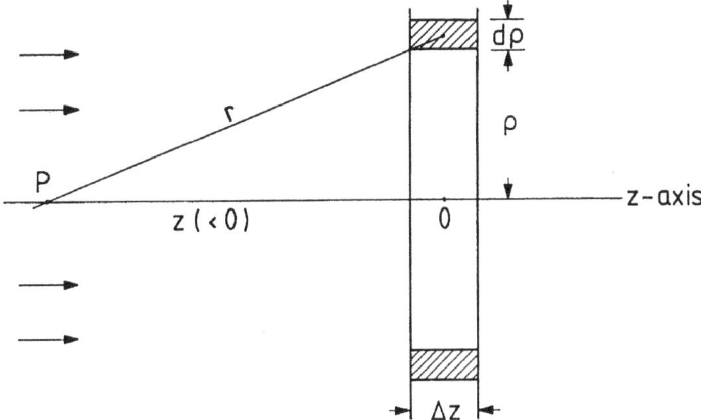

Figure 1-12. Reflection of neutrons by a slab of scatterers. The thickness Δz is such that $k\Delta z$ is small, so that the phase of the plane wave e^{ikz} (incident from the left) is nearly constant over the slab.

In practice the incident wave would be a finite beam, with an amplitude decreasing with $\varrho = (x^2 + y^2)^{1/2}$. The resulting integral for ψ_r then is well-defined. When such decrease is slow on the scale of k^{-1} (the beam is many wavelengths wide), the integral is equal to $-e^{-ikz}/ik$, and

$$\psi_r = \frac{2\pi n \Delta z b}{ik} e^{-ikz}. \tag{71}$$

This is a reflected wave, with reflection amplitude equal to the factor multiplying e^{-ikz}.

We will now show that the reflection amplitude due to a thin slab of thickness Δz and effective potential V is (to lowest order in V)

$$r_1 = \left(\frac{\Delta z}{2ik}\right)\frac{2mV}{\hbar^2}. \tag{72}$$

(What follows is heuristic; a rigorous proof is given in Chapter 3.) Consider the effect of a potential hump, or well, which is small in extent in comparison with the wavelength. Seen on the scale of the wavelength, the hump appears as a spike, and its main effect is to create a change of slope in the wavefunction: on integrating (42) (at normal incidence) across the hump, we have

$$\psi'(z_2) - \psi'(z_1) = -\int_{z_1}^{z_2} dz \frac{2m}{\hbar^2}[\mathscr{E} - V(z)]\,\psi(z). \tag{73}$$

For wavelengths long compared to the extent of the hump, ψ is nearly constant over its effect, so when $z_2 - z_1$ is small compared to $k^{-1} = \hbar(2m\mathscr{E})^{-1/2}$, and the hump is centred on the origin,

$$\psi'(z_2) - \psi'(z_1) \simeq \frac{2m}{\hbar^2}\psi(0)\int_{z_1}^{z_2} dz\, V(z). \tag{74}$$

(The zero of energy has been chosen so that V goes to zero on either side of the hump). From (47) the left side is equal to $ik(t - 1 + r) + 0(k^2)$. The assumption that ψ remains nearly constant from z_1 to z_2 also implies $1 + r \simeq \psi(0) \simeq t$. Thus (74) gives

$$r \simeq \left(\frac{1}{2ik}\right)\frac{2m}{\hbar^2}\int_{z_1}^{z_2} dz\, V(z). \tag{75}$$

For V constant inside the hump (of extent Δz), and zero outside, this reduces to (72).

We can now give an expression for the effective potential of a collection of scatterers: (71) and (72) together imply that this is

$$V = 4\pi\left(\frac{\hbar^2}{2m}\right) nb. \tag{76}$$

The scattered waves interfere constructively to give a reflected and a transmitted wave, *as if* the medium were completely homogeneous and acted on the particles with a potential given by (76). We have considered normal incidence; at oblique incidence the constructive interference of the spherically diverging waves from the scatterers within the slab is in the specular and straight-through directions.

We now turn to the more complex question of electromagnetic radiation interacting with the atoms or molecules in a thin slab of material. We will calculate the field at P in front of a slab, as in Figure 1-12. The incident electric field propagates in the z direction, and is taken to be polarized along the x direction,

$$\mathbf{E} = \mathrm{e}^{i(kz-\omega t)}\,(E_0, 0, 0). \tag{77}$$

When the wavelength is large compared to atomic size, each atom radiates predominantly as a dipole. For a given atom with dipole $\mathbf{p}$, oscillating at the impressed angular frequency ω, the electric field at $\mathbf{r} = r\hat{\mathbf{r}}$ from the atom is (see for example Jackson (1962), Section 9.2)

$$\mathbf{E} = \frac{\mathrm{e}^{ikr}}{r}\left\{k^2(\hat{\mathbf{r}} \times \mathbf{p}) \times \hat{\mathbf{r}} + [3(\hat{\mathbf{r}}\cdot\mathbf{p})\hat{\mathbf{r}} - \mathbf{p}]\left(\frac{1}{r^2} - \frac{ik}{r}\right)\right\}, \tag{78}$$

where $k = \omega/c$. The far field (given by the first term) is a spherically diverging wave, with $\mathbf{E}$ transverse to $\mathbf{r}$. We do not omit the near field, since we do not wish to assume that $kr \gg 1$. All dipoles are taken to lie along the direction of the incident electric field, and to have the same strength αE_0, where α is the polarizability of an atom:

$$\mathbf{p} = \mathrm{e}^{-i\omega t}\,(\alpha E_0, 0, 0). \tag{79}$$

The point P is at $(0, 0, z)$, with $z < 0$. The contribution to the electric field at P from a dipole at $(x, y, 0)$ is then

$$E_x = \frac{\alpha E_0\,\mathrm{e}^{i(kr-\omega t)}}{r^3}\left\{k^2(y^2 + z^2) + (1 - ikr)\left(\frac{3x^2}{r^2} - 1\right)\right\}, \tag{80}$$

with E_y and E_z odd in x and thus integrating to zero when we sum over the dipolar fields. Thus the net field at P due to all the dipoles (of number density n) in the thin slab is, on changing to the cylindrical coordinates ϱ and ϕ and integrating over ϕ,

$$\begin{aligned} E_x^{\text{dipoles}} &= \mathrm{e}^{-i\omega t}\,\alpha E_0 n\Delta z\pi \int_0^\infty \mathrm{d}\varrho\,\varrho\,\frac{\mathrm{e}^{ikr}}{r^3}\left\{k^2(\varrho^2 + 2z^2) + (1 - ikr)\left(\frac{3\varrho^2}{r^2} - 2\right)\right\} \\ &= \mathrm{e}^{-i\omega t}\,\alpha E_0 n\Delta z\pi \int_{-z}^\infty \mathrm{d}r\,\frac{\mathrm{e}^{ikr}}{r^2}\left\{k^2(r^2 + z^2) + (1 - ikr)\left(1 - \frac{3z^2}{r^2}\right)\right\}. \end{aligned} \tag{81}$$

The first term is an integral of the form (70); the others may be obtained from it by integration by parts. The result is

$$E_x^{\text{dipoles}} = \mathrm{e}^{-i(kz+\omega t)}\,2\pi ik\alpha n\Delta z E_0. \tag{82}$$

The reflection amplitude for the slab is the coefficient of $E_0\,\mathrm{e}^{-i(kz+\omega t)}$ in (82). We compare this with the result analogous to (72) for the reflection amplitude due to a thin slab of dielectric constant ε,

$$r_1 = \frac{i}{2}k\Delta z(\varepsilon - 1). \tag{83}$$

Thus the effective dielectric constant of a slab of atoms of polarizability α and number density n is

$$\varepsilon \simeq 1 + 4\pi\alpha n. \tag{84}$$

We have neglected the effect of the dipolar fields on each other. When these are taken into account, the resulting dielectric constant for a uniform medium becomes

$$\varepsilon = \frac{1 + \frac{8}{3}\pi\alpha n}{1 - \frac{4}{3}\pi\alpha n}. \tag{85}$$

This expression is known as the Clausius–Mossotti or Lorentz–Lorenz formula (Lorentz 1909). The result (84) is the first-order term in the αn expansion of (85). The form of (85), with $n = n(z)$, remains valid with a high degree of accuracy in a stratified medium of polarizable atoms (Castle and Lekner 1980, Lekner 1983).

1-6 A look ahead

In the preceding sections we have introduced the definitions and basic equations for the reflection of electromagnetic, particle and acoustic compressional waves by planar stratified media. The remainder of the book is written almost entirely in electromagnetic notation; a translation of the main results into particle-wave language is made in the Appendix. Here we preview Chapters 2 to 13, stating and discussing some of their results and techniques in order to give the reader a feel for the structure and content of the book.

Chapter 2 contains both general results, true for reflection and transmission at any transition between two uniform media, and some specific results for exactly solvable profiles. Among the general results are the conservation law

$$q_1(1 - |r_{12}|^2) = q_2|t_{12}|^2, \tag{86}$$

and reciprocity relations such as

$$r_{21} = -\frac{t_{12}}{t_{12}^*} r_{12}^* \tag{87}$$

and

$$q_2 t_{12} = q_1 t_{21}. \tag{88}$$

The conservation law (86), which holds for real q_1 and q_2 and in the absence of absorption within the interface, represents conservation of energy in the electromagnetic case, and conservation of probability density current in the particle case. The relation (87) holds under the same conditons, and implies that the reflectance $R = |r|^2$ is the same from either side of incidence on a nonabsorbing interface. The relation (88) is more general, being valid also in the presence of absorption within the interface. It implies the equality of the transmittances $T_{12} = (q_2/q_1)|t_{12}|^2$ and $T_{21} = (q_1/q_2)|t_{21}|^2$, representing the energy or particle flux through the inhomogeneity, for incidence from medium 1 or from medium 2. (When the polarization subscripts s and p are omitted, the relation quoted is understood to be valid for either wave.)

For inhomogeneities extending from z_1 to z_2, with $\varepsilon = \varepsilon_1$ for $z \leqslant z_1$ and $\varepsilon = \varepsilon_2$ for $z \geqslant z_2$, the s wave reflection and transmission amplitudes may be expressed in terms of the values and derivatives of two linearly independent solutions F and G of (7) within $z_1 \leqslant z \leqslant z_2$, evaluated at z_1 and z_2:

$$r_s = e^{2iq_1z_1} \times \frac{q_1q_2(F_1G_2 - G_1F_2) + iq_1(F_1G_2' - G_1F_2') + iq_2(F_1'G_2 - G_1'F_2) - (F_1'G_2' - G_1'F_2')}{q_1q_2(F_1G_2 - G_1F_2) + iq_1(F_1G_2' - G_1F_2') - iq_2(F_1'G_2 - G_1'F_2) + (F_1'G_2' - G_2'F_2')}, \tag{89}$$

$$t_s = \frac{e^{i(q_1z_1 - q_2z_2)}\, 2iq_1(F_2G_2' - G_2F_2')}{q_1q_2(F_1G_2 - G_1F_2) + iq_1(F_1G_2' - G_1F_2') - iq_2(F_1'G_2 - G_1'F_2) + (F_1'G_2' - G_1'F_2')}. \tag{90}$$

Similar expressions can be written down for the p polarization. These results are useful for specific profiles for which the solutions are known functions, such as the Airy functions for the linear profile, and the Bessel functions for the exponential profile. General results may also be deduced from (89) and (90), for example that $r_s \to -1$ and $t_s \to 0$ at grazing incidence, and that r_s and t_s tend to the Fresnel values (15) as $\Delta z = z_2 - z_1$ tends to zero. From the p polarization expressions one finds that $r_p \to 1$ and $t_p \to 0$ at grazing incidence. Thus r_p/r_s always moves in the complex plane from $+1$ at normal incidence to -1 at grazing incidence, and the number of ellipsometric Brewster angles (or *principal angles*), where Re $(r_p/r_s) = 0$, is odd (1, 3, 5, . . .).

Chapter 2 also lists the exact solutions for three dielectric function profiles which are solvable for both the s and p polarizations, and another (the important hyperbolic tangent profile) which is solvable for the s wave only. Two other cases which are solvable for the s wave case, the $\mathrm{sech}^2(z/a)$ and the linear profile, are discussed in Sections 4-3 and 5-2 respectively, where their solution is relevant to the problem at hand.

Chapter 3 treats the reflection of long waves, that is those whose wavelength is large compared to the thickness of the reflecting inhomogeneity. The long wave results are obtained from perturbation theories, which in turn derive from exact integral and integro-differential equations obeyed by the s and p waves. For example, from the perturbation theory for the s wave one finds that the reflection amplitude, to second order in the interface thickness, is given by

$$r_s = r_{s0} + \frac{2q_1\omega^2/c^2}{(q_1 + q_2)^2}\left\{i\lambda_1 - 2q_2\lambda_2 - \frac{\omega^2/c^2}{q_1 + q_2}\lambda_1^2\right\} + \ldots, \tag{91}$$

where the λ_n are integrals of dimension (length)^n,

$$\lambda_n = \int_{-\infty}^{\infty} dz\, [\varepsilon(z) - \varepsilon_0(z)]z^{n-1}. \tag{92}$$

In (92), ε is the dielectric function profile under consideration, and ε_0 is the step dielectric function defined in (11), which has the reflection amplitude r_{s0} given in (13). The integrals λ_n depend on the relative positioning of the actual profile ε and the step profile ε_0. A theory which calculates reflection amplitudes as a perturbation series about a reference profile (here ε_0) must obtain results for observables, such as $|r_s|^2$, which are invariant to the relative positioning of the actual and reference profile. If $r = r_0 + r_1 + r_2 + \ldots,$

$$R = |r|^2 = |r_0|^2 + 2\,\mathrm{Re}\,(r_0^* r_1) + \{|r_1|^2 + 2\,\mathrm{Re}\,(r_0^* r_2)\} + \ldots, \tag{93}$$

and we see from (91) that the first order term r_1 is imaginary in the absence of absorption or total internal reflection for the s wave. (The same is true for the p wave, as is shown in Section 3-4.) Then there is no term in R of first order in the interface thickness, for either polarization. The second order term is given by the expression within the braces in (93); from (91) we have

$$R_s = \left(\frac{q_1 - q_2}{q_1 + q_2}\right)^2 - \frac{4q_1 q_2 \omega^4/c^4}{(q_1 + q_2)^4}\, i_2 + \ldots, \tag{94}$$

where the *integral invariant* i_2 is given by

$$i_2 = 2(\varepsilon_1 - \varepsilon_2)\lambda_2 - \lambda_1^2. \tag{95}$$

(The subscript 2 denotes dimensionality (length)^2.) The integrals λ_1 and λ_2 which enter into r_s and R_s depend on the relative positioning of the actual and reference profiles, but the combination of integrals which comprise i_2 is invariant with respect to the choice of positioning. Similar results are obtained for the observables r_p/r_s and $R_p = |r_p|^2$:

$$r_{s0}\left(\frac{r_p}{r_s}\right) = r_{p0} - \frac{2iQ_1}{(Q_1 + Q_2)^2}\frac{K^2}{\varepsilon_1\varepsilon_2}\mathscr{I}_1 + \ldots, \tag{96}$$

$$R_p = \left(\frac{Q_1 - Q_2}{Q_1 + Q_2}\right)^2 - \frac{4Q_1 Q_2}{\varepsilon_1\varepsilon_2(Q_1 + Q_2)^4}\left\{\frac{\omega^4}{c^4} i_2 - \frac{\omega^2}{c^2}K^2\left[j_2 + \left(\frac{1}{\varepsilon_1} + \frac{1}{\varepsilon_2}\right)i_2\right] + \frac{K^2}{\varepsilon_1\varepsilon_2}[(\varepsilon_1 + \varepsilon_2)j_2 - \mathscr{I}_1^2]\right\} + \ldots, \tag{97}$$

where j_2 is another second order invariant, and the first order invariant $\mathscr{I}_1$ is defined by

$$\mathscr{I}_1 = \int_{-\infty}^{\infty} dz\, \frac{(\varepsilon_1 - \varepsilon)(\varepsilon - \varepsilon_2)}{\varepsilon} = \int_{-\infty}^{\infty} dz \left\{\varepsilon_1 + \varepsilon_2 - \frac{\varepsilon_1\varepsilon_2}{\varepsilon} - \varepsilon\right\}. \tag{98}$$

These results show that, in the long wave limit, the observables R_s, R_p and r_p/r_s take *universal form*. The integral invariants $\mathscr{I}_1$, i_2 and j_2 depend on the profile shape and extent only. All frequency and angular dependence is contained in the coefficients of $\mathscr{I}_1$, i_2 and j_2, and is the same for all non-singular profiles. (The degenerate case $\varepsilon_1 = \varepsilon_2$ requires special consideration for the ellipsometric ratio r_p/r_s, however.)

Equations (94), (96) and (97) illustrate how theory answers the question "what information can be obtained from a given experiment?" From (94) we see that measurement of the s reflectivity in the long wave case can determine only one characteristic of the interface, the invariant i_2. Experimental data at different angles of incidence give no new information (we are assuming that the interface has no roughness, and the absence of absorption in the interface or substrate), merely the opportunity to reduce the uncertainty in i_2. The same is true for ellipsometry to lowest order in the interface thickness: one parameter (the invariant $\mathscr{I}_1$) can be determined, at any angle of incidence. The p wave reflectance (97) carries more information, because the direction of the electric field relative to the interface changes with the angle of incidence. In principle, the values of $\mathscr{I}_1^2$, i_2 and j_2 may be determined by intensity measurements at a mimimum of three angles of incidence.

The long wave results described above were obtained from perturbation theory, the perturbation being the deviation of the actual profile $\varepsilon(z)$ from the step profile $\varepsilon_0(z)$. The simplest example of a perturbation theory expression for the reflection amplitude is that for reflection by a film between like media:

$$r_s^{\text{pert}} = -\frac{\omega^2/c^2}{2iq_0}\int_{-\infty}^{\infty} \mathrm{d}z\,(\varepsilon - \varepsilon_0)\,\mathrm{e}^{2iq_0 z}. \tag{99}$$

Here q_0 is the common value of q_1 and q_2; ε_0 is likewise the common value of ε_1 and ε_2. The normal incidence, thin film version of this result has been used in Section 1-5 (equations (72) and (83)). Note that r_s^{pert} diverges at grazing incidence (as $q_0 \to 0$). This is unphysical: for passive media the reflection amplitude must stay within the unit circle, and in fact we saw that the exact r_s tends to -1 at grazing incidence. This troublesome divergence at grazing incidence remains in higher order perturbation expressions, but is removed by the variational theory developed in *Chapter 4*. The simplest trial function, $\psi_0 = \mathrm{e}^{iq_0 z}$, leads to the variational expression

$$r_s^{\text{var}} = \frac{-\dfrac{\omega^2/c^2}{2iq_0}\lambda(2q_0)}{1 + \dfrac{\omega^2/c^2}{2iq_0}\dfrac{\sigma(2q_0)}{\lambda(2q_0)}}, \tag{100}$$

where $\lambda(2q_0)$ is the Fourier integral in (99), and $\sigma(2q_0)$ is a double integral defined in Chapter 4. The variational result (100) is not divergent at grazing incidence; in fact it tends to the correct value of -1 as $q_0 \to 0$, since the integrals λ and σ have the property that $\sigma(0) = \lambda^2(0)$. Further, r_s^{var} is correct to second order in the film thickness, whereas r_s^{pert} is not. These properties are shared by the variational expressions, derived in Chapter 4, for s and p wave reflection amplitudes between unlike media.

Non-linear, first order differential equations (of the Riccati type) for the reflection amplitudes are derived in *Chapter 5*. Two kinds of equations are used: those for a quantity $\varrho(z)$ which tends to $r\,\mathrm{e}^{-2iq_1 z}$ as $z \to -\infty$, and those for $r(z)$, tending to r in the same limit. For the s wave, the respective equations are

$$\varrho' + 2iq\varrho = \frac{q'}{2q}(1 - \varrho^2), \tag{101}$$

$$r' = \frac{q'}{2q}(\mathrm{e}^{2i\phi} - r^2\,\mathrm{e}^{-2i\phi}), \tag{102}$$

where primes denote differentiation with respect to z, and the phase integral ϕ is defined by

$$\phi(z) = \int^{z} d\zeta \, q(\zeta). \tag{103}$$

The corresponding equations for the p wave reflection amplitudes have Q'/Q instead of q'/q on the right-hand side. From (101) it is shown in Section 5-4 that R_s has the Fresnel reflectivity as an upper bound for all monotonic profiles:

$$R_s \leqslant R_{s0} = \left(\frac{q_1 - q_2}{q_1 + q_2}\right)^2. \tag{104}$$

A similar bound holds for R_p when $Q(z) = q(z)/\varepsilon(z)$ is monotonic.

Integration of (102) from $z = -\infty$ to $+\infty$ gives

$$r_s = -\int_{-\infty}^{\infty} dz \, \frac{q'}{2q} (e^{2i\phi} - r^2 \, e^{-2i\phi}). \tag{105}$$

The $r(z)$ on the right-hand side is the reflection amplitude of a profile truncated at z. If the reflection is weak, one can get an approximate expression for r_s by omitting the term proportional to r^2 on the right. This is the weak reflection or Rayleigh approximation,

$$r_s^R = -\int_{-\infty}^{\infty} dz \, \frac{q'}{2q} \, e^{2i\phi}. \tag{106}$$

The corresponding approximation in the p wave case is

$$r_p^R = \int_{-\infty}^{\infty} dz \, \frac{Q'}{2Q} \, e^{2i\phi}. \tag{107}$$

At normal incidence both (106) and (107) reduce to

$$r_n^R = -\tfrac{1}{2}\int_{-\infty}^{\infty} dz \, \frac{n'}{n} \, e^{2i\phi_n}, \qquad \phi_n(z) = \frac{\omega}{c}\int^{z} d\zeta \, n(\zeta), \tag{108}$$

where $n = \varepsilon^{1/2}$ is the refractive index. If one makes the further (and drastic) approximations of replacing $2n$ by $n_1 + n_2$ and ϕ_n by $(\omega/c)(n_1 + n_2)z = (k_1 + k_2)z$, (108) simplifies to

$$r_n \simeq -\frac{1}{n_1 + n_2}\int_{-\infty}^{\infty} dz \, \frac{dn}{dz} \, e^{i(k_1+k_2)z}. \tag{109}$$

Expressions closely related to (109) have been used by Buff, Lovett and Stillinger (1965) and by Huang and Webb (1969) in the analysis of reflection from the diffuse interface of a binary mixture.

The Rayleigh approximation works very well when the reflection is weak, but fails near grazing incidence. The Rayleigh approximation (106) and the long wave limiting form (94) are compared in Figure 1-13 with the exact reflectivity for the hyperbolic tangent profile

$$\varepsilon(z) = \tfrac{1}{2}(\varepsilon_1 + \varepsilon_2) - \tfrac{1}{2}(\varepsilon_1 - \varepsilon_2) \tanh z/2a. \tag{110}$$

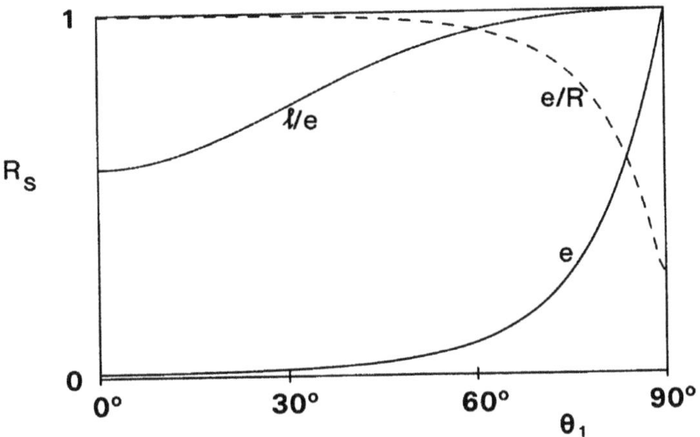

Figure 1-13. Reflectivity of the s wave off the tanh profile (110), for $\varepsilon_1 = 1$, $\varepsilon_2 = (4/3)^2$, $(\omega/c)a = 0.2$. For this value of $(\omega/c)a$ the distance in which the dielectric function changes over 80% of its range (from $(9\varepsilon_1 + \varepsilon_2)/10$ to $(\varepsilon_1 + 9\varepsilon_2)/10$) is about one seventh of the wavelength of the incident radiation. The full curve (e) is the exact reflectivity (111), the dashed curve (e/R) gives the ratio of the exact to the Rayleigh reflectivities, and the dot–dashed curve (l/e) gives the ratio of the long-wave limiting form (94) to the exact value.

For this profile the phase integral can be evaluated analytically (see Section 6-4), $i_2 = (\pi^2/3)(\varepsilon_1 - \varepsilon_2)^2 a^2$ from Table 3-1, and the exact reflectivity is (from Section 2-5)

$$R_s = \left\{\frac{\sinh \pi(q_1 - q_2)a}{\sinh \pi(q_1 + q_2)a}\right\}^2. \tag{111}$$

The reflection of short waves, that is those whose wavelength is small compared to the thickness of the interface, is discussed in *Chapter 6*. In the short wave limit the reflection properties approach the behaviour of classical particles, which are either totally reflected or not reflected. In the absence of turning points (zeros of $q^2(z)$), approximate solutions of (7) are the Liouville–Green functions

$$E_+ = \left(\frac{q_1}{q}\right)^{1/2} e^{i\phi}, \qquad E_- = \left(\frac{q_2}{q}\right)^{1/2} e^{-i\phi} \tag{112}$$

(ϕ is the phase integral defined in (103)). A perturbation theory based on a Green's function constructed from $E_\pm$ gives the first order reflection amplitude

$$r_s^{(1)} = -\frac{1}{2}\int_{-\infty}^{\infty} d\phi\, e^{2i\phi}\,\{\gamma - \gamma^2/4i\}, \tag{113}$$

where

$$\gamma = \frac{dq/dz}{q^2} = \frac{dq}{q\,d\phi} \tag{114}$$

is a dimensionless function which must be small for the short wave approximation to hold. The perturbation theory result is closely related to the Rayleigh approximation (which is the first term of another perturbation approach, the Bremmer series discussed in Section 6-5), as can be seen by writing (106) in the form

$$r_s^R = -\frac{1}{2}\int_{-\infty}^{\infty} d\phi\, \gamma\, e^{2i\phi}. \tag{115}$$

Unlike the long wave case, the reflection properties of short waves depend on the detail of the reflecting inhomogeneity, and do not take a universal form. For example: in the case of the profiles of finite range, which have a discontinuity in slope at the endpoints z_1 and z_2, both $r_s^{(1)}$ and r_s^R give

$$r_s = \frac{1}{4i} e^{i(\phi_1+\phi_2)} \{\gamma_1 e^{-i\Delta\phi} - \gamma_2 e^{i\Delta\phi}\} + \ldots, \tag{116}$$

where ϕ_1 and ϕ_2 are the values of the phase integral at z_1 and z_2, $\Delta\phi = \phi_2 - \phi_1$, and . . . denotes that exponentially small terms have been omitted. (The function γ changes in value from 0 to γ_1 at z_1, and from γ_2 to 0 at z_2). A similar result holds for the p wave:

$$r_p = \frac{1}{4i} e^{i(\phi_1+\phi_2)} \{\gamma_1 \cos 2\theta_1 e^{-i\Delta\phi} - \gamma_2 \cos 2\theta_2 e^{i\Delta\phi}\} + \ldots. \tag{117}$$

The s reflectivity is thus an oscillatory function of $\Delta\phi$, and decays as the inverse square of the vacuum wavenumber ω/c:

$$R_s = \tfrac{1}{16}\{\gamma_1^2 + \gamma_2^2 - 2\gamma_1\gamma_2 \cos 2\Delta\phi\} + \ldots. \tag{118}$$

(The p reflectivity has the same form, with $\gamma \cos 2\theta$ replacing γ.) This oscillatory/power-law behaviour is characteristic of profiles with discontinuities in slope or higher order derivatives. Profiles with no such discontinuities, such the hyperbolic tangent, show exponential decrease with ω/c in the short wave limit. Approximations such as (116) and (117), and the Rayleigh approximation, fail at grazing incidence, and in the presence of turning points. When there is a single turning point ($q^2 \leqslant 0$ for $z \geqslant z_0$, say) there is total reflection. For the s wave

$$r_s = e^{i\delta_s}, \qquad \delta_s \simeq 2\int_0^{z_0} dz\, q(z) - \pi/2, \tag{119}$$

the phase decrement $\pi/2$ being universal for smooth profiles. In the case of two turning points ($q^2 < 0$ for $z_1 \leqslant z \leqslant z_2$), the classically forbidden region $q^2 < 0$ is tunnelled through by a portion of the wave. The transmission amplitude then varies approximately as $\exp(-2\Delta\Phi)$, where $\Delta\Phi$ is the increment in the imaginary part of the phase integral between the turning points:

$$\Delta\Phi = \int_{z_1}^{z_2} dz\, |q(z)|. \tag{120}$$

Reflection from anisotropic systems is considered in *Chapter 7*. In the simplest case where the system retains azimuthal symmetry about the normal to the interface, it is characterized by two dielectric functions $\varepsilon_x(z, \omega)$ and $\varepsilon_z(z, \omega)$, which give the response of the system to electric field components respectively parallel and perpendicular to the interface. The resolution of electromagnetic waves into s and p components remains valid in this case, with the equations to be satisfied modified from (7) and (20) to

$$\frac{d^2E}{dz^2} + \left(\varepsilon_x \frac{\omega^2}{c^2} - K^2\right) E = 0, \tag{121}$$

$$\frac{d}{dz}\left(\frac{1}{\varepsilon_x}\frac{dB}{dz}\right) + \left(\frac{\omega^2}{c^2} - \frac{K^2}{\varepsilon_z}\right) B = 0. \tag{122}$$

Equation (121) has the same form as (7), with ε_x replacing ε, but (122) differs from the isotropic case, since it contains both ε_x and ε_z. There are corresponding changes in the p wave reflection amplitude, and in r_p/r_s. The ellipsometric ratio, for example, still takes the form (96) in the long wave case, but the invariant $\mathscr{I}_1$ is now given by

$$\mathscr{I}_1 = \int_{-\infty}^{\infty} dz \left\{ \varepsilon_1 + \varepsilon_2 - \frac{\varepsilon_1 \varepsilon_2}{\varepsilon_z} - \varepsilon_x \right\} \tag{123}$$

(this applies to the case of an anisotropic thin film between isotropic media 1 and 2). For reflection at a sharp boundary between an isotropic medium 1 and an anisotropic medium characterized by ε_x and ε_z, there is transparency for the p polarization at

$$\theta_B = \arctan \left\{ \frac{\varepsilon_x(\varepsilon_x - \varepsilon_1)}{\varepsilon_1(\varepsilon_z - \varepsilon_1)} \right\}^{1/2}. \tag{124}$$

In the case of an anisotropic film, characterized by $\varepsilon_x(z)$ and $\varepsilon_z(z)$, on a uniform anisotropic substrate characterized by ε_{2x} and ε_{2z}, the form (96) is still valid, with ε_{2z} replacing ε_2 in the factor multiplying $\mathscr{I}_1$ and

$$\mathscr{I}_1 = \int_{-\infty}^{\infty} dz \left\{ \frac{\varepsilon_1^2 - \varepsilon_{2x}\varepsilon_{2z}}{\varepsilon_1 - \varepsilon_{2x}} - \frac{\varepsilon_1 - \varepsilon_{2z}}{\varepsilon_1 - \varepsilon_{2x}} \varepsilon_x - \frac{\varepsilon_1 \varepsilon_{2z}}{\varepsilon_z} \right\}. \tag{125}$$

The effect of absorption (the dissipation of electromagnetic energy within the medium) is discussed in *Chapter 8*. Absorption is included phenomenologically in the Maxwell equations by means of a complex dielectric function, $\varepsilon = \varepsilon_r + i\varepsilon_i$. This simple change has far-reaching consequences for reflection properties. In the case of reflection at the sharp surface of an absorbing medium (a metal, for example), the Fresnel equations (13) and (31) retain their form, but now $q_2 = q_r + iq_i$ and $Q_2 = Q_r + iQ_i = (q_r + iq_i)/(\varepsilon_r + i\varepsilon_i)$, where

$$\left(\frac{cq_r}{\omega}\right)^2 = \tfrac{1}{2}\{\varepsilon_r - \varepsilon_1 \sin^2\theta_1 + [(\varepsilon_r - \varepsilon_1 \sin^2\theta_1)^2 + \varepsilon_i^2]^{1/2}\}. \tag{126}$$

$$\frac{cq_i}{\omega} = \frac{\varepsilon_i/2}{cq_r/\omega}. \tag{127}$$

The ellipsometric ratio r_p/r_s no longer has the real axis as its trajectory, but lies within the upper half of the unit circle:

$$\frac{r_p}{r_s} = \frac{q_1^2(q_r^2 + q_i^2) - K^4 + 2iq_1q_iK^2}{(q_1q_r + K^2)^2 + q_1^2q_i^2}. \tag{128}$$

Some of the general results derived in Chapter 2 still hold, notably the fact that $r_s \to -1$ and $r_p \to 1$ at grazing incidence, and the implication that there is an odd number of angles of incidence at which $\mathrm{Re}(r_p/r_s) = 0$. The reciprocity relation (88) also holds, and thus the transmittance of an absorbing system is independent of the direction of propagation of the radiation.

Zero reflection is not possible off an absorbing medium with a sharp boundary, for either polarization. If however a dielectric layer is deposited on the absorber, zero reflectance is possible for both polarizations (at different angles of incidence); this interference–absorption effect thus produces reflection polarizers (Section 8-3).

A thin absorbing film on a transparent substrate always decreases the transmittance, but reflectance can be either increased or decreased, depending on the polarization and whether $\varepsilon_1 < \varepsilon_2$ or $\varepsilon_1 > \varepsilon_2$. For example, the s reflectivity to first order in the film thickness is given by

$$R_s = \left(\frac{q_1 - q_2}{q_1 + q_2}\right)^2 - \frac{4q_1(q_1 - q_2)}{(q_1 + q_2)^3}\frac{\omega^2}{c^2}\int_{-\infty}^{\infty} \mathrm{d}z\, \varepsilon_i(z) + \ldots. \tag{129}$$

The form (96) for the ellipsometric ratio is still valid, with $\mathscr{I}_1$ complex:

$$\mathscr{I}_1 = \int_{-\infty}^{\infty} \mathrm{d}z \left(\varepsilon_1 + \varepsilon_2 - \frac{\varepsilon_1\varepsilon_2\varepsilon_r}{\varepsilon_r^2 + \varepsilon_i^2} - \varepsilon_r\right) + i\int_{-\infty}^{\infty} \mathrm{d}z \left[\frac{\varepsilon_1\varepsilon_2}{\varepsilon_r^2 + \varepsilon_i^2}\right]\varepsilon_i. \tag{130}$$

An important and dramatic effect due to absorption is that of *attenuated total reflection*, discussed in Section 8-6. An absorbing layer (typically a metal film) deposited between two dielectrics can turn a total reflection configuration into one whose p reflectance is small at resonance, and can be zero for proper choice of thickness of metal film and angle of incidence. This phenomenon is an interference-attenuation effect, associated with the resonant excitation of electromagnetic surface waves at a metal–dielectric interface.

Chapter 9 deals with the inversion of reflectance and ellipsometric data to obtain the parameters of the reflector. For example, if the real and imaginary parts ϱ_r and ϱ_i of r_p/r_s are measured at angle of incidence θ_1, and the interface is assumed to be sharply defined, the real and imaginary parts of ε may be found from

$$\frac{\varepsilon_r + i\varepsilon_i}{\varepsilon_1} = \sin^2\theta_1 + \sin^2\theta_1 \tan^2\theta_1 \frac{(1 - \varrho_r^2)^2 - 4\varrho_i^2 + 4i(1 - \varrho_r^2)\varrho_i}{[(1 - \varrho_r)^2 + \varrho_i^2]^2}. \tag{131}$$

If a model reflection amplitude is constructed as a function of wavevector component in medium 1, and analytically continued to negative q_1 via $r(-q_1) = r^*(q_1)$, an explicit inversion is possible (Section 9-3) to obtain the dielectric function profile which would give this reflection amplitude in the Rayleigh approximation. In the s wave case the result is

$$\frac{\varepsilon(x)}{\varepsilon_1} \simeq \sin^2\theta_1 + \cos^2\theta_1 \exp\left(-4\int_{-\infty}^{2x} \mathrm{d}y\, F_s(y)\right), \tag{132}$$

where $x = q_1^{-1}\int^z \mathrm{d}\zeta\, q(\zeta)$ and F_s is the Fourier transform of r_s:

$$F_s(y) = \frac{1}{2\pi}\int_{-\infty}^{\infty} \mathrm{d}q_1\, \mathrm{e}^{-iq_1 y}\, r_s(q_1). \tag{133}$$

The reflection of pulses, and of finite beams, is considered in *Chapter 10*. We find that nearly monochromatic pulses reflect (in the first approximation) without change of shape, with a time delay Δt determined by the frequency variation of the phase of the reflection amplitude:

$$\Delta t = \frac{\mathrm{d}\delta}{\mathrm{d}\omega}. \tag{134}$$

(The derivative is to be evaluated at the dominant angular frequency of the pulse.) For example: total reflection at normal incidence has $r_n = e^{i\delta_n}$, and the short wave limiting form is found from (119) to be

$$\delta_n = 2\frac{\omega}{c}\int_0^{z_0} dz\, n(z, \omega) - \pi/2, \tag{135}$$

where $n(z, \omega)$ is the refractive index, and z_0 is the turning point determined by $n(z_0, \omega) = 0$. (This formula applies to reflection from the ionosphere, for example, in which case $n^2 = \varepsilon \simeq 1 - \omega_p^2/\omega^2$, where ω_p is the plasma angular frequency, proportional to the square root of ionospheric electron density.) From (134) and (135) we find that the time delay is the same as for pulse travel to the turning point and back at the ground velocity $u(z, \omega) = d\omega/dk$, where $k = n\omega/c$:

$$\Delta t = 2\int_0^{z} \frac{dz}{u} = \frac{2}{c}\int_0^{z_0} dz \left[n + \omega \frac{\partial n}{\partial \omega} \right]. \tag{136}$$

Pulses are built up from waves of differing frequencies. Bounded beams can be regarded as superpositions of plane waves of differing directions of propagation. Just as the reflection of pulses is determined by the frequency dependence of the reflection amplitude, the reflection of beams depends on the angular dependence of the reflection amplitude. There is a lateral shift on reflection of a beam of radiation,

$$\Delta x = -\frac{d\delta}{dK}, \tag{137}$$

where K is the lateral component of the wavevector ($K = k_x$), and the derivative is to be evaluated at the dominant value of K for the incident beam. A particularly interesting case is total reflection at a sharp boundary. For $\theta_1 > \theta_c$ the phase of the s wave reflection amplitude is

$$\delta_s = -2\arctan\frac{|q_2|}{q_1}, \tag{138}$$

and (137) leads to the beam shift

$$\Delta x_s = \frac{2K}{q_1|q_2|} = \frac{\lambda_1}{\pi}\frac{\tan\theta_1}{(\sin^2\theta_1 - \sin^2\theta_c)^{1/2}}, \tag{139}$$

where λ_1 is the wavelength $2\pi/k_1 = 2\pi c/n_1\omega$ in the first medium. This beam shift is divergent at the critical angle (where $q_2 = 0$), and in fact the formula (139) fails there, since (137) is derived on the assumption of a slow variation of the phase shift with angle. In practice (139) works well to close proximity of the critical angle, as discussed in Section 10-3. In Section 10-2 we show that the $|q_2|$ singularity in the phase shift at $\theta_1 = \theta_c^+$ is universal for nonabsorbing profiles.

Chapter 11 deals with reflection from rough surfaces. A planar stratified surface, no matter how diffuse, gives specular reflection of an incident plane wave, but rough surfaces scatter as well as reflect. The Rayleigh criterion for negligible roughness is (Section 11-1)

$$qh \ll 1, \tag{140}$$

where q (as always) is the normal component of the wavevector, and h is a measure of the variation in the height of the surface. When (140) is satisfied the surface will reflect specularly. According to the Rayleigh criterion, for given roughness and angle of incidence long waves may be reflected specularly and short waves diffusely, or for given roughness and wavelength there may be diffuse scattering near normal incidence and specular reflection near grazing incidence. Chapter 11 treats the reflection from corrugated surfaces (diffraction gratings), from liquid metal and liquid dielectric surfaces (scattering by thermally excited surface waves), in both cases using the methods of Rayleigh, and gives an outline of the application of the Helmholtz theorem to the scattering by rough surfaces (the Kirchhoff or surface integral method).

Matrix methods are developed in *Chapter 12.* Any stratified medium may be approximated by a set of uniform layers. The matrix methods connect, via a two-by-two matrix, the coefficients of either the two independent solutions, or the field amplitude and its derivative, at the entry and exit points of a layer. In the latter case these matrix relations are, for a uniform layer between z_n and z_{n+1},

$$\begin{pmatrix} E_{n+1} \\ D_{n+1} \end{pmatrix} = \begin{pmatrix} \cos \delta_n & \dfrac{\sin \delta_n}{q_n} \\ -q_n \sin \delta_n & \cos \delta_n \end{pmatrix} \begin{pmatrix} E_n \\ D_n \end{pmatrix} \tag{141}$$

$$\begin{pmatrix} B_{n+1} \\ C_{n+1} \end{pmatrix} = \begin{pmatrix} \cos \delta_n & \dfrac{\sin \delta_n}{Q_n} \\ -Q_n \sin \delta_n & \cos \delta_n \end{pmatrix} \begin{pmatrix} B_n \\ C_n \end{pmatrix} \tag{142}$$

for the p wave (here $C = \mathrm{d}B/\varepsilon \mathrm{d}z$, $Q_n = q_n/\varepsilon_n$). For a profile approximated by N uniform layers, the reflection and transmission properties are determined by the profile matrix, which is a product of N matrices such as those in (141) or (142). If the elements of the profile matrix for the s polarization are s_{ij}, for example, the reflection and transmission amplitudes for an interface between media a and b are

$$r_s = \mathrm{e}^{2i\alpha} \frac{q_a q_b s_{12} + s_{21} + iq_a s_{22} - iq_b s_{11}}{q_a q_b s_{12} - s_{21} + iq_a s_{22} + iq_b s_{11}}, \tag{143}$$

$$t_s = \mathrm{e}^{i(\alpha-\beta)} \frac{2iq_a}{q_a q_b s_{12} - s_{21} + iq_a s_{22} + iq_b s_{11}}. \tag{144}$$

(Here $\alpha = q_a z_1$ and $\beta = q_b z_{N+1}$, z_1 and z_{N+1} being the boundaries of the inhomogeneity.) In the absence of absorption the matrix elements are real; the matrix formulation, and the results (143) and (144), are valid in the presence of absorption also. The matrices in (141) and (142) are unimodular (have unit determinant); this fact simplifies the treatment of periodically stratified media (Section 12-3), which in turn has important application to the multilayer dielectric mirrors (Section 12-4).

Numerical methods are discussed in *Chapter 13.* These are based on matrices derived from those in (141) and (142): an arbitrary layer extending from a to

b can be represented, to second order in the layer thickness, by the s matrix

$$\begin{pmatrix} 1 - \int_a^b \mathrm{d}z\, q^2(z)(b-z) & b-a \\ -\int_a^b \mathrm{d}z\, q^2(z) & 1 - \int_a^b \mathrm{d}z\, q^2(z)(z-a) \end{pmatrix} \tag{145}$$

(This result, and a similar one for the p matrix, are derived in Section 12-5.) A given interface is approximated by a set of layers within which the dielectric function $\varepsilon(z)$, and thus also $q^2(z)$, vary linearly with z. The matrix methods can be applied without modification to total reflection and tunnelling; reflection and transmission through absorbing layers requires computation with complex matrix elements, the formalism being otherwise unaltered. Wavefunctions within the stratification may be obtained as a by-product of the profile matrix calculation.

Finally, the *Appendix* translates some of the main results of the book from electromagnetic to particle wave notation. Readers interested in the reflection and transmission of particles satisfying the Schrödinger equation may find it useful to read the Appendix next, since simpler s wave results, which are in one-to-one correspondence to the particle results, are interspersed in the main text with the interesting but more difficult matter relating to the electromagnetic p wave.

References

References quoted in the text

A. Fresnel (1823) "Mémoire sur la loi des modifications que la réflexion imprime a la lumière polarisée", Mémoires de l'Académie **11**, 393–433.

P. G. Bergmann (1946) "The wave equation in a medium with a variable index of refraction", J. Acoust. Soc. Amer. **17**, 329–333.

G. Green (1838) "On the reflexion and refraction of sound", Trans. Camb. Phil. Soc. **6**, 403–412.

E. Fermi (1950) "Nuclear physics" (Notes compiled by J. Orear, A. H. Rosenfeld, and R. A. Schluter), University of Chicago Press, pp. 201–202.

J. D. Jackson (1962) "Classical electrodynamics", John Wiley and Sons.

H. A. Lorentz (1909/1952) "The theory of electrons", Dover Publications, p. 145.

C. G. Darwin (1924) "The optical constants of matter", Trans. Camb. Phil. Soc. **23**, 137–167.

D. R. Hartree (1928) "The propagation of electromagnetic waves in a stratified medium", Proc. Camb. Phil. Soc. **25**, 97–120.

P. J. Castle and J. Lekner (1980) "Variation of the local field through the liquid-vapour interface", Physica **101A**, 99–11.

J. Lekner (1983) "Anisotropy of the dielectric function within a liquid-vapour interface", Molecular Physics **49**, 1385–1400.

F. P. Buff, R. A. Lovett, and F. H. Stillinger (1965) "Interfacial density profile for fluids in the critical region", Phys. Rev. Lett. **15**, 621–623.

J. S. Huang and W. W. Webb (1969) "Diffuse interface in a critical fluid mixture", J. Chem. Phys. **50**, 3677–3693.

Books and review articles dealing with aspects of reflection in stratified media

P. Drude (1902/1959) "Theory of optics", Dover publications.

J. W. S. Rayleigh (1912) "On the propagation of waves through a stratified medium, with special reference to the question of reflection", Proc. Roy. Soc. **86**, 207–266.

F. Abelès (1950) "Recherches sur la propagation des ondes électromagnétiques sinusoidales dans les milieux stratifiés. Application aux couches minces". Annales de Physique **5**, 596–640.
A. Vašíček (1960) "Optics of thin films", North-Holland.
J. Heading (1962) "An introduction to phase-integral methods", Methuen.
J. R. Wait (1962) "Electromagnetic waves in stratified media", Pergamon.
V. L. Ginzburg (1964) "The propagation of electromagnetic waves in plasmas", Pergamon.
M. Born and E. Wolf (1965) "Principles of optics" (3rd ed.), Pergamon.
N. Fröman and P. O. Fröman (1965) "JWKB approximation", North-Holland.
L. D. Landau and E. M. Lifshitz (1965) "Quantum mechanics", Pergamon.
R. Jacobsson (1966) "Light reflection from films of continuously varying refractive index", Progress in Optics **5**, 247–286.
I. Tolstoy (1973) "Wave propagation", McGraw-Hill.
J. Heading (1975) "Ordinary differential equations, theory and practice", Elek Science, London.
Z. Knittl (1976) "Optics of thin films", Wiley.
H. Levine (1978) "Unidirectional wave motions" North-Holland.
L. M. Brekhovskikh (1980) "Waves in layered meida", (2nd ed.) Academic Press.
B. L. N. Kennett (1983) "Seismic wave propagation in stratified media", Cambridge.
K. G. Budden (1985) "The propagation of radio waves", Cambridge.

A detailed discussion of all aspects of ellipsometry can be found in

R. M. A. Azzam and N. M. Bashara (1977) "Ellipsometry and polarized light".
See also R. H. Muller (1972) "Principles of ellipsometry", Advances in Electrochemistry and Electrochemical Engineering, vol. 9, Interscience.

Sections 1-1 and 1-3 are based in part on

J. Lekner (1982) "Reflection of long waves by interfaces", Physica **112A**, 544–556.

The polarization of light by reflection from glass at 56° incidence, and from water at 53°, was discovered by Malus in 1808. In 1814 Brewster found, as the result of extensive experiments, that "the index of refraction is the tangent of the angle of polarization". See Chapter 24 of

D. Brewster (1853) "A treatise on optics", Longman, Brown and Green.

The explanation of near transparency of a general interface to the p wave given in Section 1-2, and of the similar effect in the reflection of acoustic waves (Section 1-4), is based on

J. Lekner (1982) "Second-order ellipsometric coefficients", Physica **113A**, 506–520.

The result (Section 1-5), for the reflection of electromagnetic waves in terms of the superposition of fields radiated by oscillating dipoles within a material slab, is an adaptation of a similar calculation for the transmitted wave, given by

M. Sargent, M. O. Scully, and W. E. Lamb (1974) "Laser physics", Addison-Wesley, Appendix A.

For a recent treatment of the molecular basis of reflection see

F. Hynne and R. K. Bullough (1984) "The scattering of light. I. The optical response of a finite molecular fluid", Phil. Trans. Roy. Soc. Lond. **A312**, 251–293.

2

Exact results

We shall first derive general results, that is those valid for an arbitrary interfacial profile (Sections 2-1 to 2-3); some of these results will be restricted to non-absorbing interfaces and substrates. The remainder of the chapter is devoted to important special profiles for which the reflection amplitude may be obtained exactly. Both the general and the specific exact results are useful in testing approximate theories and numerical calculations.

The title of this chapter is not intended to imply that all other results in the book are approximate: for example in Chapter 3 we shall be deriving exact integral equations, and results which are exact (and general) to second order in the interface thickness.

2-1 Comparison identities, and conservation and reciprocity laws

In sections 1-1 and 1-2 we saw that, for planar stratified media, the s and p wave equations may be put in the form

$$\frac{d^2\psi}{dz^2} + q^2\psi = 0, \tag{1}$$

with the reflection and transmission amplitudes defined in terms of the asymptotic forms of ψ at large negative and positive z:

$$e^{iq_1 z} + r\,e^{-iq_1 z} \leftarrow \psi \rightarrow t\,e^{iq_2 z}. \tag{2}$$

For the s wave $\psi = E$ and $q_E^2 = \varepsilon\omega^2/c^2 - K^2$. For the p wave, $\psi = b = (\varepsilon_1/\varepsilon)^{1/2}B$, and q^2 is given by (1.22):

$$q_b^2 = q_E^2 + \frac{1}{2\varepsilon}\frac{d^2\varepsilon}{dz^2} - \frac{3}{4}\left(\frac{1}{\varepsilon}\frac{d\varepsilon}{dz}\right)^2. \tag{3}$$

For the s wave, the r and t of (2) are r_s and t_s as defined in (1.10); for the p wave, from (1.25), $r = -r_p$ and $t = t_p$.

Let $\tilde{\psi}$ be the solution for another dielectric function profile $\tilde{\varepsilon}(z)$, which has the same limiting values ε_1 and ε_2 as $\varepsilon(z)$:

$$\frac{d^2\tilde{\psi}}{dz^2} + \tilde{q}^2\tilde{\psi} = 0, \qquad e^{iq_1z} + \tilde{r}\,e^{-iq_1z} \leftarrow \tilde{\psi} \rightarrow \tilde{t}\,e^{iq_2z}. \tag{4}$$

We can obtain *comparison identities* relating r and $\tilde{r}$, and t and $\tilde{t}$, from the differential equations and their boundary conditions. The relation between r and $\tilde{r}$ is obtained by multiplying the wave equation for $\tilde{\psi}$ by ψ, the wave equation for ψ by $\tilde{\psi}$, and subtracting. We get

$$\frac{d}{dz}\left(\psi\frac{d\tilde{\psi}}{dz} - \tilde{\psi}\frac{d\psi}{dz}\right) = (q^2 - \tilde{q}^2)\psi\tilde{\psi}. \tag{5}$$

We now integrate from a point z_1 deep inside medium 1, to a point z_2 deep inside medium 2 (i.e., z_1 and z_2 are such that ψ and $\tilde{\psi}$ have attained their asymptotic forms (2) and (4)). The integral of the left-hand side of (5) gives $2iq_1(\tilde{r} - r)$, all dependence on z_1 and z_2 cancelling out. Thus

$$r = \tilde{r} - \frac{1}{2iq_1}\int_{-\infty}^{\infty} dz\,(q^2 - \tilde{q}^2)\psi\tilde{\psi}, \tag{6}$$

where we have replaced z_1 by $-\infty$ and z_2 by $+\infty$.

Another comparison identity may be obtained, in the same way, for ψ and $\tilde{\psi}^*$ (the complex conjugate of $\tilde{\psi}$):

$$2iq_1(1 - r\tilde{r}^*) - 2iq_2t\tilde{t}^* = \int_{-\infty}^{\infty} dz\,(q^2 - \tilde{q}^2)\psi\tilde{\psi}^*. \tag{7}$$

On setting $\tilde{\varepsilon} = \varepsilon$ (and thus $\tilde{q} = q$), we obtain

$$q_1(1 - |r|^2) = q_2|t|^2. \tag{8}$$

The energy density of a plane electromagnetic wave is proportional to $\varepsilon|E|^2$, and the speed to $1/\sqrt{\varepsilon}$; thus the intensity is proportional to $\sqrt{\varepsilon}|E|^2$. The amount of energy in the primary wave which is incident on a unit area of the interface in unit time is proportional to $\sqrt{\varepsilon_1}\cos\theta_1$. The energy carried away by the reflected wave is proportional to $\sqrt{\varepsilon_1}\cos\theta_1|r|^2$, and that carried by the transmitted wave is proportional to $\sqrt{\varepsilon_2}\cos\theta_2|t|^2$. Thus (8) expresses the law of conservation of energy. The geometry leading to the factors $\cos\theta_1$ and $\cos\theta_2$ is shown in Figure 2-1.

In quantum mechanics (8) expresses the conservation of the probability density current; the conservation of energy is guaranteed, since (1.40) [which leads to (1) on the substitution $\Psi = e^{iKx}\psi(z)$] is the Schrödinger equation for an energy eigenstate. The probability current density is

$$\mathbf{J} = \frac{\hbar}{2im}(\Psi^*\nabla\Psi - \Psi\nabla\Psi^*) = \frac{\hbar}{m}\,\mathrm{Im}\,(\Psi^*\nabla\Psi), \tag{9}$$

and has (for our geometry) zero y component, and

$$J_x = \frac{\hbar K}{m}. \tag{10}$$

As ψ takes the asymptotic forms specified in (2), the z component of $\mathbf{J}$ takes the limiting values

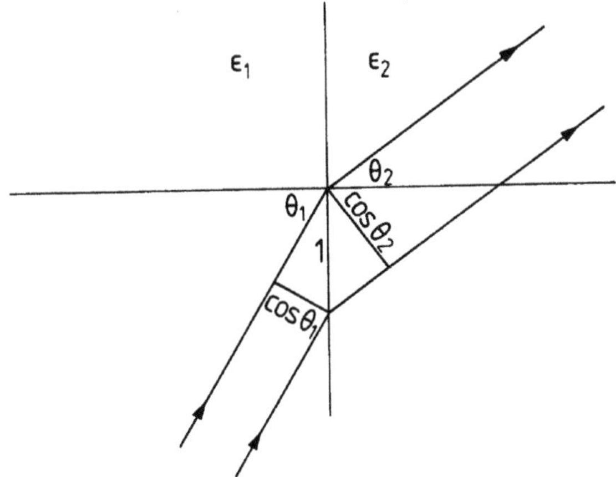

Figure 2-1. A strip of unit width on the boundary is illuminated by a beam of width $\cos\theta_1$; the transmitted wave is a beam of width $\cos\theta_2$ (the reflected beam is not shown).

$$\frac{\hbar q_1}{m}(1 - |r|^2) \leftarrow J_z \rightarrow \frac{\hbar q_2}{m}|t|^2. \tag{11}$$

Thus the form of the wavefunction guarantees conservation of probability density current tangential to the (smooth) interface, while (8) and (11) demonstrate the conservation of the normal component.

The relation (8) is valid only for real q: the reality of $\tilde{q}$ was assumed in obtaining (7). The extension of (8) to include absorption will be discussed in Chapter 8; here we comment only on the case of total internal reflection, when $q_2 = i|q_2|$. Then $|r|^2 = 1$ (as we shall see in the next section), so that the left-hand side of (8) is zero. But $|t|^2$ is not zero in general: for example $|t_s|^2$ takes the value $4q_1^2/(q_1^2 + |q_2|^2)$ for the step profile. When total internal reflection occurs, there is no energy (or particle flux) propagated into the second medium, and t becomes the coefficient of an exponentially decaying wave: it is no longer a transmission amplitude in the true sense.

Further comparison identities may be obtained by comparing waves incident from opposite directions. We define ψ_{21} as a solution of (1) with asymptotic forms

$$t_{21}\,\mathrm{e}^{-iq_1z} \leftarrow \psi_{21} \rightarrow \mathrm{e}^{-iq_2z} + r_{21}\,\mathrm{e}^{iq_2z}. \tag{12}$$

The previously used ψ and $\tilde{\psi}$, with asymptotics given by (2) and (4), will now be referred to as ψ_{21} and $\tilde{\psi}_{21}$ to emphasize that these represent motions originating in medium 1, and partially transmitted into medium 2. On comparing $\tilde{\psi}_{12}$ with ψ_{21}, we find

$$2i(q_2\tilde{t}_{12} - q_1t_{21}) = \int_{-\infty}^{\infty} \mathrm{d}z\,(q^2 - \tilde{q}^2)\psi_{21}\tilde{\psi}_{12}. \tag{13}$$

When $\tilde{q} = q$, this gives (for real q_1, q_2, but possibly complex q within the interface)

$$q_2t_{12} = q_1t_{21}, \tag{14}$$

so that (13) can be written in a form similar to (6):

$$t_{12} = \tilde{t}_{12} - \frac{1}{2iq_2}\int_{-\infty}^{\infty} \mathrm{d}z\,(q^2 - \tilde{q}^2)\psi_{21}\tilde{\psi}_{12}. \tag{15}$$

Comparing ψ_{12} with $\tilde{\psi}_{21}$ gives a relation like (15), with $\psi_{12}\tilde{\psi}_{21}$ in the integrand. Finally, comparing ψ_{21} with ψ_{12}^* gives

$$-2i(q_2 r_{21}\tilde{t}_{12}^* + q_1 t_{21}\tilde{r}_{12}^*) = \int_{-\infty}^{\infty} \mathrm{d}z\,(q^2 - \tilde{q}^2)\psi_{21}\tilde{\psi}_{12}^*. \tag{16}$$

On setting $\tilde{q} = q$, we find

$$q_2 r_{21} t_{12}^* + q_1 r_{12}^* t_{21} = 0, \tag{17}$$

which together with (14) shows that

$$r_{21} = -\frac{t_{12}}{t_{12}^*} r_{12}^*. \tag{18}$$

Thus the $1 \rightarrow 2$ and $1 \leftarrow 2$ reflection amplitudes have equal absolute value. The relations (17) and (18) are valid only in the absence of absorption or total internal reflection.

The result $|r_{12}|^2 = |r_{21}|^2$, implied by (18), is remarkable in juxtaposition with our intuitive picture of particles going up or down a potential gradient (as in Figure 1-8). In the semiclassical limit, one might expect stronger reflection for particle waves moving uphill than for those going downhill. In fact the reflectivity is exactly the same in the two cases, unless there is total internal reflection.

Some of the above relations were given (in a restricted form) by Stokes (1849), using the idea of reversing the wavemotions, as illustrated in Figure 2-2.

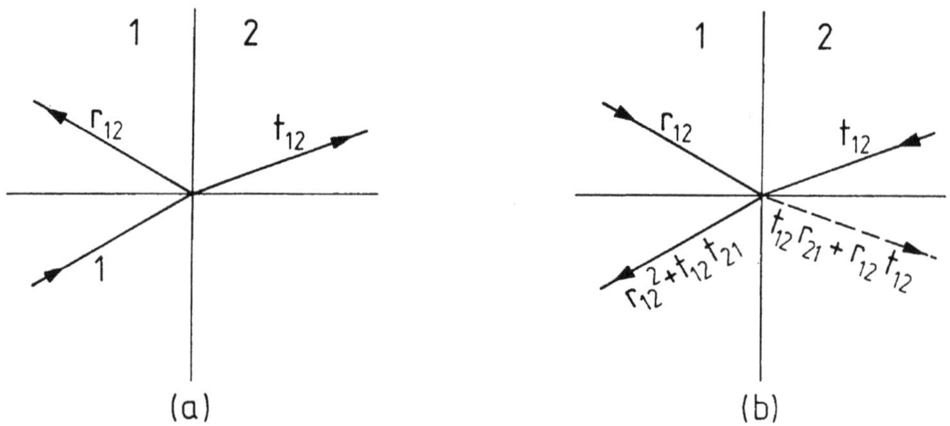

Figure 2-2. Stokes' "principle of reversion". Reversing the wavemotions, according to Stokes, will lead to a retracing of the previous motion.

In part (a) of Figure 2-2, an incident wave of unit amplitude is split into a reflected and a transmitted wave, with amplitudes r_{12} and t_{12}. Stokes argued that on reversing the wavemotions, the situation in part (b) results. On comparing the resultant amplitudes with those of part (a), one gets $r_{12}^2 + t_{12}t_{21} = 1$ and $r_{21} = -r_{12}$. The first of these, together with (14), gives $q_1(1 - r_{12}^2) = q_2 t_{12}^2$, which agrees with (8) when r_{12} and t_{12} are real. The second agrees with (18), under the same condition. Thus Stokes' results are valid when the phases associated with r and t are restricted to 0 and π. A slight modification of Stokes' argument produces the general results: we replace his "reversion" by time reversal plus complex conjugation. The effect, shown in Fig 2-3, is to give the correct analytical forms of the reversed beams.

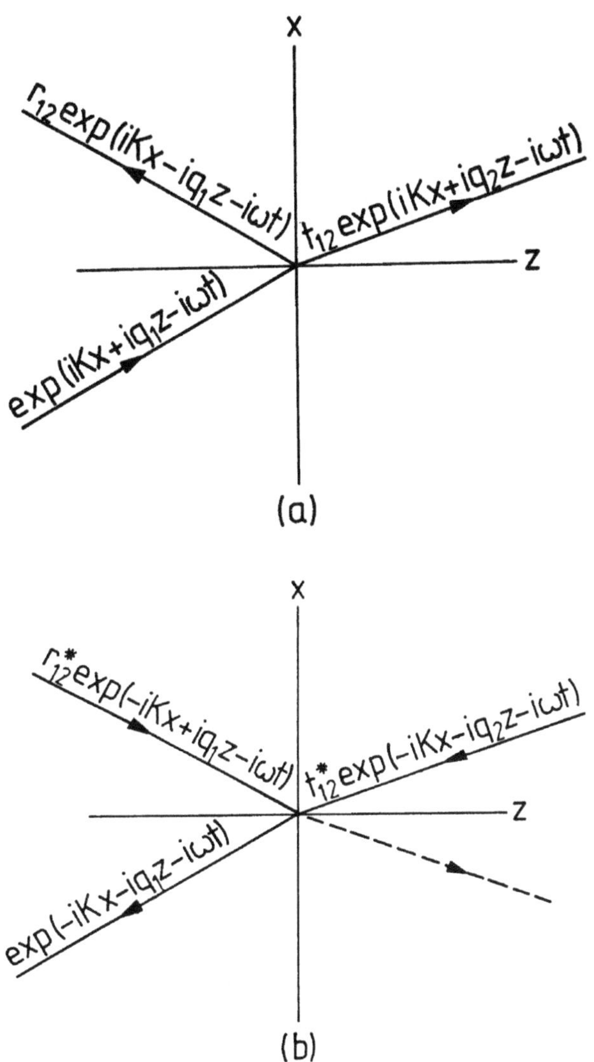

Figure 2-3 Time reversal plus complex conjugation applied to the waveform in (a) gives the waveforms in (b).

The arguments used above give

$$r_{12}^* r_{12} + t_{12}^* t_{21} = 1, \tag{19}$$

and

$$r_{12}^* t_{12} + t_{12}^* r_{21} = 0. \tag{20}$$

We now have agreement: (19) with (14) reproduces (8), and (20) is the same as (18).

Other comparison identities can be obtained: for the p wave one analogous to (6) will be used in Section 3-4, and another analogous to (15) will be derived and used in Section 8-5.

2-2 General expressions for r_s and r_p.

Consider an interface for which $\varepsilon = \varepsilon_1$ for $z \leqslant z_1$, and $\varepsilon = \varepsilon_2$ for $z \geqslant z_2$; the thickness $\Delta z = z_2 - z_1$ of the non-uniform region can be large. By a limiting process, the results (to be derived below) for such finite-ranged interfaces can be extended to continuously varying interfaces. For example: for the hyperbolic tangent dielectric function

$$\varepsilon(z) = \tfrac{1}{2}(\varepsilon_1 + \varepsilon_2) - \tfrac{1}{2}(\varepsilon_1 - \varepsilon_2) \tanh [(z - z_0)/2a], \tag{21}$$

one can take $z_1 = z_0 - \Delta z/2$, $z_2 = z_0 + \Delta z/2$, and by making $\Delta z/a$ large enough, any desired accuracy can be achieved.

For the s wave, $\mathbf{E} = (0, E(z)\, \mathrm{e}^{i(Kx-\omega t)}, 0)$, where $E(z)$ satisfies

$$\frac{\mathrm{d}^2 E}{\mathrm{d}z^2} + q^2 E = 0. \tag{22}$$

This second-order linear differential equation has two linearly independent solutions (for arbitrary form of $\varepsilon(z)$). We call these $F(z)$ and $G(z)$ in the region $z_1 \leqslant z \leqslant z_2$. Then

$$E(z) = \begin{cases} \mathrm{e}^{iq_1 z} + r_s\, \mathrm{e}^{-iq_1 z} & (z < z_1) \\ \alpha F(z) + \beta G(z) & (z_1 \leqslant z \leqslant z_2) \\ t_s\, \mathrm{e}^{iq_2 z} & (z > z_2). \end{cases} \tag{23}$$

When $\varepsilon(z)$ has no delta-function singularities (or worse), as we assume here, E and $\mathrm{d}E/\mathrm{d}z$ are continuous. Continuity of E and $\mathrm{d}E/\mathrm{d}z$ at z_1 and z_2 gives us four linear equations in the four unknowns r_s, t_s, α, β; namely

$$\begin{aligned} \mathrm{e}^{iq_1 z_1} + r_s\, \mathrm{e}^{-iq_1 z_1} &= \alpha F_1 + \beta G_1. \\ iq_1(\mathrm{e}^{iq_1 z_1} - r_s\, \mathrm{e}^{-iq_1 z_1}) &= \alpha F_1' + \beta G_1'. \\ \alpha F_2 + \beta G_2 &= t_s\, \mathrm{e}^{iq_2 z_2}, \\ \alpha F_2' + \beta G_2' &= iq_2 t_s\, \mathrm{e}^{iq_2 z_2}. \end{aligned} \tag{24}$$

We have written F_1 for $F(z_1)$, F_1' for $\mathrm{d}F/\mathrm{d}z$ evaluated at z_1, etc. Solving for r_s we find

$$r_s = \mathrm{e}^{2iq_1 z_1} \times \frac{q_1 q_2(F_1 G_2 - G_1 F_2) + iq_1(F_1 G_2' - G_1 F_2') + iq_2(F_1' G_2 - G_1' F_2) - (F_1' G_2' - G_1' F_2')}{q_1 q_2(F_1 G_2 - G_1 F_2) + iq_1(F_1 G_2' - G_1 F_2') - iq_2(F_1' G_2 - G_1' F_2) + (F_1' G_2' - G_1' F_2')}. \tag{25}$$

The corresponding expression for the transmission amplitude is

$$t_s = \times \frac{e^{i(q_1 z_1 - q_2 z_2)} 2iq_1(F_2G_2' - G_2F_2')}{q_1q_2(F_1G_2 - G_1F_2) + iq_1(F_1G_2' - G_1F_2') - iq_2(F_1'G_2 - G_1'F_2) + (F_1'G_2' - G_1'F_2')}. \tag{26}$$

The bilinear form in the numerator of t_s is the Wronskian, W, of the solutions F and G of (22), and is a constant:

$$W = FG' - GF', \qquad W' = FG'' - GF'', \tag{27}$$

and $F'' = -q^2F$, $G'' = -q^2G$ from (22); thus W' is zero and W is independent of z.

Some general properties of r_s and t_s follow directly from the forms of (25) and (26), for arbitrary nonsingular profiles.

We first note that in the absence of absorption (ε real), F and G may be chosen to be real. This is because they are solutions of a linear differential equation with real coefficients: if F is a complex solution of (22), so is F^* and therefore so is $F^* + F$, which is real. Henceforth, F and G are taken to be real functions, unless otherwise indicated.

Total internal reflection occurs when $q_2^2 \leqslant 0$ (i.e., when $\sin^2\theta_1 \geqslant \varepsilon_2/\varepsilon_1$, as discussed in Section 1-3); when $q_2 = i|q_2|$, the wave in medium 2 decays exponentially as $e^{-|q_2|z}$. Thus no *propagation* of the wave into medium 2 takes place, and we expect the reflection amplitude to lie on the unit circle. This is true, as may be verified directly from (25), which takes the form $e^{2iq_1z_1}(-f + ig)/(f + ig)$, with f and g real, when $q_2 = i|q_2|$.

The first equality in (27) may be regarded as the first order linear differential equation for G:

$$G' - \frac{F'}{F}G = \frac{W}{F}. \tag{28}$$

This has the solution

$$G(z) = WF(z)\int^z d\zeta/F^2(\zeta). \tag{29}$$

For non-zero W, this is a second solution, linearly independent of F, of the differential equation (22). Let $\int_1^2$ denote $\int_{z_1}^{z_2} dz/F^2(z)$, and set $W = 1$. Then

$$\begin{aligned} F_1G_2 - G_1F_2 &= F_1F_2\int_1^2, \\ F_1G_2' - G_1F_1' &= F_1/F_2 + F_1F_2'\int_1^2, \\ F_1'G_2 - G_1'F_2 &= -F_2/F_1 + F_1'F_2\int_1^2, \\ F_1'G_2' - G_1'F_1' &= \frac{F_1'}{F_2} - \frac{F_2'}{F_1} + F_1'F_2'\int_1^2. \end{aligned} \tag{30}$$

As Δz tends to zero, the first and last bilinear forms tend to zero, and the second and third tend to $+1$ and -1, respectively. Thus the reflection and transmission amplitudes for an arbitrary non-singular profile of extent Δz approach the step profile values (1.15) as $\Delta z \to 0$.

For passive media, we must have $|r|^2 \leqslant 1$. After some algebra, the physical requirement that $|r_s|^2 \leqslant 1$ reduces to $W(z_1)W(z_2) \geqslant 0$, where W is the Wronskian $FG' - GF'$. We have seen that $W(z_1) = W(z_2)$, so this condition is satisfied identically.

The conservation law $q_1(1 - |r|^2) = q_2|t|^2$ proved in the last section may be verified directly from (25) and (26). As in the result $|r_s|^2 \leqslant 1$, the proof involves the identity

$$(F_1G_2 - G_1F_2)(F_1'G_2' - G_1'F_2') - (F_1G_2' - G_1F_2')(F_1'G_2 - G_1'F_2) = W^2. \tag{31}$$

We now turn to the reciprocity laws (14) and (18), which relate t_{21} to t_{12} and r_{21} to t_{12} and r_{12} (the suffix s will be dropped for the moment). The results (25) and (26) may be rewritten as

$$r_{12} = e^{2iq_1z_1}\frac{N}{D}, \tag{32}$$

$$t_{12} = e^{i(q_1z_1-q_2z_2)}\frac{2iq_1W}{D}, \tag{33}$$

where N is the numerator of (25), and D is the denominator common to (25) and (26). The corresponding results for r_{21} and t_{21} follow from the continuity of E and dE/dz at z_1 and z_2, where E now has the forms outside the interval $[z_1, z_2]$ as given by (12). We find

$$r_{21} = e^{-2iq_2z_2}\frac{N^*}{D}, \tag{34}$$

$$t_{21} = e^{i(q_1z_1-q_2z_2)}\frac{2iq_2W}{D}. \tag{35}$$

The reciprocity law (14) follows on comparison of (33) and (35); note that the reality of F and G (and thus the lack of absorption within the interface) is not needed. Both media 1 and 2 must be nonabsorbing, however, for the asymptotic forms (2) and (12) to be valid.

The reciprocity law (18) follows from (32) to (34); in this case the reality of F and G has been assumed in cancelling W with W^* and in writing the numerator of r_{21} as N^*, the general expression being

$$N_{21} = q_1q_2(F_1G_2 - G_1F_2) - iq_1(F_1G_2' - G_1F_2') - iq_2(F_1'G_2 - G_2'F_2) - (F_1'G_2' - G_1'F_2'). \tag{36}$$

We now turn to the p wave reflection and transmision amplitudes. For the standard geometry defined in Chapter 1, $\mathbf{B} = (0, e^{i(Kx-\omega t)}B(z), 0)$ where $B(z)$ satisfies

$$\frac{\mathrm{d}}{\mathrm{d}z}\left(\frac{1}{\varepsilon}\frac{\mathrm{d}B}{\mathrm{d}z}\right)+\left(\frac{\omega^2}{c^2}-\frac{K^2}{\varepsilon}\right)B \;=\; 0. \tag{37}$$

We will derive general expresions for r_p and t_p, analogous to the results for r_s and t_s. Let $C(z)$ and $D(z)$ be two linearly independent solutions of (37) within the interval $[z_1, z_2]$. Then

$$B(z) \;=\; \begin{cases} \mathrm{e}^{iq_1 z} - r_p\,\mathrm{e}^{-iq_1 z} & z < z_1 \\ \gamma C(z) + \delta D(z) & z_1 \leqslant z \leqslant z_2 \\ \left(\dfrac{\varepsilon_2}{\varepsilon_1}\right)^{1/2} t_p\,\mathrm{e}^{iq_2 z} & z > z_1. \end{cases} \tag{38}$$

The form of (37) shows that $\mathrm{d}B/\varepsilon\mathrm{d}z$ must be continuous (discontinuity would give rise to a delta-function term, which we assume to be absent from $\varepsilon(z)$). From the continuity of B and $\mathrm{d}B/\varepsilon\mathrm{d}z$ at z_1 and z_2 we obtain four equations in the four unknowns r_p, t_p, γ and δ. These are, for ε continuous at z_1 and z_2,

$$\begin{aligned} \mathrm{e}^{iq_1 z_1} - r_p\,\mathrm{e}^{-iq_1 z_1} \;&=\; \gamma C_1 + \delta D_1\,, \\ iq_1(\mathrm{e}^{iq_1 z_1} + r_p\,\mathrm{e}^{-iq_1 z_1}) \;&=\; \gamma C_1' + \delta D_1', \\ \gamma C_2 + \delta D_2 \;&=\; \left(\frac{\varepsilon_2}{\varepsilon_1}\right)^{1/2} t_p\,\mathrm{e}^{iq_2 z_2}, \\ \gamma C_2' + \delta D_2' \;&=\; \left(\frac{\varepsilon_2}{\varepsilon_1}\right)^{1/2} t_p iq_2\,\mathrm{e}^{iq_2 z_2}. \end{aligned} \tag{39}$$

Solving for r_p and t_p, we find

$$-r_p \;=\; \mathrm{e}^{2iq_1 z_1} \times \frac{q_1q_2(C_1D_2-D_1C_2)+iq_1(C_1D_2'-D_1C_2')+iq_2(C_1'D_2-D_1'C_2)-(C_1'D_2'-D_1'C_2')}{q_1q_2(C_1D_2-D_1C_2)+iq_1(C_1D_2'-D_1C_2')-iq_2(C_2'D_2-D_1'C_2)+(C_1'D_2'-D_1'C_2')}, \tag{40}$$

$$\left(\frac{\varepsilon_2}{\varepsilon_1}\right)^{1/2} t_p \;=\; \frac{\mathrm{e}^{i(q_1z_1-q_2z_2)}\,2iq_1(C_2D_2' - D_2C_2')}{q_1q_2(C_1D_2-D_1C_2)+iq_1(C_1D_2'-D_1C_2')-iq_2(C_1'D_2-D_1'D_2)+(C_1'D_2'-D_1'C_2')}. \tag{41}$$

All of the general properties derived in Section 1-1 may be verified for the p wave. The proofs are as for the s wave, with a slight difference in the case of the reciprocity relations (14) and (18), which we will make explicit. We write (40) and (41) as

$$-r_{12} = e^{-2iq_1 z_1}\frac{N_{12}}{D}, \tag{42}$$

$$\left(\frac{\varepsilon_2}{\varepsilon_1}\right)^{1/2} t_{12} = e^{i(q_1 z_1 - q_2 z_2)}\frac{2iq_1 W_2}{D}, \tag{43}$$

where N_{12} is the denominator of (40), D is the denominator common to (40) and (41), and $W_2 = C_2 D_2' - D_2 C_2'$ is the Wronskian at z_2 of the pair of solutions of (37). (The Wronskian for (37) is not independent of z, as we shall see shortly.) The corresponding expressions for a wave incident from medium 2 are

$$-r_{21} = e^{-2iq_2 z_2}\frac{N_{21}}{D}, \tag{44}$$

$$\left(\frac{\varepsilon_1}{\varepsilon_2}\right)^{1/2} t_{21} = e^{i(q_1 z_1 - q_2 z_2)}\frac{2iq_2 W_1}{D}, \tag{45}$$

where

$$N_{21} = q_1 q_2(C_1 D_2 - D_1 C_2) - iq_1(C_1 D_2' - D_1 C_2')$$
$$- iq_2(C_1' D_2 - D_1' C_2) - (C_1' D_2' - D_1' C_2'). \tag{46}$$

From (37), the Wronskian $W = CD' - DC'$ has the derivative

$$W' = CD'' - DC'' = \frac{\varepsilon'}{\varepsilon} W, \tag{47}$$

and so W is proportional ε, or W/ε is constant. The relation $q_2 t_{12} = q_1 t_{21}$ follows from (43) and (45) on using

$$\frac{W_1}{\varepsilon_1} = \frac{W_2}{\varepsilon_2}, \tag{48}$$

which we have just proved. The relation (18) or (20) follows from (42) to (44) provided $N_{21} = N_{12}^*$ (it thus holds in the absence of absorption within the interface).

On using the identity

$$(C_1 D_2 - D_1 C_2)(C_1' D_2' - D_1' C_2')$$
$$- (C_1 D_2' - D_1 C_2')(C_1' D_2 - D_1' C_2) = W_1 W_2, \tag{49}$$

we find that $|r_p|^2 \leqslant 1$ provided $(W_1/\varepsilon_1)(W_2/\varepsilon_2) \geqslant 0$, which follows from (48). Finally, the conservation law $q_1(1 - |r_p|^2 = q_2|t_p|^2$ follows from (40) and (41) on using (48) and (49).

2-3 Reflection at grazing incidence, and the existence of a Brewster angle

We will show that $r_s \to -1$ and $r_p \to +1$ at grazing incidence, exactly and without ambiguity of phase. A direct consequence is that a Brewster angle (as defined in polarization modulation ellipsometry) always exists. These results hold for interfaces with arbitrary dielectric function profiles, for internal as well as external reflections, and in the presence of adsorption within the reflecting layer or its substrate. In Chapter 7 we shall see that the results also hold for those anisotropic media for which the s and p wave characterization is adequate.

At grazing incidence, $\theta_1 \to \pi/2$ and $q_1 = \sqrt{\varepsilon_1}(\omega/c)\cos\theta_1 \to 0$. The functions F and G in the general expression (25) also depend on angle of incidence, through $q^2(z) = \varepsilon(z)\omega^2/c^2 - K^2$, where $K = \sqrt{\varepsilon_1}(\omega/c)\sin\theta_1 = \sqrt{\varepsilon_2}(\omega/c)\sin\theta_2$. Thus at grazing incidence $q^2(z_1) = \varepsilon_1\omega^2/c^2 - K^2 \to 0$, but this does not imply singular behaviour of F or G through (22). On letting $q_1 \to 0$ in (25), we find that $r_s \to -1$.

Note that there is usually an arbitrariness of the phase of a reflection amplitude, associated with the arbitrariness of choice of the origin of coordinates. For example, the reflection amplitude of a step profile located at z_1 is given by (1.15):

$$r_{s0} = e^{2iq_1z_1}\frac{q_1 - q_2}{q_1 + q_2}, \tag{50}$$

and carries the origin-dependent phase $2q_1z_1$. But as $q_1 \to 0$, this phase arbitrariness disappears. We have just shown this to be true for all profiles: at grazing incidence the reflection amplitude is known in magnitude *and* in phase. The incident and reflected waves are then both moving parallel to the interface, and there is no motion perpendicular to the interface to give rise to a phase shift associated with the path difference $2z_1$ between the incident and reflected waves.

Note that the reality of F and G, or of q_2, has not been assumed. Thus there is total reflection at grazing incidence, with reversal of the electric field, even in the presence of absorption and irrespective of the sign of $\varepsilon_1 - \varepsilon_2$.

A similar result holds for the p wave. On letting $q_1 \to 0$ in (40), we find $r_p \to +1$. This result, together with (1.27) shows that again the electric field is reversed on reflection at grazing incidence.

Thus the reflected electric fields of both the s and p waves are exactly out of phase with the incident electric fields, whether the reflecting surface is metallic or dielectric, sharp or diffuse, and for internal as well as external reflection. It follows that Lloyd's mirror experiment should produce diffraction fringes, with destructive interference at the mirror's edge, under these very general conditions. This is in accord with experiment (Jenkins and White (1950), Sections 13.8 and 28.10).

The convention in use throughout this book, and established in Chapter 1, has $r_p = r_s$ at normal incidence, where the s and p waves are physically indistinguishable. Thus the ratio $r_p/r_s = +1$ at normal incidence, and tends to -1 at grazing incidence. At general incidence r_p/r_s is a complex number, with no ambiquity of phase, since in taking the ratio one cancels out the arbitrary phase factors associated with the choice of origin. The ratio r_p/r_s is measured by ellipsometry. In polarization modulation ellipsometry (Jasperson and Schnatterly 1969, Beaglehole 1980), it is experimentally most convenient to measure $\mathrm{Im}(r_p/r_s)$ at the angle where $\mathrm{Re}(r_p/r_s) = 0$ (often called the principal angle). The vanishing of the real part of r_p/r_s is one of several possible operational definitions of the Brewster angle (other possibilities are locations of minima of $|r_p|^2$ or of $|r_p/r_s|^2$). For the step profile, with r_s and r_p given by (50) and

$$r_{p0} = -e^{2iq_1z_1}\frac{Q_1 - Q_2}{Q_1 + Q_2}, \tag{51}$$

all these definitions reduce to the Brewster angle (1.33), determined by $Q_1 = Q_2$, that is, purely in terms of the dielectric functions of media 1 and 2.

The question arises as to whether an ellipsometric Brewster angle, defined by the location of Re $(r_p/r_s) = 0$, always exists. The answer is *yes*: we have seen that r_p/r_s moves in the complex plane from the point $+1$ at normal incidence to the point -1 at grazing incidence, and it follows that it must cross the line Re $(r_p/r_s) = 0$ at least once (and in general an odd number of times). This is a consequence of the continuity of solutions of linear differential equations as a function of the parameters of the equations (here the parameter is the angle of incidence, appearing in the differential equations through K^2). See for example Birkhoff and Rota (1969), Sections 4 and 10 of Chapter 6. Some examples of the path of r_p/r_s as θ_1 varies are shown in Figure 2-4; a further case, illustrating triple Brewster angles, appears in the next section (Figure 2-8).

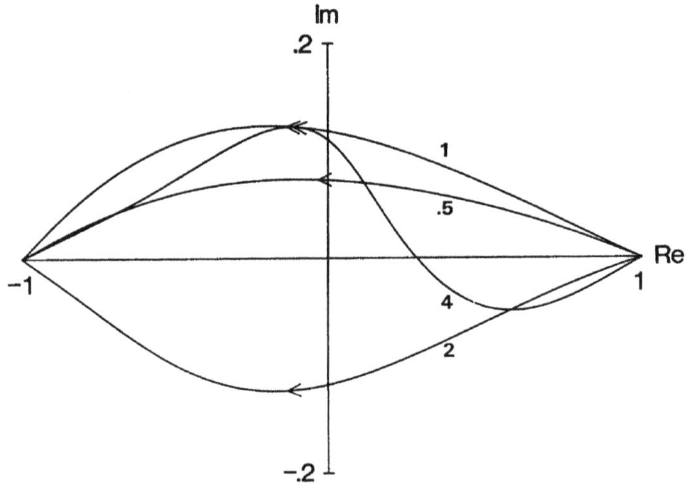

Figure 2-4. The path of r_p/r_s in the complex plane, for a uniform layer of thickness Δz. Four values of $(\omega/c)\Delta z$ (indicated on the paths) are shown. All paths begin at $+1$ and end at -1; they cut the vertical axis at the ellipsometric Brewster angle, where Re $(r_p/r_s) = 0$. The arrows indicate the direction of increasing angle of incidence; the head of each arrow is located at arctan $(\varepsilon_2/\varepsilon_1)^{1/2}$, the Brewster angle for a film of vanishing thickness. The curves are drawn for $\varepsilon_1 = 1$, $\varepsilon = (4/3)^2$ and $\varepsilon_2 = (3/2)^2$, approximating a layer of water on glass.

2-4 Reflection by a uniform layer

After the step dielectric function profile, the simplest and most commonly occurring is the two-step profile, representing a uniform layer between media 1 and 2 (Figure 2-5).

In the interval $z_1 \leqslant z' \leqslant z_2$ the s wave equation (22) has $q^2 = \varepsilon\omega^2/c^2 - K^2$, with ε constant. The solutions are thus $e^{\pm iqz}$ or cos qz, sin qz. On matching E and dE/dz

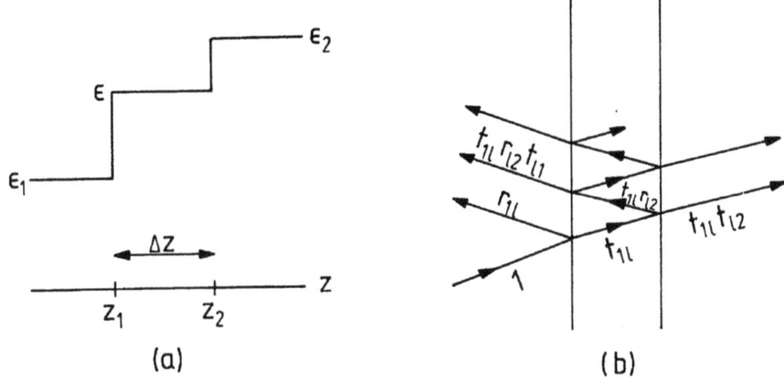

Figure 2-5. (a) The dielectric function profile of a uniform layer. (b) Schematics of the multiple reflection method for calculation of the reflection and transmission amplitudes.

at z_1 and $z_2 = z_1 + \Delta z$ to $e^{iq_1z} + r_s e^{-iq_1z}$ and $t_s e^{iq_2z}$ respectively, we find

$$r_s = e^{2iq_1z_1}\frac{q(q_1 - q_2)c + i(q^2 - q_1q_2)s}{q(q_1 + q_2)c - i(q^2 + q_1q_2)s}, \tag{52}$$

$$t_s = e^{i(q_1z_1 - q_2z_2)}\frac{2q_1q}{q(q_1 + q_2)c - i(q^2 + q_1q_2)s}, \tag{53}$$

where $c = \cos q\Delta z$ and $s = \sin q\Delta z$. Equivalently, we may substitute $F = \cos qz$, $G = \sin qz$ in (25) and (26); for these solutions of (22) we have $W = q$, $F_1G_2 - G_1F_2 = s$, $F_1G_2' - G_1F_2' = qc$, $F_1'G_2 - G_1'F_2 = -qc$, $F_1'G_1'F_2' = q^2s$.

It is instructive to consider another derivation of these results, using the multiple reflection method shown schematically in Figure 2-5(b). An incident wave of unit amplitude will produce a reflected wave of amplitude r_{1l} (this being the reflection amplitude at the step at z_1) and a transmitted wave of amplitude t_{1l} within the layer, this being the transmission amplitude at the step from medium 1 to the layer. This wave is in turn partly transmitted at z_2 (with amplitude $t_{1l}t_{l2}$), and partly reflected (amplitude $t_{1l}r_{l2}$). The reflected wave is then partly transmitted at z_1, giving a reflected wave amplitude $t_{1l}r_{l2}t_{l1}$, and partly reflected. The continuation of this process gives

$$\begin{aligned} r_s &= r_{1l} + t_{1l}t_{l1}(r_{l2} + r_{l2}r_{l1}r_{l2} + \ldots) \\ &= r_{1l} + \frac{t_{1l}t_{l1}r_{l2}}{1 - r_{l1}r_{l2}}, \end{aligned} \tag{54}$$

and

$$\begin{aligned} t_s &= t_{1l}t_{l2}(1 + r_{l2}r_{l1} + (r_{l2}r_{l1})^2 + \ldots) \\ &= \frac{t_{1l}t_{l2}}{1 - r_{l1}r_{l2}}. \end{aligned} \tag{55}$$

The various reflection and transmission amplitudes are for reflection at a single step, and can be found from (1.15), and the reciprocity relations (14) and (18):

$$r_{1l} = e^{2iq_1z_1}\frac{q_1 - q}{q_1 + q}, \quad r_{l1} = e^{-2iqz_1}\frac{q - q_1}{q + q_1}, \quad r_{l2} = e^{2iqz_2}\frac{q - q_2}{q + q_2}, \tag{56}$$

$$t_{1l} = e^{i(q_1-q)z_1}\frac{2q_1}{q_1 + q}, \quad t_{l1} = e^{i(q_1-q)z_1}\frac{2q}{q_1 + q}, \quad t_{l2} = e^{i(q-q_2)z_2}\frac{2q}{q + q_2}. \tag{57}$$

If we write $r = (q_1 - q)/(q_1 + q)$ and $r' = (q - q_2)/(q + q_2)$, r_s and t_s reduce to

$$r_s = e^{2iq_1z_1}\frac{r + r'\,e^{2iq\Delta z}}{1 + rr'\,e^{2iq\Delta z}}, \tag{58}$$

$$t_s = e^{i(q_1z_1 - q_2z_2)}\frac{(1 + r)(1 + r')\,e^{iq\Delta z}}{1 + rr'\,e^{2iq\Delta z}}, \tag{59}$$

and are readily shown to be equivalent to (52) and (53).

We see from the above equations that $|r_s|^2$ and $|t_s|^2$ are periodic functions of the thickness Δz of the film, at given ε_1, ε_2, ε and angle of incidence. The period in Δz is π/q, which increases from $\pi c/\omega\sqrt{\varepsilon}$ at normal incidence to $(\pi c/\omega)(\varepsilon - \varepsilon_1)^{-1/2}$ at grazing incidence.

Zero reflection is possible if $r' = r$ and $e^{2iq\Delta z} = -1$, and also if $r' = -r$ and $e^{2iq\Delta z} = 1$. The first of these pairs of conditions holds if $q^2 = q_1q_2$ and $2q\Delta z$ is an odd multiple of π. At normal incidence, these give the familiar characteristics of an antireflection lens coating:

$$\varepsilon^2 = \varepsilon_1\varepsilon_2 \quad \text{and} \quad \Delta z = \lambda/4,\ 3\lambda/4, \ldots \tag{60}$$

– the refractive index of the layer has to be the geometric mean of the refractive indices of the two outer media, and the thickness has to be equal to an odd multiple of a quarter wavelength (λ is the wavelength within the layer). At oblique incidence the condition $q^2 = q_1q_2$ can be satisfied only if $\varepsilon^2 < \varepsilon_1\varepsilon_2$; it then holds at

$$\theta_1 = \arcsin\{(\varepsilon_1\varepsilon_2 - \varepsilon^2)/\varepsilon_1(\varepsilon_1 + \varepsilon_2 - 2\varepsilon)\}^{1/2}.$$

The second pair of conditions holds if $q_1 = q_2$ and $2q\Delta z$ is an even multiple of π. These are equivalent to

$$\varepsilon_1 = \varepsilon_2 \quad \text{and} \quad \Delta z = \lambda/2,\ \lambda, \ldots, \tag{61}$$

where $\lambda = 2\pi/q$. At normal incidence this happens when $\varepsilon^{1/2}(\omega/c)\Delta z = n\pi$ (n an integer). At oblique incidence a uniform film between like media is perfectly transparent at the angles

$$\theta_1 = \arcsin\left\{\left(\varepsilon - \left(\frac{n\pi}{(\omega/c)\Delta z}\right)^2\right)\Big/\varepsilon_1\right\}^{1/2}.$$

An example of reflectivity at normal incidence as a function of layer thickness is shown in Figure 2-6. R_s as a function of angle of incidence will be shown together with R_p in Figure 2-7.

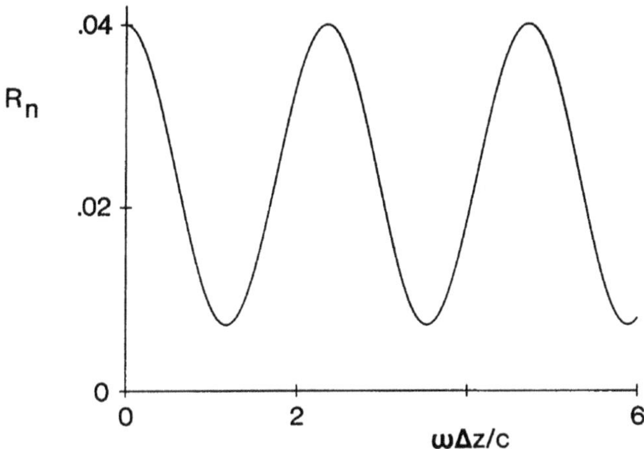

Figure 2-6. Reflectivity of a uniform layer as a function of layer thickness. R_n stands for the common value of R_s and R_p at normal incidence. The refractive index values used are $\sqrt{\varepsilon_1} = 1$, $\sqrt{\varepsilon} = 4/3$ and $\sqrt{\varepsilon_2} = 3/2$, representing a layer of water on glass.

The reflectivity in Figure 2-6 displays the periodicity mentioned above. For the case shown, the maximum of R_n is the Fresnel (zero thickness) value

$$R_{n0} = \left(\frac{\sqrt{\varepsilon_1} - \sqrt{\varepsilon_2}}{\sqrt{\varepsilon_1} + \sqrt{\varepsilon_2}}\right)^2. \tag{62}$$

In fact the uniform layer reflectivity is never greater than the Fresnel reflectivity R_{s0}, at any angle of incidence, provided ε lies between ε_1 and ε_2. This intuitively plausible result follows from the equivalence of

$$R_s \leqslant \left(\frac{q_1 - q_2}{q_1 + q_2}\right)^2 \tag{63}$$

to

$$q^4 + q_1^2 q_2^2 \leqslant q^2(q_1^2 + q_2^2), \tag{64}$$

which in turn is equivalent to

$$(\varepsilon_1 - \varepsilon)(\varepsilon - \varepsilon_2) \leqslant 0. \tag{65}$$

Note that the corresponding result for the p wave reflectivity is not true at all angles: the Fresnel or single-step reflectivity is zero at $\theta_1 = \arctan(\varepsilon_2/\varepsilon_1)^{1/2}$, at which angle the uniform layer or two-step reflectivity is not zero (in general).

The s wave reflectivity may be written in the form

$$R_s = |r_s|^2 = \frac{r^2 + 2rr' \cos 2q\Delta z + (r')^2}{1 + 2rr' \cos 2q\Delta z + (rr')^2} \tag{66}$$

(provided r and r' are real; this requires the absence of absorption within the film, and $\theta_1 < \arcsin(\varepsilon/\varepsilon_1)^{1/2}$, $\theta_1 < \arcsin(\varepsilon_2/\varepsilon_1)^{1/2}$). It thus has extrema when $\sin 2q\Delta z = 0$ (when $\cos 2a\Delta z = \pm 1$). These extrema take the values

$$R_s^+ = \left(\frac{q_1 - q_2}{q_1 + q_2}\right)^2 = R_{s0}, \qquad R_s^- = \left(\frac{q_1 q_2 - q^2}{q_1 q_2 + q^2}\right)^2. \tag{67}$$

Note that R_s^- is zero when $q^2 = q_1 q_2$, the antireflection coating condition. R_s^- is less than R_s^+ provided (64) holds, that is when ε lies between ε_1 and ε_2. When ε is outside this range, R_s^+ becomes the minimum value, and R_s^- the maximum. They are equal when ε is equal to either ε_1 or ε_2, in which case the layer is nonexistent as far as reflection is concerned.

The p wave reflection and transmission amplitudes are obtained by matching B and $dB/\varepsilon dz$ at z_1 and z_2; we find

$$-r_p = e^{2iq_1z_1} \frac{Q(Q_1 - Q_2)c + i(Q^2 - Q_1Q_2)s}{Q(Q_1 + Q_2)c - i(Q^2 + Q_1Q_2)s}, \tag{68}$$

$$\left(\frac{\varepsilon_2}{\varepsilon_1}\right)^{1/2} t_p = e^{i(q_1z_1 - q_2z_2)} \frac{2Q_1Q}{Q(Q_1 + Q_2)c - i(Q^2 + Q_1Q_2)s}, \tag{69}$$

where $Q_i = q_i/\varepsilon_i$ and $Q = q/\varepsilon$, and $c = \cos q\Delta z$, $s = \sin q\Delta z$ as before. The multiple reflection method gives the alternative forms

$$-r_p = e^{2iq_1z_1} \frac{r + r' e^{2iq\Delta z}}{r + rr' e^{2iq\Delta z}}, \tag{70}$$

$$\left(\frac{\varepsilon_2}{\varepsilon_1}\right)^{1/2} t_p = e^{i(q_1z_1 - q_2z_2)} \frac{(1 + r)(1 + r') e^{iq\Delta z}}{1 + rr' e^{2iq\Delta z}}, \tag{71}$$

where now $r = (Q_1 - Q)/(Q_1 + Q)$, $r' = (Q - Q_2)/(Q + Q_2)$. At normal incidence, $r_p = r_s$ and $t_p = t_s$.

Zero reflection occurs when $r' = r$ and $e^{2iq\Delta z} = -1$, and also when $r' = -r$ and $e^{2iq\Delta z} = 1$. The equality of r and r' holds if $Q^2 = Q_1Q_2$, which at normal incidence is equivalent to $\varepsilon^2 = \varepsilon_1\varepsilon_2$, as for s wave. The other possibility, $r' = -r$ and $e^{2iq\Delta z} = 1$, holds if $Q_1 = Q_2$ and $q\Delta z$ is an integer multiple of π (the same condition can be read off from (68)). The equality of Q_1 and Q_2 is satisfied at all angles if $\varepsilon_1 = \varepsilon_2$, or at the Brewster angle $\arctan(\varepsilon_2/\varepsilon_1)^{1/2}$ for general values of the dielectric constants ε_1, ε_2. Transparency of a uniform layer between *like media* has the same condition (61) as for the s wave. Transparency of a uniform film between unlike media at the Brewster angle for vanishing thickness, $\theta_B = \arctan(\varepsilon_2/\varepsilon_1)^{1/2}$, will occur for thicknesses such that $q\Delta z$ is an integer times π. This gives (on using 1.37)

$$\frac{\omega}{c}\Delta z = \frac{\text{integer} \times \pi}{(\varepsilon - \varepsilon_h)^{1/2}}, \qquad \varepsilon_h = \frac{\varepsilon_1\varepsilon_2}{\varepsilon_1 + \varepsilon_2}. \tag{72}$$

(Transparency at the Brewster angle is possible for non-zero thickness only if $\varepsilon > \varepsilon_h$; since the harmonic mean ε_h is always less than either ε_1 or ε_2, this is not a strong constraint.) The variation of R_p with angle is compared with that of R_s in Figure 2-7.

$R_p = |r_p|^2$ has extrema when $\cos 2q\Delta z = \pm 1$; these are

$$R_p^+ = \left(\frac{Q_1 - Q_2}{Q_1 + Q_2}\right)^2 = R_{p0}, \qquad R_p^- = \left(\frac{Q_1Q_2 - Q^2}{Q_1Q_2 + Q^2}\right)^2. \tag{73}$$

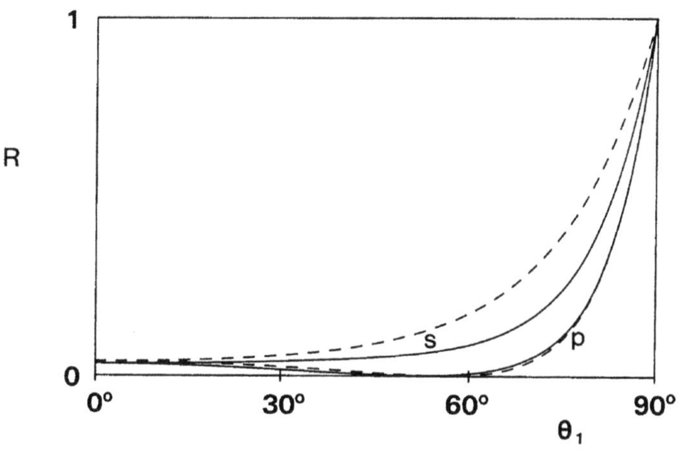

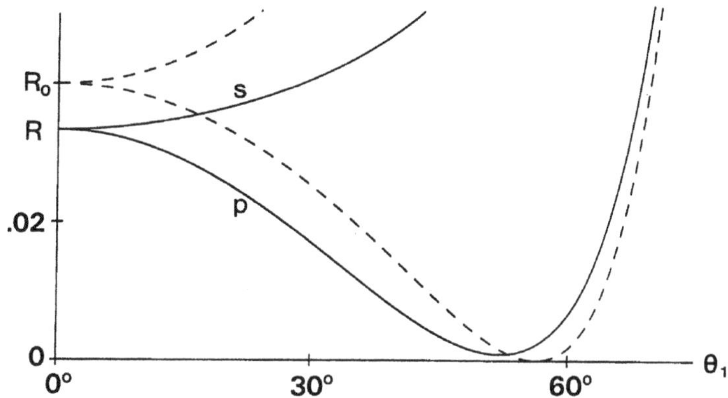

Figure 2-7. Angular variation of the s and p reflectivities, for $\varepsilon_1 = 1$, $\varepsilon_2 = (3/2)^2$, $\varepsilon = (4/3)^2$ and $(\omega/c)\Delta z = 2$. These parameters approximate a layer of water on glass, about one third of a wavelength thick. The corresponding r_p/r_s curve is one of those displayed in Figure 2-4. The dashed curves are for zero thickness of water (air/glass only).

The extrema are zero when $Q_1 = Q_2$ ($\theta_1 = \arctan(\varepsilon_2/\varepsilon_1)^{1/2} = \theta_B$), and $Q^2 = Q_1 Q_2$, respectively. The extrema are equal (to R_{p0}) when $Q = Q_1$ or Q_2, that is when $\theta_1 = \arctan(\varepsilon/\varepsilon_1)^{1/2}$ or when $\theta_1 = \arcsin(\varepsilon\varepsilon_2/\varepsilon_1(\varepsilon + \varepsilon_2))^{1/2}$. *At these two angles, the reflectivity of the p wave is independent of the thickness of the layer.* The separation of variables constant K^2 at these angles takes the values

$$K_1^2 = \left(\frac{\omega}{c}\right)^2 \frac{\varepsilon_1 \varepsilon}{\varepsilon_1 + \varepsilon}, \qquad K_2^2 = \left(\frac{\omega}{c}\right)^2 \frac{\varepsilon\varepsilon_2}{\varepsilon + \varepsilon_2}, \tag{74}$$

appropriate to Brewster – transparency at the first or second interface, respectively.

We mentioned in Section 2-3 the possibility of multiple Brewster angles. An example of triple Brewster angles, defined by the location of Re $(r_p/r_s) = 0$, is shown in Figure 2-8 for a uniform film.

When incidence is from the medium with higher dielectric function, total internal reflection occurs for $\theta_1 > \theta_c = \arcsin(\varepsilon_2/\varepsilon_1)^{1/2}$. For $\theta_1 > \theta_c$ both r_p and r_s lie on the unit circle, and so does r_p/r_s. The limiting value of r_p/r_s at grazing incidence is still -1. An example is shown in Figure 2-9.

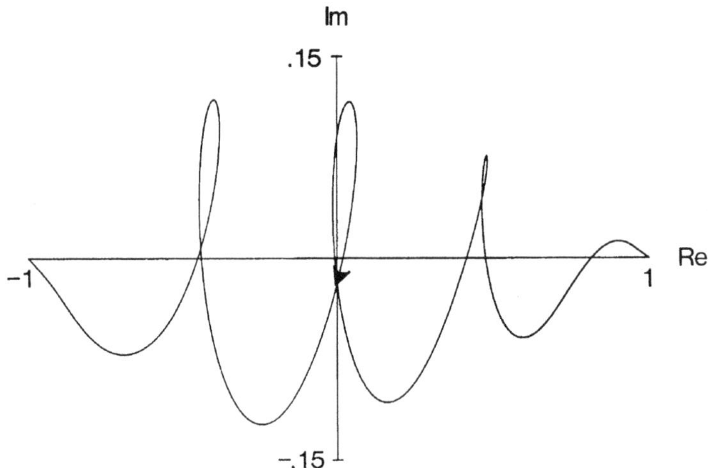

Figure 2-8. Illustration of triple Brewster angles. The curve is the locus of r_p/r_s in the complex plane, as a function of the angle of incidence. The arrow shows the direction of increasing angle of incidence; the point of the arrow is located at the zero-thickness Brewster angle, arctan $(\varepsilon_2/\varepsilon_1)^{1/2}$. The values of ε_1, ε and ε_2 are as in Figure 2-4, representing a layer of water on glass. The thickness of the film is about four wavelengths $((\omega/c)\Delta z = 27)$.

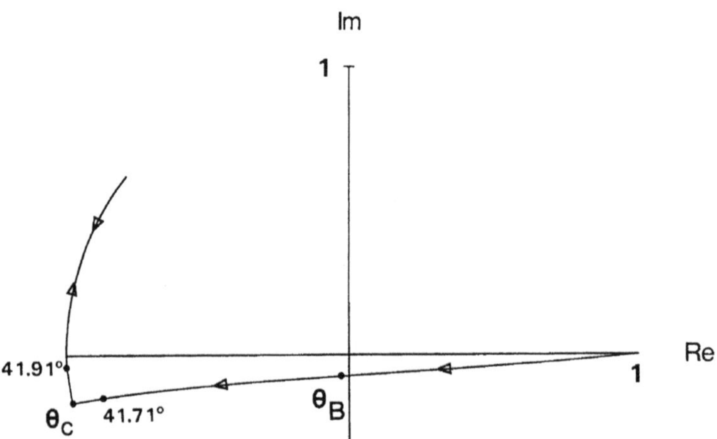

Figure 2-9. Locus of r_p/r_s in the complex plane, for light incident from glass onto a layer of water bounded by air ($\sqrt{\varepsilon_1} = 3/2$, $\sqrt{\varepsilon} = 4/3$, $\varepsilon_2 = 1$. The thickness of the uniform water layer is such that $(\omega/c)\Delta z = 1/2$. Note the rapid variation with angle near θ_c ($\simeq 41.81°$), where 0.1 degree intervals are indicated. The position of the zero thickness Brewster angle (arctan $(\varepsilon_2/\varepsilon_1)^{1/2} \simeq 33.69°$) is also shown.

2-5 Other exactly solvable profiles

It is possible to construct an infinite number of dielectric function profiles for which the reflection amplitude at normal incidence is known analytically. For a given function F, *define* $\varepsilon(\omega^2/c^2)$ as $-F''/F$ in the interval $z_1 \leqslant z \leqslant z_2$. A dielectric function so defined has the normal incidence reflection amplitude given by (25) and (30). Continuity of ε at z_1 or z_2 is not demanded. For example, $F = z^p$ gives the profile $\varepsilon(\omega^2/c^2) = p(1 - p)/z^2$; $F = e^{ikz}$ gives $\varepsilon(\omega^2/c^2) = k^2$, the uniform layer dielectric function discussed in the last section.

The latter example can be applied to oblique incidence as well, by setting $-F''/F = \varepsilon(\omega^2/c^2) - K^2$. How can one construct other solutions which are valid at oblique incidence? This was answered by Heading (1965) in the electromagnetic case. The same question has been examined in quantum mechanics, as the problem of constructing solvable potentials for the Schrödinger equation, and in acoustics (construction of solvable velocity profiles) [Bose (1964), Deavenport (1966), Vasudevan, Venkatesan and Jagannathan (1967)]. The method developed consists in transforming an equation whose solutions are known into the wave equation, and then stating the solutions of the wave equation in terms of the original equation. This systematic development has been extended to the electromagnetic p wave by Wescott (1969) [see also Heading, 1970].

We are most interested in profiles which are solvable for both the s and p waves, to which we will turn shortly. But first we give one example of a profile solvable for the s wave, which is included in the systematic development, but predates it by more than thirty years. This is the useful hyperbolic tangent profile [Eckart (1930), Epstein (1930), Landau and Lifshitz 1965, Section 25],

$$\begin{aligned}\varepsilon(z) &= \tfrac{1}{2}(\varepsilon_1 + \varepsilon_2) - \tfrac{1}{2}(\varepsilon_1 - \varepsilon_2)\tanh z/2a \\ &= \frac{\varepsilon_1 + \varepsilon_2\,\mathrm{e}^{z/a}}{1 + \mathrm{e}^{z/a}} = \frac{\varepsilon_1}{1 + \mathrm{e}^{z/a}} + \frac{\varepsilon_2}{1 + \mathrm{e}^{-z/a}}.\end{aligned} \tag{75}$$

Figure 2-10 shows this profile, and Figure 2-11 the corresponding reflectivity at normal incidence as a function of its thickness.

The s wave equation

$$\frac{\mathrm{d}^2E}{\mathrm{d}z^2} + \left(\varepsilon\frac{\omega^2}{c^2} - K^2\right)E = 0 \tag{76}$$

can be transformed to the hypergeometric differential equation by the substitutions

$$\zeta = -\mathrm{e}^{-z/a}, \quad E = \zeta^{-iq_2a}\,w(\zeta), \tag{77}$$

Figure 2-10. The hyperbolic tangent dielectric function profile (75), also known as the Fermi profile. Here $\varepsilon_1 = 1$ and $\varepsilon_2 = (4/3)^2$, representing the air–water interface at optical frequencies.

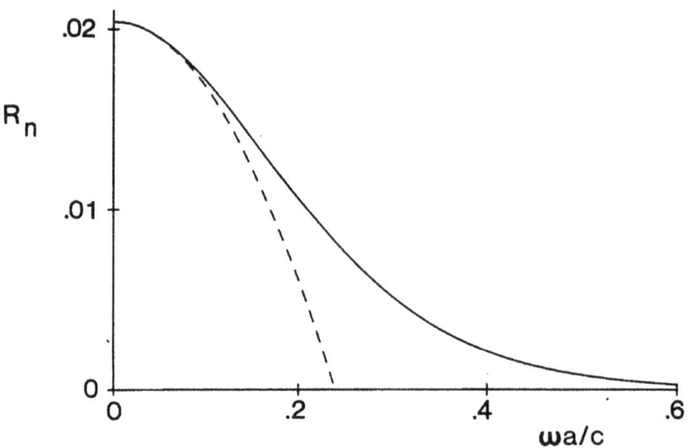

Figure 2-11. The reflectivity at normal incidence, $R_n = [\sinh \pi a(k_1 - k_2)/\sinh \pi a(k_1 + k_2)]^2$, for the hyperbolic tangent profile, as a function of interface thickness. Also shown is the leading term in the long-wave expansion: $R_n = R_{n0}[1 - (4\pi^2/3)a^2 k_1 k_2 + \ldots]$. This is an example of a general result to be derived in Chapter 3. The dielectric function values are as in Figure 2-10.

where $w(\zeta)$ tends to a constant as $\zeta \to 0$ ($z \to \infty$). The function w satisfies

$$\zeta(1 - \zeta)\frac{d^2 w}{d\zeta^2} + (1 - 2iy_2)(1 - \zeta)\frac{dw}{d\zeta} + (y_2^2 - y_1^2)w = 0, \tag{78}$$

where $y_1 = q_1 a$ and $y_2 = q_2 a$. (The references quoted above give solutions for normal incidence; the generalization to oblique incidence is given here.) The hypergeometric function

$$F(\alpha, \beta; \gamma; \zeta) = 1 + \frac{\alpha\beta}{\gamma}\frac{\zeta}{1!} + \frac{\alpha(\alpha + 1)\beta(\beta + 1)}{\gamma(\gamma + 1)}\frac{\zeta^2}{2!} + \ldots \tag{79}$$

satisfies the equation

$$\zeta(1 - \zeta)\frac{d^2 F}{d\zeta^2} + [\gamma - (\alpha + \beta + 1)\zeta]\frac{dF}{d\zeta} - \alpha\beta F = 0, \tag{80}$$

so that $w(\zeta)$ is equal to $F(i(y_1 - y_2), -i(y_1 + y_2); 1 - 2iy_2; \zeta)$. To extract the reflection amplitude we need the limiting form as $z \to -\infty$, i.e., as $\zeta \to -\infty$. This is obtained from the formula [Oberhettinger (1964), 15.3.7].

$$F(\alpha; \beta; \gamma; \zeta) = \frac{\Gamma(\gamma)\Gamma(\beta - \alpha)}{\Gamma(\beta)\Gamma(\gamma - \alpha)}(-\zeta)^{-\alpha}F(\alpha, 1 + \alpha - \gamma; 1 + \alpha - \beta; 1/\zeta)$$
$$+ \frac{\Gamma(\gamma)\Gamma(\alpha - \beta)}{\Gamma(\alpha)\Gamma(\gamma - \beta)}(-\zeta)^{-\beta}F(\beta, 1 + \beta - \gamma; 1 + \beta - \alpha; 1/\zeta), \tag{81}$$

valid for $|\arg(-\zeta)| < \pi$. As $\zeta \to -\infty$, the leading terms in (81), on using the expansion (79), give the limiting form

$$(-)^{-iq_2a}\Gamma(1-2iy_2)$$
$$\times\left\{\frac{\Gamma(-2iy_1)\,\mathrm{e}^{iq_1z}}{\Gamma(-i(y_1+y_2))\Gamma(1-i(y_1+y_2))}+\frac{\Gamma(2iy_1)\,\mathrm{e}^{-iq_1z}}{\Gamma(i(y_1-y_2))\Gamma(1+i(y_1-y_2))}\right\}\leftarrow E. \tag{82}$$

The reflection amplitude is defined as the ratio of the coefficient of e^{-iq_1z} to that of e^{iq_1z}; on using the formula

$$\Gamma(z)\Gamma(1-z) = \pi/\sin \pi z, \tag{83}$$

we find

$$r_s = -\frac{\Gamma(2iy_1)\Gamma(-i(y_1+y_2)\Gamma(-i(y_1-y_2))\sinh\pi(y_1-y_2)}{\Gamma(-2iy_1)\Gamma(i(y_1+y_2))\Gamma(i(y_1-y_2))\sinh\pi(y_1+y_2)}. \tag{84}$$

Ratios of the type $\Gamma(-iy)/\Gamma(iy)$ can be evaluated by using the infinite product representation of the gamma function (Whittaker and Watson (1927), Section 12.1)

$$\frac{1}{\Gamma(z)} = z\,\mathrm{e}^{\gamma z}\prod_{n=1}^{\infty}\left(1+\frac{z}{n}\right)\mathrm{e}^{-z/n}. \tag{85}$$

Here γ is Euler's constant .5772157 . . . . We find

$$\frac{\Gamma(-iy)}{\Gamma(iy)} = -\exp 2i\{\gamma y-\phi(y)\}, \tag{86}$$

where

$$\phi(y) = \sum_{n=1}^{\infty}\left(\frac{y}{n}-\arctan\frac{y}{n}\right). \tag{87}$$

Thus

$$r_s = \exp 2i\{\phi(2y_1)-\phi(y_1+y_2)-\phi(y_1-y_2)\}\,\frac{\sinh\pi(y_1-y_2)}{\sinh\pi(y_1-y_2)}. \tag{88}$$

The combination of ϕ functions within the braces simplifies to

$$\sum_{n=1}^{\infty}\arctan\left\{\frac{2y_1}{n}\cdot\frac{y_1^2-y_2^2}{n^2+3y_1^2+y_2^2}\right\}. \tag{89}$$

In this form it is clear that the phase of r_s is third order in the interface thickness when the profile is centred on the origin, and also that $|r_s| = 1$ when $q_2 = i|q_2|$ (total internal reflection).

We have given some detail for this model profile, since it is frequently used and has the virtue that the reflection amplitude, complete with phase, is expressible in terms of elementary functions. Another interesting feature is the relationship between *reflection at oblique incidence to that at normal incidence*. The solution at oblique incidence is obtained from that at normal incidence by replacing $k_i = \sqrt{\varepsilon_i}(\omega/c)$ by $q_i = k_i\cos\theta_i$ in the formulae above. This is a general property of dielectric function (or potential energy) profiles of the form

$$\varepsilon(z) = \tfrac{1}{2}(\varepsilon_1 + \varepsilon_2) - \tfrac{1}{2}(\varepsilon_1 - \varepsilon_2)f(z, a), \tag{90}$$

where the function f depends on parameters (such as the length a characterizing the interface thickness) which are *independent of* ε_1 *and* ε_2. When this holds, $q^2 = \varepsilon\omega^2/c^2 - K^2$ may be written as

$$q^2(z) = \tfrac{1}{2}(q_1^2 + q_2^2) - \tfrac{1}{2}(q_1^2 - q_2^2)f(z, a), \tag{91}$$

and information concerning the dielectric constants ε_1 and ε_2, the vacuum wavenumber ω/c, and the angle of incidence is contained within the normal components of the wavevector, q_1 and q_2. From (90), the function f is given by

$$f = \frac{\varepsilon_1 + \varepsilon_2 - 2\varepsilon}{\varepsilon_1 - \varepsilon_2}. \tag{92}$$

Any ε will give a function f, but only profiles which can be put in the form (90) will have f independent of ε_1 and ε_2. We shall shortly see examples of profiles which do not have this scaling property, and for which the reflection amplitude at oblique incidence cannot be obtained from the formula for normal incidence.

We now turn to profiles for which both s and p wave solutions may be obtained analytically, and concentrate on two continuous dielectric function profiles of finite range: an exponential variation with z of the refractive index or dielectric function, and a linear variation with z of the reciprocal of the refractive index. The exponential one was considered by Galejs (1961), Burman and Gould (1963), and Abelès (1964). The dielectric function is given by

$$\varepsilon(z) = \begin{cases} \varepsilon_1 & z < z_1 \\ (\varepsilon_1\varepsilon_2)^{1/2} \exp\left\{\dfrac{(z - \bar{z})}{\Delta z} \log \dfrac{\varepsilon_2}{\varepsilon_1}\right\} & z_1 \leqslant z \leqslant z_2 \\ \varepsilon_2 & z > z_2 \end{cases} \tag{93}$$

where $\bar{z} = (z_1 + z_2)/2$ and $\Delta z = z_2 - z_1$. A simpler but less symmetric form for ε is $\varepsilon_1 \exp[(z - z_1)/a]$, where $a = \Delta z/\log(\varepsilon_2/\varepsilon_1)$. Figure 2-12 shows an exponential profile representing the air–water interface at optical frequencies.

A change from z to a dimensionless independent variable proportional to the refractive index,

$$u = 2a\frac{\omega}{c}\sqrt{\varepsilon} \equiv 2ka, \tag{94}$$

transforms the s wave equation into Bessel's equation

$$\frac{d^2E}{du^2} + \frac{1}{u}\frac{dE}{du} + \left(1 - \frac{(2Ka)^2}{u^2}\right)E = 0. \tag{95}$$

The general solution within $[z_1, z_2]$ is $\alpha J_s(u) + \beta Y_s(u)$, with $s = 2Ka$. The order s of the Bessel functions depends on the angle of incidence (it is proportional to $\sin\theta_1$); both s and u are proportional to the interface thickness. The s wave reflection and transmission amplitudes may be obtained from (25) and (26), with $F(z) = J_s(u)$ and $G(z) = Y_s(u)$.

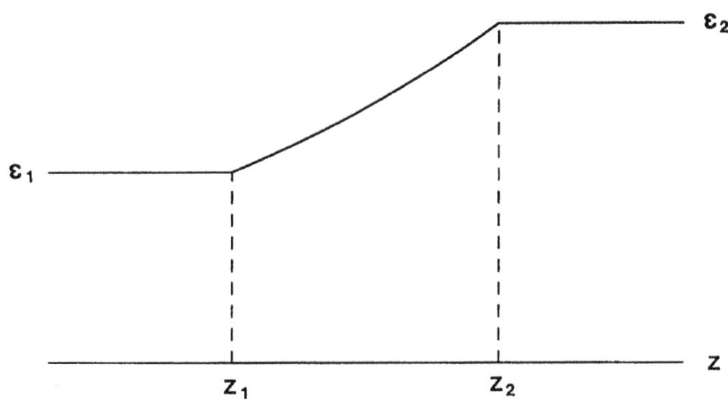

Figure 2-12 The exponential function profile, equation (93), with $\sqrt{\varepsilon_1} = 1$, $\sqrt{\varepsilon_2} = 4/3$.

The p wave equation reads, in the u variable,

$$\frac{d^2B}{du^2} - \frac{1}{u}\frac{dB}{du} + \left(1 - \frac{(2Ka)^2}{u^2}\right)B = 0, \tag{96}$$

and is satisfied by $\alpha u J_p(u) + \beta u Y_p(u)$, where $p^2 = (2Ka)^2 + 1$. The reflection amplitude may be found from (40), with $C(z) = uJ_p(u)$ and $D(z) = uY_p(u)$. It is useful to work in terms of the cross products

$$\begin{aligned} A_\nu &= J_\nu(u_1)Y_\nu(u_2) - Y_\nu(u_1)J_\nu(u_2), \\ B_\nu &= J_\nu(u_1)Y_\nu'(u_2) - Y_\nu(u_1)J_\nu'(u_2), \\ C_\nu &= J_\nu'(u_1)Y_\nu(u_2) - Y_\nu'(u_1)J_\nu(u_2), \\ D_\nu &= J_\nu'(u_1)Y_\nu'(u_2) - Y_\nu'(u_1)J_\nu'(u_2), \end{aligned} \tag{97}$$

where the primes denote differentiation with respect to u. The p wave reflection amplitude then reads

$$\begin{aligned} -r_p &= e^{2iq_1z_1} \\ &\times \frac{q_1q_2A_p + iq_1k_2(B_p + A_p/u_2) + iq_2k_1(C_p + A_p/u_1) - k_1k_2(D_p + B_p/u_1 + C_p/u_2 + A_p/u_1u_2)}{q_1q_2A_p + iq_1k_2(B_p + A_p/u_2) - iq_2k_1(C_p + A_p/u_1) + k_1k_2(D_p + B_p/u_1 + C_p/u_2 + A_p/u_1u_2)}. \end{aligned} \tag{98}$$

The s wave result is

$$r_s = e^{2iq_1z_1}\frac{q_1q_2A_s + iq_1k_2B_s + iq_2k_1C_s - k_1k_2D_s}{q_1q_2A_s + iq_1k_2B_s - iq_2k_1C_s + k_1k_2D_s}. \tag{99}$$

At normal incidence $s = 0$ and $p = 1$. On using the identities [compare Olver (1964), 9.1.32, 33]

$$\begin{aligned} &A_1 = D_0, \quad B_1 + A_1/u_2 = -C_0, \quad C_1 + A_1/u_1 = -B_0, \\ &D_1 + B_1/u_1 + C_1/u_2 + A_1/u_1u_2 = A_0, \end{aligned} \tag{100}$$

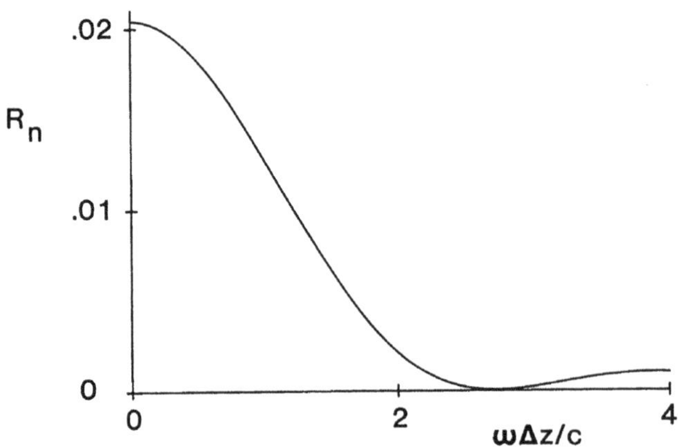

Figure 2-13. Normal incidence reflectivity for the exponential profile, as a function of the interface thickness. The values of ε_1 and ε_2 are as in Figure 2-12. The first minimum is at $(\omega/c)\Delta z \simeq 2.71$; for the same ε_1 and ε_2 the similar Rayleigh profile (Figures 2-14 and 15) has its first zero at $(\omega/c)\Delta z \simeq 2.73$.

we find that both the reflection amplitudes at normal incidence reduce to

$$r_n = e^{2ik_1z_1}\frac{A_0 + iB_0 + iC_0 - D_0}{A_0 + iB_0 - iC_0 + D_0}. \tag{101}$$

The corresponding reflectivity is

$$R_n = |r_n|^2 = \frac{A_0^2 + B_0^2 + C_0^2 + D_0^2 - 8/\pi^2 u_1 u_2}{A_0^2 + B_0^2 + C_0^2 + D_0^2 + 8/\pi^2 u_1 u_2}. \tag{102}$$

In obtaining (102) we have used the identity (49), and the fact that the Wronskian $J_\nu(u)Y_\nu'(u) - J_\nu'(u)Y_\nu(u)$ is equal to $2/\pi u$. The reflectivity at normal incidence as a function of interfacial thickness is shown in Figure 2-13.

The second dielectric function profile for which a solution is known for both the s and p waves was first considered by Rayleigh (1880) (for normal incidence only), and a solution for general incidence of both polarizations was given by Burman and Gould (1963). For the Rayleigh profile the reciprocal of the refractive index varies linearly with distance between the interfacial boundaries z_1 and z_2; as usual we have $\varepsilon = \varepsilon_1$ for $z < z_2$ and $\varepsilon = \varepsilon_2$ for $z > z_2$. This profile is shown in Figure 2-14.

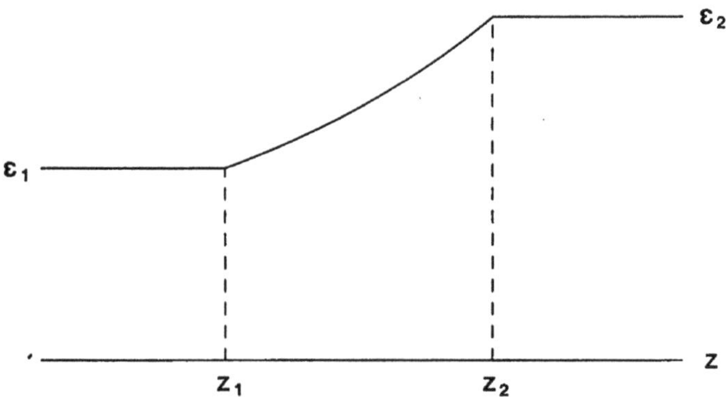

Figure 2-14. Dielectric function $\varepsilon(z)$ for the Rayleigh profile. The values $\varepsilon_1 = 1$ and $\varepsilon_2 = (4/3)^2$ approximate air on the left and water on the right (at optical frequencies).

Since $\varepsilon^{-1/2}$ varies linearly with z, it will be useful to work in terms of this function, which we will call η:

$$\varepsilon^{-1/2}(z) = \eta(z) = \bar{\eta} + (z - \bar{z})\frac{\Delta\eta}{\Delta z}. \tag{103}$$

Here $\Delta\eta = \eta_2 - \eta_1 = \varepsilon_2^{-1/2} - \varepsilon_1^{-1/2}$, $\Delta z = z_2 - z_1$, $\bar{\eta} = (\eta_1 + \eta_2)/2$, and $\bar{z} = (z_1 + z_2)/2$. At normal incidence the s wave equation becomes, on changing the independent variable from z to η,

$$\frac{d^2E}{d\eta^2} + (\tfrac{1}{4} - \nu^2)\frac{E}{\eta^2} = 0, \tag{104}$$

where

$$\nu^2 = \tfrac{1}{4} - \left(\frac{\omega}{c}\frac{\Delta z}{\Delta\eta}\right)^2. \tag{105}$$

This equation has the power-law solutions $E_\pm = \eta^{\frac{1}{2}\pm\nu}$, and the reflection amplitude can be found from (25):

$$r_n = e^{2ik_1z_1}\frac{\left(\dfrac{\varepsilon_1}{\varepsilon_2}\right)^\nu - 1}{2\nu\left[\left(\dfrac{\varepsilon_1}{\varepsilon_2}\right)^\nu + 1\right] - 2i\dfrac{\omega}{c}\dfrac{\Delta z}{\Delta\eta}\left[\left(\dfrac{\varepsilon_1}{\varepsilon_2}\right)^\nu - 1\right]}. \tag{106}$$

The reflectivity at normal incidence takes different forms according as $\left(\dfrac{\omega}{c}\dfrac{\Delta z}{\Delta\eta}\right)^2$ is less or greater than $\frac{1}{4}$. When $\left(\dfrac{\omega}{c}\dfrac{\Delta z}{\Delta\eta}\right)^2 < \frac{1}{4}$, ν is real and

$$\begin{aligned} R_n &= \frac{\left[\left(\dfrac{\varepsilon_1}{\varepsilon_2}\right)^\nu - 1\right]^2}{4\nu^2\left[\left(\dfrac{\varepsilon_1}{\varepsilon_2}\right)^\nu + 1\right]^2 + 4\left(\dfrac{\omega}{c}\dfrac{\Delta z}{\Delta\eta}\right)^2\left[\left(\dfrac{\varepsilon_1}{\varepsilon_2}\right)^\nu - 1\right]^2} \\ &= \frac{\left[\left(\dfrac{\varepsilon_1}{\varepsilon_2}\right)^\nu - 1\right]^2}{\left[\left(\dfrac{\varepsilon_1}{\varepsilon_2}\right)^\nu + 1\right]^2 - 16\left(\dfrac{\omega}{c}\dfrac{\Delta z}{\Delta\eta}\right)^2\left(\dfrac{\varepsilon_1}{\varepsilon_2}\right)^\nu}. \end{aligned} \tag{107}$$

For $\left(\dfrac{\omega}{c}\dfrac{\Delta z}{\Delta\eta}\right)^2 > \frac{1}{4}$, $\nu = i|\nu|$, and

$$R_n = \frac{\sin^2\left(\frac{1}{2}|\nu|\log\dfrac{\varepsilon_1}{\varepsilon_2}\right)}{4|\nu|^2 + \sin^2\left(\frac{1}{2}|\nu|\log\dfrac{\varepsilon_1}{\varepsilon_2}\right)}. \tag{108}$$

At $v = 0$ these two forms take the common value

$$R_n(v = 0) = \frac{\left(\log \frac{\varepsilon_1}{\varepsilon_2}\right)^2}{16 + \left(\log \frac{\varepsilon_1}{\varepsilon_2}\right)^2}. \tag{109}$$

We note from (108) that the reflectivity is zero whenever $|v| \log (\varepsilon_1/\varepsilon_2) = 2n\pi$ ($n = 1, 2, \ldots$), that is when

$$\frac{\omega}{c}\frac{\Delta z}{\Delta \eta} = \pm\left\{\frac{1}{4} + \left(\frac{2n\pi}{\log (\varepsilon_1/\varepsilon_2)}\right)^2\right\}^{1/2}. \tag{110}$$

The reflectivity at normal incidence as a function of interface thickness is shown in Figure 2-15.

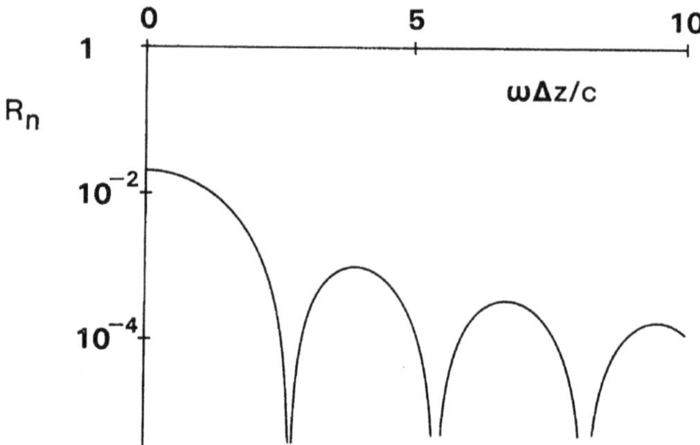

Figure 2-15. Normal incidence reflectivity for the Rayleigh profile, as a function of the interface thickness. The values $\varepsilon_1 = 1$, $\varepsilon_2 = (4/3)^2$ have been used, representing air–water interface, in common with Figure 2-13.

The exponential profile, which also has discontinuities in slope at its boundaries and has a similar shape, has minima at points approximated by (110) (see the caption to Figure 2-13). The reflectivities for both profiles show oscillatory decay with increasing thickness. The uniform layer discussed in the last section, which has discontinuities in value at the boundaries, has its reflectivity strictly periodic in the thickness. In contrast, the hyperbolic tangent profile, which is continuous in value and in all its derivatives, shows a monotonic decrease in reflectivity with interface thickness.

For oblique incidence, the s wave equation (22) for the Rayleigh profile becomes, in the η variable,

$$\frac{d^2E}{d\eta^2} + \left[\frac{\frac{1}{4} - v^2}{\eta^2} - \left(K\frac{\Delta z}{\Delta \eta}\right)^2\right]E = 0, \tag{111}$$

and has solutions proportional to $\eta^{1/2}$ times a Bessel function of order v and imaginary argument $\pm iK(\Delta z/\Delta\eta)\eta$. Thus r_s may be obtained from the general formula given in Section 2-2.

The p wave is most conveniently discussed in terms of the variable $b = (\varepsilon_1/\varepsilon)^{1/2} B$, which satisfies (1.22):

$$\frac{d^2 b}{dz^2} + \left[q^2 - \varepsilon^{1/2} \frac{d^2 \varepsilon^{-1/2}}{dz^2} \right] b = 0. \tag{112}$$

Since $\varepsilon^{-1/2}$ is linear for z for the Rayleigh profile, the E and b equations are the same, *except* at the end-points z_1 and z_2. There, because of the discontinuity in the slope of $\varepsilon^{-1/2}$, the equation for b contains additional delta-function terms:

$$\varepsilon^{1/2} \frac{d^2 \varepsilon^{-1/2}}{dz^2} = \frac{\Delta\eta}{\Delta z} \left\{ \frac{1}{\eta_1} \delta(z - z_1) - \frac{1}{\eta_2} \delta(z - z_2) \right\}. \tag{113}$$

As a consequence, db/dz is discontinuous at z_1 and z_2. Within the interface, b has the same Bessel function solutions as the s wave. Expressions for r_s and r_p, graphs of r_s, r_p and r_p/r_s, and a comparison with theory for the reflection of long waves (to be discussed in the next chapter), are given in Lekner (1982).

We have concentrated on general results, and on the reflection by four special profiles, three of which are solvable for both the s and p waves. Discussion of other special profiles (which are solvable for the s wave only) may be found in Sections 4-3 and 5-2, in Heading (1965), and in the texts listed in the references for this section. Other exact and general results will appear elsewhere in this book, in particular Chapters 3, 4, 5, 6, 8 and 12.

References

References quoted in text

G. G. Stokes (1849) "On the perfect blackness of the central spot in Newton's rings, and on the verification of Fresnel's formulae for the intensities of the reflected and refracted rays", Cambridge and Dublin Mathematical Journal **4**, 1–15.

F. A. Jenkins and H. E. White (1950) "Fundamentals of optics", McGraw-Hill, Sections 13.8 and 28.10.

S. N. Jasperson and S. E. Schnatterly (1969). "An improved method for high reflectivity ellipsometry based on a new polarization modulation technique", Rev. Sci. Instr. **40**, 761–767.

D. Beaglehole (1980) "Ellipsometric study of the surface of simple liquids", Physica **100B**, 163–174.

G. Birkhoff and G. -C. Rota (1969) "Ordinary differential equations", Blaisdell (Waltham, Massachusetts).

J. Heading (1965) "Refractive index profiles based on the hypergeometric equation and the confluent hypergeometric equation", Proc. Camb. Phil. Soc. **61**, 897–913.

A. K. Bose (1964) "A class of solvable potentials", Nuovo Cimento **32**, 679–688.

R. L. Deavenport (1966) "A normal mode theory of an underwater acoustic duct by means of Green's function", Radio Science **1**, 709–724.

R. Vasudevan, K. Venkatesan, and G. Jagannathan (1967) "Construction of solvable potentials and some aspects of regularization of singular potentials", Nuovo Cimento (suppl.) **5**, 621–643.

B. S. Westcott (1969) "Exact solutions for vertically polarized electromagnetic waves in horizontally stratified media", Proc. Camb. Phil. Soc. **66**, 675–684.

J. Heading (1970) "The equality of the moduli of certain ratios occurring in the connexion formulae of solutions of some transcendental differential equations", Proc. Camb. Phil. Soc. **67**, 347–361.

C. Eckart (1930) "The penetration of a potential barrier by electrons", Phys. Rev. **35**, 1303–1309.

P. S. Epstein (1930) "Reflection of waves in an inhomogeneous absorbing medium", Proc. Nat. Acad. Sci. **16**, 627–637.

L. D. Landau and E. M. Lifshitz (1965) "Quantum mechanics", Pergamon.

E. T. Whittaker and G. N. Watson (1927) "A course of modern analysis", Cambridge.

F. Oberhettinger (1964) "Hypergeometric functions", Chapter 15 of Handbook of Mathematical Functions, edited by M. Abramowitz and I. A. Stegun, NBS Applied Mathematics Series No. 55.

J. G. Galejs (1961) "ELF waves in the presence of exponential ionospheric conductivity profiles", IRE Trans. Ant. Prop. **9**, 554–562.

R. Burman and R. N. Gould (1963) "On the propagation of vertically polarized electromagnetic waves in a horizontally stratified medium", J. Atm. Terr. Phys. **25**, 543–549.

F. Abelès (1964) "Optical properties of inhomogeneous films", NBS Misc. Publ. **265**, 41–58.

F. W. J. Olver (1964) "Bessel functions of integer order" (chapter 9 of "Handbook of mathematical functions", edited by M. Abramowitz and I. A. Stegun, NBS Appl. Math. Series **55**).

J. W. S. Rayleigh (1880) "On the reflection of vibrations at the confines of two media between which the transition is gradual", Proc. Lond. Math. Soc. **11**, 51–56.

J. Lekner (1982) "Exact reflection amplitudes for the Rayleigh profile", Physica **116A**, 235–247.

The derivation of conservation laws and reciprocity relations from comparison identities (Section 2-1) is based on

J. Lekner (1982) "Reflection of long waves by interfaces", Physica **112A**, 544–556.

For alternative derivations, see Landau and Lifshitz (1965), Section 25 and

J. Heading (1975) "Ordinary differential equations, theory and practice", Elek Science, Chapter 4.

Extension of Stokes' idea of reversing the wavemotions is discussed by Šantavý and Knittl:

U. Šantavý (1961) "On the reversibility of light beams in conducting media", Optica Acta **8**, 301–307;
Z. Knittl (1962) "The principle of reversibility and thin film optics", Optica Acta **9**, 33–45;
Z. Knittl (1976) "Optics of thin films", Wiley, Chapter 6.

Sections 2-2 and 2-3 are based on the author's paper on the Rayleigh profile (quoted above) and on

J. Lekner (1985) "Reflection at oblique incidence and the existence of a Brewster angle", J. Opt. Soc. Amer. **A2**, 186–188.

The uniform layer between two bulk media (Section 2-4) is considered (for example), in

M. Born and E. Wolf (1965) "Principles of optics" (3rd edition), Pergamon, Section 1.6.4, and
R. M. Azzam and N. M. Bashara (1977) "Ellipsometry and polarized light", North Holland.

Exactly solvable profiles (Section 2-5) are listed in the texts by

K. G. Budden (1961) "Radio waves in the ionosphere", Cambridge;

V. L. Ginzburg (1964) "The propagation of electromagnetic waves in plasmas", Pergamon;

J. R. Wait (1970) "Electromagnetic waves in stratified media", (second edition), Pergamon; and

L. M. Brekhovskikh (1980) "Waves in layered media" (second edition), Academic Press.

The manufacture of solvable profiles from basic transcendental equations is also considered in the text on differential equations by Heading (1975) (quoted above); Section 9.2.

3

Reflection of long waves

We have seen in Section 2-2 that the reflection amplitudes of an arbitrary profile tend to the Fresnel values as the thickness Δz of the profile tends to zero. An equivalent limit to consider is that of reflection by a profile of fixed extent, as the wavelength increases. We might expect the reflection amplitudes to be well represented, in the long wave limit, by the first few terms of a series in the ratio of the interface thickness to the wavelength. This expectation turns out to be essentially correct, with the coefficient of a given power of $(\omega/c)\Delta z$ depending on the angle of incidence, as well as on the profile characteristics.

In the long wave limit, a given dielectric function profile reflects as a step profile, plus a small correction which depends on the deviation of the profile from a single step. The long wave theory treats the deviation as a *perturbation*; the perturbation theory of reflection is developed in Section 3-1 (for any type of perturbation), and then used to obtain the long wave expansion in the following sections.

3-1 Integral equation and perturbation theory for the *s* wave

The results of this section hold for the electromagnetic *s* wave, and for Schrödinger particle waves. We wish to express ψ, the solution of

$$\frac{d^2\psi}{dz^2} + +q^2\psi = 0, \quad e^{iq_1 z} + r\,e^{-iq_1 z} \leftarrow \psi \rightarrow t\,e^{iq_2 z}, \tag{1}$$

in terms of a known function ψ_0, the solution of

$$\frac{d^2\psi_0}{dz^2} + q_0^2\psi_0 = 0, \quad e^{iq_1 z} + r_0\,e^{-iq_1 z} \leftarrow \psi_0 \rightarrow t_0\,e^{iq_2 z}. \tag{2}$$

Note that q and q_0 share the same asymptotic values, q_1 and q_2: the reference dielectric function $\varepsilon_0(z)$ (or potential $V_0(z)$) must be chosen to tend to the same limits ε_1 and ε_2 as $\varepsilon(z)$ (or the same V_1 and V_2 as $V(z)$). Write $q^2 = q_0^2 + \Delta q^2$, and $\psi = \psi_0 + \psi_1 + \psi_2 + \ldots$, a series in powers of Δq^2. From (1) and (2), the correction to ψ_0 of nth order in Δq^2 satisfies the equation

$$\frac{d^2\psi_n}{dz^2} + q_0^2\psi_n = -\Delta q^2\psi_{n-1}, \tag{3}$$

for $n = 1, 2, \ldots$. To solve (3) we need to construct a Green's function $G(z, \zeta)$ which satisfies

$$\frac{\partial^2 G}{\partial z^2} + q_0^2(z)G = \delta(z - \zeta), \tag{4}$$

and has the appropriate asymptotic behaviour to make each ψ_n take the limiting forms given by (1), with $r = r_n$ and $t = t_n$. When such a Green's function has been constructed, the corrections $\psi_1, \psi_2, \ldots$ are given by

$$\psi_n(z) = -\int_{-\infty}^{\infty} d\zeta\ \Delta q^2(\zeta)G(z, \zeta)\psi_{n-1}(\zeta), \tag{5}$$

so that one can solve for ψ_1 in terms of the known ψ_0, for ψ_2 in terms of ψ_1, and so on. An equivalent formulation is to write (1) as an integral equation, using (4):

$$\psi(z) = \psi_0(z) - \int_{-\infty}^{\infty} d\zeta\ \Delta q^2(\zeta)G(z, \zeta)\psi(\zeta). \tag{6}$$

The sequence (5) is then obtained by iteration of (6). [The reader not familiar with Green's functions may verify that (6) solves (1) by operating on both sides with $\partial^2/\partial z^2 + q_0^2(z)$, and using (4).]

The above is for *any* reference q_0 and ψ_0. For long waves, the natural choice for q_0 and ψ_0 are the functions corresponding to the step dielectric function profile

$$\begin{aligned} \varepsilon_0(z) &= \begin{cases} \varepsilon_1 & (z < 0) \\ \varepsilon_2 & (z > 0) \end{cases} \\ &= \tfrac{1}{2}(\varepsilon_1 + \varepsilon_2) - \tfrac{1}{2}(\varepsilon_1 - \varepsilon_2)\ \mathrm{sgn}\ (z), \end{aligned} \tag{7}$$

for which q_0 takes the values q_1 and q_2 for $z \lessgtr 0$, and

$$\psi_0(z) = \begin{cases} e^{iq_1 z} + r_0\, e^{-iq_1 z} & (z < 0) \\ t_0\, e^{iq_2 z} & (z > 0), \end{cases} \tag{8}$$

where (from the continuity of ψ_0 and $d\psi_0/dz$ at $z = 0$)

$$r_0 = \frac{q_1 - q_2}{q_1 + q_2}, \qquad t_0 = \frac{2q_1}{q_1 + q_2} = 1 + r_0. \tag{9}$$

The appropriate Green's function $G(z, \zeta)$ must satisfy (4), and be such that ψ (as given by (6)) have the asymptotic forms of (1). We see from (4) that G is built up from the functions $e^{\pm iq_0 z}$, and that $\partial G/\partial z$ must have a unit discontinuity along the line $z = \zeta$. For each of $\zeta > 0$ and $\zeta < 0$, G has three different analytic forms, the dividing lines being at $z = 0$, and $z = \zeta$. The continuity of G at both boundaries, and the respective continuity and discontinuity of $\partial G/\partial z$ at $z = 0$ and ζ, impose conditions on the coefficients. When the coefficients are evaluated, we find the six analytic forms

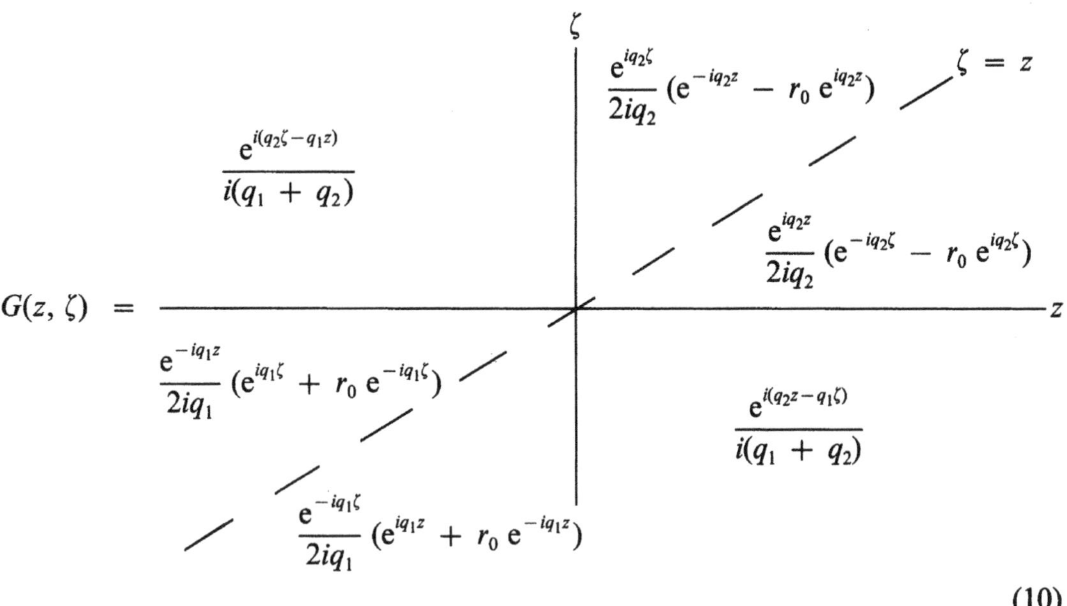

(10)

From (10) and (6) we find the analytic form of ψ as $z \to -\infty$:

$$\psi(z) \to e^{iq_1z} + r_0\,e^{-iq_1z} - e^{-iq_1z}\left\{\frac{1}{2iq_1}\int_{-\infty}^{0} d\zeta\;\Delta q^2(\zeta)\,(e^{iq_1\zeta} + r_0\,e^{-q_1\zeta})\psi(\zeta)\right.$$
$$\left. + \frac{1}{i(q_1 + q_2)}\int_0^{\infty} d\zeta\;\Delta q^2(\zeta)\,e^{iq_2\zeta}\,\psi(\zeta)\right\}. \tag{11}$$

On comparing (11) with the asymptotic form in (1), we identify the reflection amplitude as r_0 minus the expression within the braces. The nth order term r_n is therefore

$$r_n = \frac{i}{2q_1}\int_{-\infty}^{0} d\zeta\;\Delta q^2(\zeta)(e^{iq_1\zeta} + r_0\,e^{-iq_1\zeta})\psi_{n-1}(\zeta)$$
$$+ \frac{i}{q_1 + q_2}\int_0^{\infty} d\zeta\;\Delta q^2(\zeta)\,e^{iq_2\zeta}\,\psi_{n-1}(\zeta). \tag{12}$$

In particular, the first order term r_1 is given by

$$r_1 = \frac{i}{2q_1}\int_{-\infty}^{0} d\zeta\;\Delta q^2(\zeta)(e^{iq_1\zeta} + r_0\,e^{-iq_1\zeta})^2$$
$$+ \frac{i}{q_1 + q_2}(1 + r_0)\int_0^{\infty} d\zeta\;\Delta q^2(\zeta)\,e^{2iq_2\zeta}. \tag{13}$$

When $\varepsilon_1 = \varepsilon_2$ (and q_1, q_2 take the common value q_0) r_1 takes the simple form

$$r_1 = \frac{i}{2q_0}\int_{-\infty}^{\infty} d\zeta\;\Delta q^2(\zeta)\,e^{2iq_0\zeta}, \tag{14}$$

from which the reflection formulae used in Section 1-5 may be obtained.

The general expression for r implicit in (11) may be put into a simpler form by using (8) and (9):

$$r = r_0 - \frac{1}{2iq_1}\int_{-\infty}^{\infty} d\zeta\ \Delta q^2(\zeta)\psi(\zeta)\psi_0(\zeta). \tag{15}$$

This formula can be obtained directly from the comparison identity (2.6), but without the perturbation theory derived above is capable of giving only r_1.

We have developed the perturbation series in terms of Δq^2. Since $q^2(z) = \varepsilon(z)\omega^2/c^2 - K^2$,

$$\Delta q^2 = \frac{\omega^2}{c^2}(\varepsilon - \varepsilon_0), \tag{16}$$

and the perturbation is independent of the angle of incidence. This simple result holds only for the s and particle waves; the more complex p wave perturbation theory has terms in K^2, as we shall see in Section 3-4.

3-2 The s wave to second order in the interface thickness

We see from (16) that the perturbation Δq^2 is small, for arbitrary $\Delta\varepsilon$, in the long wavelength limit. In this section we will obtain corrections to r_0 in terms of integrals over the difference $\varepsilon - \varepsilon_0$ between the actual and the step profile. It will be convenient to define the integrals λ_n, of dimension (length)^n, as

$$\lambda_n = \int_{-\infty}^{\infty} dz\ [\varepsilon(z) - \varepsilon_0(z)]z^{n-1}. \tag{17}$$

Consider the first order (in Δq^2) expression for r_1, as given by (13). The integrals in (13) contain a factor $\Delta q^2 = (\omega^2/c^2)(\varepsilon - \varepsilon_0)$, assumed to be of short range, by which we mean that the λ_n converge for all n. (This includes profiles such as the hyperbolic tangent defined in (2.75), for which $\varepsilon - \varepsilon_0$ tends to zero exponentially in z as $z \to \pm\infty$.) Let the range of $\varepsilon - \varepsilon_0$ be characterized by a length Δz, an interface thickness. In the long wave limit, $(\omega/c)\Delta z$ is a small dimensionless parameter, as are $q_1\Delta z$ and $q_2\Delta z$. We expand the factors $e^{iq_1z} + r_0\,e^{-iq_1z}$ and e^{iq_2z} in (12) and (13) in powers of q_iz, to obtain an expansion in the above small parameters. For r_1 we have, to second order in the interface thickness,

$$r_1 = \frac{2iq_1\omega^2/c^2}{(q_1 + q_2)^2}\{\lambda_1 + 2iq_2\lambda_2 + \dots\}. \tag{18}$$

The corresponding expression for r_2 is

$$r_2 = \frac{-2q_1\omega^4/c^4}{(q_1 + q_2)^3}\lambda_1^2 + \dots, \tag{19}$$

obtained from (12) and the result

$$\psi_1(0) = \frac{2iq_1\omega^2/c^2}{(q_1 + q_2)^2}\lambda_1 + \dots. \tag{20}$$

The leading term in r_3 is of third order in Δz:

$$r_3 = \frac{2iq_1\omega^6/c^6}{(q_1 + q_2)^4}\lambda_3 + \dots, \tag{21}$$

and can thus be omitted from the second order expression, which reads

$$r = r_0 + \frac{2iq_1\omega^2/c^2}{(q_1 + q_2)^2}\left\{\lambda_1 + 2iq_2\lambda_2 + \frac{i(\omega^2/c^2)\lambda_1^2}{q_1 + q_2}\right\} + \dots. \tag{22}$$

We now need a notation to distinguish between terms which are nth order in Δq^2, and those which are nth order in Δz. We write the latter as r_{sn} (for s wave reflection amplitude component which is nth order in Δz); r_{s0} and r_0 are equal, but from $n = 1$ onwards r_n and r_{sn} are different. In the Δz expansion we have

$$r_{s1} = \frac{2iq_1\omega^2/c^2}{(q_1 + q_2)^2}\lambda_1, \tag{23}$$

$$r_{s2} = \frac{-2q_1\omega^2/c^2}{(q_1 + q_2)^2}\left\{2q_2\lambda_2 + \frac{\omega^2/c^2}{q_1 + q_2}\lambda_1^2\right\}. \tag{24}$$

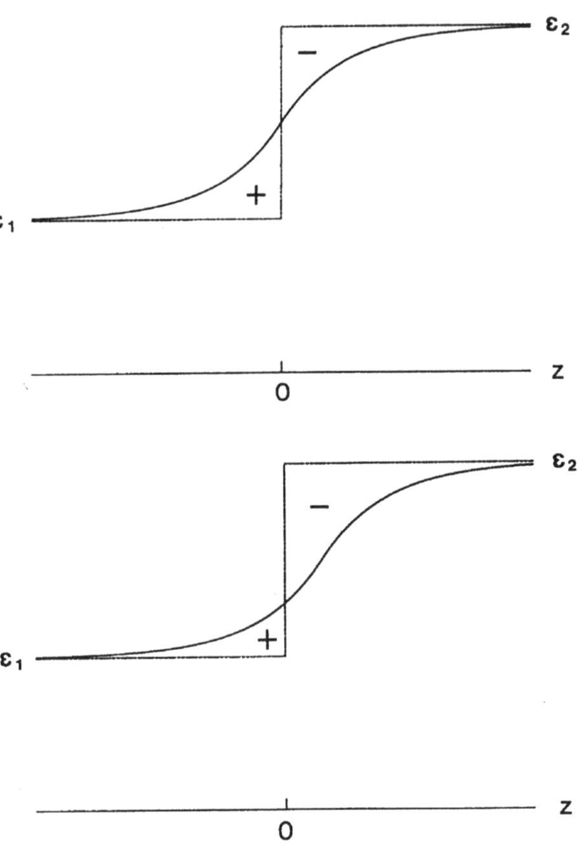

Figure 3-1. Dependence of the integral $\lambda_1 = \int_{-\infty}^{\infty} dz(\varepsilon - \varepsilon_0)$ on the relative positioning of ε and ε_0. The sign of the contributions is indicated; the magnitude of the positive and negative contributions is equal to the area enclosed between the actual dielectric function profile and the reference step profile. In the upper diagram the areas are equal, and $\lambda_1 = 0$; such positioning is always possible when $\varepsilon(z)$ is real and $\varepsilon_1 \neq \varepsilon_2$. The profile drawn is the double exponential, defined in Section 3-6, equation (69).

The s wave reflection amplitude is thus determined to second order in the interface thickness in terms of the two integrals λ_1 and λ_2 over the difference between the given profile and the step profile. These integrals depend on the choice of relative positioning of $\varepsilon(z)$ and $\varepsilon_0(z)$, as shown in Figure 3-1.

The result obtained by theory for any observable, such as $|r_s|^2$, must be independent of an arbitrary choice made in calculating that observable. It follows that

$$\begin{aligned} |r_s|^2 &= |r_{s0} + r_{s1} + r_{s2} + \ldots|^2 \\ &= |r_{s0}|^2 + 2 \operatorname{Re}(r_{s0}^* r_{s1}) + \{|r_{s1}|^2 + 2 \operatorname{Re}(r_{s0}^* r_{s2})\} + \ldots \end{aligned} \tag{25}$$

must be invariant with respect to the choice of relative positioning of $\varepsilon(z)$ and $\varepsilon_0(z)$. Consider for example the case where ε is real, and also q_2 is real (thus excluding total internal reflection). Then r_{s0} and r_{s2} are both real, while r_{s1} is purely imaginary. The lowest order correction to r_{s0}^2 is then the second order term within the braces in (25). From (23) and (24), this is equal to

$$\{|r_{s1}|^2 + 2r_{s0}r_{s2}\} = -\frac{4q_1q_2\omega^4/c^4}{(q_1+q_2)^4}[2(\varepsilon_1 - \varepsilon_2)\lambda_2 - \lambda_1^2]. \tag{26}$$

We thus expect the combination $2(\varepsilon_1 - \varepsilon_2)\lambda_2 - \lambda_1^2$ to be invariant with respect to the relative positioning of ε and ε_0. This turns out to be true, as we will show in the next section.

3-3 Integral invariants

Let $\varepsilon(z)$ be a function with asymptotic values $\varepsilon(-\infty) = \varepsilon_1$ and $\varepsilon(+\infty) = \varepsilon_2$, and $\varepsilon_0(z)$ the step function taking values ε_1 for $z < 0$ and ε_2 for $z > 0$. Consider the dependence of the integral

$$\lambda_{n+1}(s) = \int_{-\infty}^{\infty} \mathrm{d}z\,[\varepsilon(z - s) - \varepsilon_0(z)]z^n \tag{27}$$

on the shift parameter s. (In changing s we change the relative positioning of the two profiles.) We have

$$\begin{aligned} \lambda_{n+1}(s) &= \int_{-\infty}^{\infty} \mathrm{d}z[\varepsilon(z) - \varepsilon_0(z) + \varepsilon_0(z) - \varepsilon_0(z+s)](z+s)^n \\ &= \int_{-\infty}^{\infty} \mathrm{d}z[\varepsilon(z) - \varepsilon_0(z)](z+s)^n + (\varepsilon_1 - \varepsilon_2)\int_{-s}^{0} \mathrm{d}z(z+s)^n \\ &= \int_{-\infty}^{\infty} \mathrm{d}z[\varepsilon(z) - \varepsilon_0(z)]\left[z^n + \binom{n}{1} z^{n-1}s + \cdots + s^n\right] + (\varepsilon_1 - \varepsilon_2)\frac{s^{n+1}}{n+1} \\ &= \lambda_{n+1}(0) + \binom{n}{1}\lambda_n(0)s + \cdots + \lambda_1(0)s^n + (\varepsilon_1 - \varepsilon_2)\frac{s^{n+1}}{n+1}. \end{aligned} \tag{28}$$

The shift-dependence of the first three integrals is

$$\begin{aligned} \lambda_1(s) &= \lambda_1(0) + (\varepsilon_1 - \varepsilon_2)s, \\ \lambda_2(s) &= \lambda_2(0) + \lambda_1(0)s + (\varepsilon_1 - \varepsilon_2)\frac{s^2}{2}, \\ \lambda_3(s) &= \lambda_3(0) + 2\lambda_2(0)s + \lambda_1^2(0)s^2 + (\varepsilon_1 - \varepsilon_2)\frac{s^3}{3}. \end{aligned} \tag{29}$$

Note that

$$2(\varepsilon_1 - \varepsilon_2)\lambda_2(s) - \lambda_1^2(s) = 2(\varepsilon_1 - \varepsilon_2)\lambda_2(0) - \lambda_1^2(0), \tag{30}$$

so that

$$i_2 = 2(\varepsilon_1 - \varepsilon_2)\lambda_2 - \lambda_1^2 \tag{31}$$

is invariant with respect to the relative positioning of ε and ε_0. Similarly

$$i_3 = 3(\varepsilon_1 - \varepsilon_2)^2\lambda_3 - 6(\varepsilon_1 - \varepsilon_2)\lambda_2\lambda_1 + 2\lambda_1^3 \tag{32}$$

is an integral invariant. There is an infinite hierarchy of such invariants (assuming the existence of λ_n for all n). The general formula has been given by Lekner (1984).

The invariance of i_2 was suggested on physical grounds, and follows in a straightforward way from the analysis given above. Nevertheless, the form (31) is not obviously invariant, and it is useful to express i_2 in a form which *is* manifestly invariant:

$$i_2 = -\int_{-\infty}^{\infty} dz_1 \int_{-\infty}^{\infty} dz_2[\varepsilon(z_1) - \varepsilon_0(z_1 - z_2)][\varepsilon(z_2) - \varepsilon_0(z_2 - z_1)]. \tag{33}$$

This form immediately suggests how other integral invariants may be generated: multiplication of the integrand in (33) by $f(z_1 - z_2)$ will produce integral invariants, for any f (subject only to the convergence of the integral). An example is $f = e^{ikz}$, which produces another set of invariants via its expansion in powers of k (Lekner (1984), Appendix A).

Any single-valued function of a step function is also a step function. The invariants developed above thus have endless generalizations. For the p wave, we will need invariants arising out of integrals over the difference between the reciprocals of ε and ε_0. We define, in parallel with (27),

$$\Lambda_{n+1}(s) = \varepsilon_1\varepsilon_2 \int_{-\infty}^{\infty} dz \left[\frac{1}{\varepsilon_0(z)} - \frac{1}{\varepsilon(z - s)}\right] z^n.$$

Their transformation properties are the same as those of the λ's: for example

$$\begin{aligned} \Lambda_1(s) &= \Lambda_1(0) + (\varepsilon_1 - \varepsilon_2)s, \\ \Lambda_2(s) &= \Lambda_2(0) + \Lambda_1(0)s + (\varepsilon_1 - \varepsilon_2)\frac{s^2}{2}. \end{aligned} \tag{35}$$

Thus

$$I_2 = 2(\varepsilon_1 - \varepsilon_2)\Lambda_2 - \Lambda_1^2 \tag{36}$$

is an integral invariant, etc. We note also that (from (29) and (35)) the difference $\mathscr{I}_1 = \Lambda_1 - \lambda_1$ is an invariant. This invariance is made explicit by means of the identity

$$\mathscr{I}_1 = \Lambda_1 - \lambda_1 = \varepsilon_1\varepsilon_2 \int_{-\infty}^{\infty} dz \left(\frac{1}{\varepsilon_0} - \frac{1}{\varepsilon}\right) - \int_{-\infty}^{\infty} dz\,(\varepsilon - \varepsilon_0)$$
$$= \int_{-\infty}^{\infty} dz\, \frac{(\varepsilon_1 - \varepsilon)(\varepsilon - \varepsilon_2)}{\varepsilon}. \tag{37}$$

The next mixed invariant is

$$\mathscr{I}_2 = \varepsilon_1\varepsilon_2 \int_{-\infty}^{\infty} dz_1 \int_{-\infty}^{\infty} dz_2 [\varepsilon(z_1) - \varepsilon_0(z_1 - z_2)] \left[\frac{1}{\varepsilon(z_2)} - \frac{1}{\varepsilon_0(z_2 - z_1)}\right]$$
$$= (\varepsilon_1 - \varepsilon_2)(\lambda_2 + \Lambda_2) - \lambda_1\Lambda_1$$
$$= \tfrac{1}{2}(i_2 + I_2 + \mathscr{I}_1^2). \tag{38}$$

Finally, we shall need another invariant, related to $\mathscr{I}_2$, which enters in the expressions for $|r_p|^2$ and r_p/r_s to second order. This is

$$j_2 = \int_{-\infty}^{\infty} dz_1 \int_{-\infty}^{\infty} dz_2\, \varepsilon_0(z_2 - z_1)[\varepsilon(z_1) - \varepsilon_0(z_1 - z_2)] \left[\frac{1}{\varepsilon(z_2)} - \frac{1}{\varepsilon_0(z_2 - z_1)}\right]. \tag{39}$$

3-4 $|r_p|^2$ and r_p/r_s to second order

The last section developed the tools which enable us to complete the characterization of the reflection of electromagnetic waves to second order in the interface thickness. For the p wave, $\mathbf{B} = (0, e^{i(Kx-\omega t)}\, B(z), 0)$, with

$$\frac{d}{dz}\left(\frac{1}{\varepsilon}\frac{dB}{dz}\right) + \left(\frac{\omega^2}{c^2} - \frac{K^2}{\varepsilon}\right) B = 0, \quad e^{iq_1 z} - r_p\, e^{-iq_1 z} \leftarrow B \rightarrow \left(\frac{\varepsilon_2}{\varepsilon_1}\right)^{1/2} t_p\, e^{iq_2 z}. \tag{40}$$

A perturbation theory for the p wave will be sketched below. For the first order results, however, the full perturbation theory with its Green's function is not needed. We can obtain the first order results directly from a comparison identity based on (40) and its companion for B_0:

$$\frac{d}{dz}\left(\frac{1}{\varepsilon_0}\frac{dB_0}{dz}\right) + \left(\frac{\omega^2}{c^2} - \frac{K^2}{\varepsilon_0}\right) B_0 = 0, \quad e^{iq_1 z} - r_{p0}\, e^{-iq_1 z} \leftarrow B_0 \rightarrow \left(\frac{\varepsilon_2}{\varepsilon_1}\right)^{1/2} t_{p0}\, e^{iq_2 z}. \tag{41}$$

This is (Lekner, 1982, equation (46)),

$$r_p = r_{p0} + \frac{1}{2iQ_1}\int_{-\infty}^{\infty} dz \left\{\left(\frac{1}{\varepsilon_0} - \frac{1}{\varepsilon}\right) K^2 BB_0 + (\varepsilon - \varepsilon_0) CC_0\right\}, \tag{42}$$

where $Q_1 = q_1/\varepsilon_1$, $C = dB/\varepsilon dz$ and $C_0 = dB_0/\varepsilon_0\, dz$. (Note that C_0 is continuous at the discontinuity of ε_0, while dB_0/dz is not.) To first order in the interface thickness, it suffices to replace BB_0 by $B_0^2(0)$ and CC_0 by $C_0^2(0)$, where (from (1.28) and (1.31))

$$B_0(0) = \frac{2Q_1}{Q_1 + Q_2}, \qquad C_0(0) = \frac{2iQ_1Q_2}{Q_1 + Q_2}. \tag{43}$$

We obtain

$$r_{p1} = \frac{-2iQ_1}{(Q_1 + Q_2)^2}\left\{\frac{K^2\Lambda_1}{\varepsilon_1\varepsilon_2} - Q_2^2\lambda_1\right\}. \tag{44}$$

This result and (23) are together sufficient to determine r_p/r_s to first order (in contrast to the expressions for $|r_s|^2$ and $|r_p|^2$, which have no first order terms when ε and q_2 are real). We have

$$\begin{aligned}\frac{r_p}{r_s} &= \frac{r_{p0} + r_{p1} + r_{p2} + \ldots}{r_{s0} + r_{s1} + r_{s2} + \ldots}\\ &= \frac{r_{p0}}{r_{s0}}\left\{1 + \left(\frac{r_{p1}}{r_{p0}} - \frac{r_{s1}}{r_{s0}}\right) + \left(\frac{r_{p2}}{r_{p0}} - \frac{r_{p1}r_{s1}}{r_{p0}r_{s0}} + \frac{r_{s1}^2}{r_{s0}^2} - \frac{r_{s2}}{r_{s0}}\right) + \ldots\right\}.\end{aligned} \tag{45}$$

The first order term is given by (for $\varepsilon_1 \neq \varepsilon_2$)

$$r_{s0}\left(\frac{r_p}{r_s}\right)_1 = r_{p0}\left(\frac{r_{p1}}{r_{p0}} - \frac{r_{s1}}{r_{s0}}\right) = -\frac{2iQ_1K^2}{(Q_1 + Q_2)^2}\frac{\mathscr{I}_1}{\varepsilon_1\varepsilon_2} \tag{46}$$

(on using $r_{s0} = (q_1 - q_2)/(q_1 + q_2)$, $-r_{p0} = (Q_1 - Q_2)/(Q_1 + Q_2)$). The special case of reflection by a thin film between like media (when $\varepsilon_1 = \varepsilon_2$) will be discussed in the next section.

For the second order term in r_p/r_s and $|r_p|^2$ we need r_{p2}. This requires a perturbation theory for (40) based on (41), which is more complex than the perturbation theory for the s wave. We will give an outline here; further details may be found in Lekner (1982) and (1984). One constructs a Green's function $G(z, \zeta)$ satisfying

$$\frac{\partial}{\partial z}\left(\frac{1}{\varepsilon_0}\frac{\partial G}{\partial z}\right) + \left(\frac{\omega^2}{c^2} - \frac{K^2}{\varepsilon_0}\right)G = \delta(z - \zeta), \tag{47}$$

and incorporating the required boundary conditions. This is (cf. (10))

$$G(z,\zeta) = \begin{cases} \dfrac{e^{iq_2\zeta}}{2iQ_2}\left(e^{-iq_2z} + r_{p0}\,e^{iq_2z}\right) & 0 < z < \zeta \\[2ex] \dfrac{e^{iq_2z}}{2iQ_2}\left(e^{-iq_2\zeta} + r_{p0}\,e^{iq_2\zeta}\right) & 0 < \zeta < z \\[2ex] \dfrac{e^{i(q_2\zeta - q_1z)}}{i(Q_1 + Q_2)} & z < 0 < \zeta \\[2ex] \dfrac{e^{i(q_2z - q_1\zeta)}}{i(Q_1 + Q_2)} & \zeta < 0 < z \\[2ex] \dfrac{e^{-iq_1z}}{2iQ_1}\left(e^{iq_1\zeta} - r_{p0}\,e^{-iq_1\zeta}\right) & z < \zeta < 0 \\[2ex] \dfrac{e^{-iq_1\zeta}}{2iQ_1}\left(e^{iq_1z} - r_{p0}\,e^{-iq_1z}\right) & \zeta < z < 0 \end{cases} \tag{48}$$

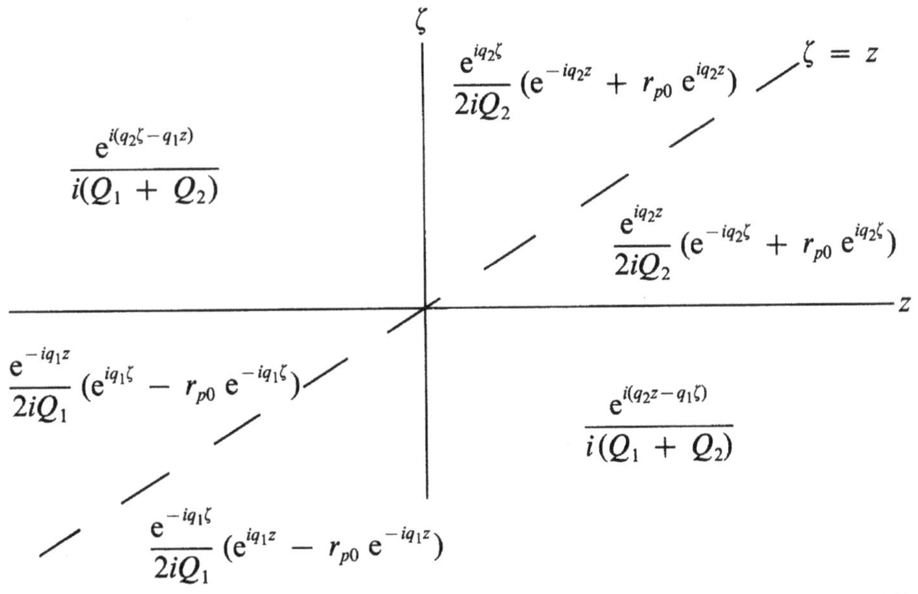

$B(z)$ satisfies an integro-differential equation

$$B(z) = B_0(z) - \int_{-\infty}^{\infty} d\zeta \left\{ \left(\frac{1}{\varepsilon_0(\zeta)} - \frac{1}{\varepsilon(\zeta)} \right) K^2 B(\zeta) G(z, \zeta) + (\varepsilon(\zeta) - \varepsilon_0(\zeta)) C(\zeta) \frac{1}{\varepsilon_0(\zeta)} \frac{\partial G}{\partial \zeta} \right\} \tag{49}$$

(from which (42) may be obtained by taking the limit $z \to -\infty$). Considerable work is required to extract the second order term r_{p2}. The result for $|r_p|^2$ to second order in the interface thickness is, for real ε and q_2,

$$|r_p|^2 = r_{p0}^2 - \frac{4Q_1Q_2}{\varepsilon_1\varepsilon_2(Q_1 + Q_2)^4} \left\{ \frac{\omega^4}{c^4} i_2 - \frac{\omega^2}{c^2} K^2 \left[j_2 + \left(\frac{1}{\varepsilon_1} + \frac{1}{\varepsilon_2} \right) i_2 \right] + \frac{K^4}{\varepsilon_1\varepsilon_2} [(\varepsilon_1 + \varepsilon_2) j_2 - \mathscr{I}_1^2] \right\} + \dots \tag{50}$$

This is to be compared with the simpler expression for the s wave:

$$|r_s|^2 = r_{s0}^2 - \frac{4q_1q_2\omega^4/c^4}{(q_1 + q_2)^4} i_2 + \dots \tag{51}$$

The p wave is more complicated, because the orientation of the electric field relative to the interface changes with the angle of incidence, while for the s polarization the electric field is always parallel to the interface. Measurement of R_s for an interface in the long wave limit can give the value of one invariant, i_2, irrespective of the angle of incidence. Measurement of R_p, at a minimum of three angles, can give the values of i_2, j_2 and $\mathscr{I}_1^2$.

As we saw in (46), the invariant $\mathscr{I}_1$ (complete with sign) can be found from the ellipsometric determination of r_p/r_s to first order in the interface thickness. The expression for r_p/r_s to second order is, for $\varepsilon_1 \neq \varepsilon_2$,

$$r_{s0}\left(\frac{r_p}{r_s}\right) = r_{p0} - \frac{2iQ_1K^2/\varepsilon_1\varepsilon_2}{(Q_1 + Q_2)^2} \mathscr{I}_1 + \frac{2Q_1(K^2/\varepsilon_1\varepsilon_2)^2}{(Q_1 + Q_2)^3} \mathscr{I}_1^2 + \frac{2Q_1Q_2K^2}{(Q_1 + Q_2)^2} \frac{j_2 - \left(\frac{1}{\varepsilon_1} + \frac{1}{\varepsilon_2} \right) i_2}{\varepsilon_1 - \varepsilon_2} + \dots \tag{52}$$

Note that the first and second order terms vanish at normal incidence ($K = 0$), as they must since there $r_p/r_s = 1$, identically.

We saw in Section 1-2 that an interface is expected to be nearly transparent to the p wave at the angle of incidence defined by $Q_1 = Q_2$, namely at the Brewster angle $\theta_B = \arctan(\varepsilon_2/\varepsilon_1)^{1/2}$. In Section 2-3 we proved the existence of at least one ellipsometric Brewster angle, θ_B', defined as the angle at which the real part of r_p/r_s is zero. We can obtain an expression for the difference $\Delta\theta_B = \theta_B' - \theta_B$ in the long wave limit. Assume absorption to be absent (real ε), in which case $\mathscr{I}_1$, i_2 and j_2 are real; and take $\theta_1 < \theta_c$, so q_2 and Q_2 are real. Then from (52) we find that (to second order in the interface thickness), Re (r_p/r_s) is zero when the expression

$$r_{p0} + \frac{2Q_1(K^2/\varepsilon_1\varepsilon_2)}{(Q_1 + Q_2)^3}\left\{\frac{K^2}{\varepsilon_1\varepsilon_2}\mathscr{I}_1^2 + \frac{\varepsilon_1\varepsilon_2 Q_2(Q_1 + Q_2)}{\varepsilon_1 - \varepsilon_2}\left[j_2 - \left(\frac{1}{\varepsilon_1} + \frac{1}{\varepsilon_2}\right)i_2\right]\right\}$$

is zero. Thus $\Delta\theta_B$ is second order in the interface thickness. On using (1.36) and (1.37) we find

$$\Delta\theta_B = \frac{(\varepsilon_1\varepsilon_2)^{5/2}\omega^2/c^2}{\varepsilon_1(\varepsilon_1^2 - \varepsilon_2^2)^2}\left[j_2 - \left(\frac{1}{\varepsilon_1} + \frac{1}{\varepsilon_2}\right)i_2 - \frac{1}{2}\left(\frac{1}{\varepsilon_1} - \frac{1}{\varepsilon_2}\right)\mathscr{I}_1^2\right] + \dots \quad (53)$$

3-5 Reflection by a thin film between like media

When media 1 and 2 are the same (as for a soap film in air), the formulae given above for $|r_s|^2$ and $|r_p|^2$ take a simpler form, and the formula for r_p/r_s takes a different form, since both r_{s0} and r_{p0} are now zero. Let ε_0 now denote the common value of ε_1 and ε_2, q_0 the common value of q_1 and q_2, etc. The integrals λ_1 and Λ_1 are now separately invariant (translating the constant functions ε_0 or ε_0^{-1} makes no difference), and the second order invariants i_2 and j_2 take the values

$$i_2 = -\lambda_1^2, \qquad j_2 = -\frac{2\lambda_1\Lambda_1}{\varepsilon_0}. \quad (54)$$

The reflectivities to second order in the interface thickness are now

$$|r_s|^2 = \frac{\omega^4/c^4}{4q_0^2}\lambda_1^2 + \dots \quad (55)$$

$$|r_p|^2 = \frac{\omega^4/c^4}{4q_0^2}\lambda_1^2\left\{1 - \frac{2(cK/\omega)^2}{\varepsilon_0}\left[\frac{\Lambda_1}{\lambda_1} + 1\right] + \frac{(cK/\omega)^4}{\varepsilon_0^2}\left[4\frac{\Lambda_1}{\lambda_1} + \left(\frac{\mathscr{I}_1}{\lambda_1}\right)^2\right]\right\} + \dots \quad (56)$$

The ellipsometric ratio r_p/r_s cannot be found from (45), since now r_{s0} and r_{p0} are zero. We have

$$\frac{r_p}{r_s} = \frac{r_{p1} + r_{p2} + \dots}{r_{s1} + r_{s2} + \dots} = \frac{r_{p1}}{r_{s1}} + \frac{r_{p2}r_{s1} - r_{s2}r_{p1}}{r_{s1}^2} + \dots, \quad (57)$$

the leading term now being zero order in the interface thickness:

$$\frac{r_{p1}}{r_{s1}} = \cos^2\theta_0 - \frac{\Lambda_1}{\lambda_1}\sin^2\theta_0. \quad (58)$$

This ratio correctly tends to unity as θ_0, the angle of incidence and refraction, tends to zero. When λ_1 and Λ_1 have the same sign, as would normally be the case, there is an angle at which the film is transparent to the p wave (to lowest order in the film thickness). This angle is an analogue of the Brewster angle arctan $(\varepsilon_2/\varepsilon_1)^{1/2}$ for unlike media. From (57), the thin film p wave transparency occurs at

$$\theta_0 = \arctan\left(\frac{\lambda_1}{\Lambda_1}\right)^{1/2} = \arctan\left\{\frac{\int_{-\infty}^{\infty} dz\,(\varepsilon - \varepsilon_0)/\varepsilon_0}{\int_{-\infty}^{\infty} dz\,(\varepsilon - \varepsilon_0)/\varepsilon}\right\}^{1/2}. \tag{59}$$

The above results may be compared to the exact formulae for a uniform layer of dielectric constant ε, with waves incident from (and transmitted into) a medium with dielectric constant ε_0. From (2.52) and (2.68),

$$r_s = e^{2iq_0z_1}\frac{i(q^2 - q_0^2)\tau}{2qq_0 - i(q^2 + q_0^2)\tau}, \tag{60}$$

and

$$-r_p = e^{2iq_0z_1}\frac{i(Q^2 - Q_0^2)\tau}{2QQ_0 - i(Q^2 + Q_0^2)\tau}, \tag{61}$$

where $q^2 = \varepsilon(\omega^2/c^2) - K^2$, $Q = q/\varepsilon$, and $\tau = \tan q\Delta z$. For the uniform film of thickness Δz, $\lambda_1 = (\varepsilon - \varepsilon_0)\Delta z$ and $\Lambda_1 = (\varepsilon_0/\varepsilon)\lambda_1$, and (55) and (56) agree with the reflectivities obtained from (60) and (61) to second order in Δz. The ratio r_p/r_s may be written as

$$\frac{r_p}{r_s} = \left[1 - \left(\frac{cK}{\omega}\right)^2\left(\frac{1}{\varepsilon} + \frac{1}{\varepsilon_0}\right)\right]\left\{\frac{1 - \frac{i}{2}\left(\frac{q}{q_0} + \frac{q_0}{q}\right)\tau}{1 - \frac{i}{2}\left(\frac{Q}{Q_0} + \frac{Q_0}{Q}\right)\tau}\right\}. \tag{62}$$

The square bracket in (62) is equal to $\cos^2\theta_0 - (\varepsilon_0/\varepsilon)\sin^2\theta_0$, and is zero at $\theta_0 = \arctan\,(\varepsilon/\varepsilon_0)^{1/2}$, which is the same as the Brewster angle for a sharp boundary between ε_0 and an *infinite* medium ε. It follows that a uniform film between like media is always transparent to the p wave at the same angle, irrespective of its thickness. A non-uniform film, on the other hand, is transparent at θ_0 given by (59) only to the lowest order in the film thickness.

There is a difficulty associated with the reflectivity formulae (55) and (56) which is shared by the first order perturbation expressions

$$r_s^{\text{pert}} = \frac{i\omega^2/c^2}{2q_0}\int_{-\infty}^{\infty} dz\,[\varepsilon(z) - \varepsilon_0]\,e^{2iq_0z}, \tag{63}$$

$$r_p^{\text{pert}} = \frac{1}{2iQ_0}\int_{-\infty}^{\infty} dz\left\{\left(\frac{1}{\varepsilon_0} - \frac{1}{\varepsilon}\right)K^2 - (\varepsilon - \varepsilon_0)Q_0^2\right\}e^{2iq_0z}. \tag{64}$$

((63) is the same as (14), and (64) may be obtained from (42), as will be discussed in Chapter 4). The difficulty occurs at grazing incidence, when $\theta_0 \to \pi/2$ and $q_0 = \sqrt{\varepsilon_0}(\omega/c)\cos\theta_0 \to 0$, and all these expressions are *divergent*. This divergence is unphysical (r_s and r_p must stay within the unit circle), and persists in higher order perturbation theory. One of the virtues of the variational expressions for r_s and r_p derived in Chapter 4 is that these are no longer divergent at grazing incidence, and in fact take the correct limiting values -1 and $+1$ as θ tends to $\pi/2$ (cf Section 2-3).

The reader may be puzzled as to how an expression like (55), which we have stated to be exact to second order in the interface thickness, could be divergent (and wrong) at grazing incidence. This puzzle is resolved by considering the order of the two limiting processes (the long wave limit, and the limit of $\theta_0 \to \pi/2$). An illustration of the application of these two limits is provided by the profile

$$\varepsilon(z) = \varepsilon_0 + \Delta\varepsilon \operatorname{sech}^2(z/a), \tag{65}$$

which will be discussed in detail in Section 4-3. The s wave reflectivity depends on two dimensionless parameters,

$$\alpha = \Delta\varepsilon(\omega/c)^2 a^2, \qquad \beta = q_0 a, \tag{66}$$

and when $\alpha > -1/4$ it is given by

$$R_s = \frac{\cos^2\left[\frac{\pi}{2}(1 + 4\alpha)^{1/2}\right]}{\cos^2\left[\frac{\pi}{2}(1 + 4\alpha)^{1/2}\right] + \sinh^2(\pi\beta)}. \tag{67}$$

When α and β are both small compared to $1/\pi$, this takes the form $\alpha^2/(\alpha^2 + \beta^2)$. For fixed (small) α, and $\beta \to 0$ (fixed interface thickness and angle of incidence tending to $\pi/2$), $R_s \to \alpha^2/\alpha^2 = 1$. For $\alpha/\beta \to 0$, which is the long wave limit $(\omega/c)a \to 0$ with θ_0 fixed, $R_s \to \alpha^2/\beta^2 = (\Delta\varepsilon(\omega^2/c^2)a/q_0)^2$, and is correctly given by (55), since $\lambda_1 = 2a\Delta\varepsilon$ for this profile.

3-6 Six profiles and their integral invariants

We have seen that, to second order in the interface thickness, the s wave reflectivity is characterized by one integral invariant (i_2), while the p wave reflectivity and r_p/r_s are characterized by the invariants $\mathscr{I}_1$, i_2 and j_2. Knowledge of these three invariants is thus sufficient to determine the reflection properties to this order. Conversely, reflection and ellipsometric studies with wavelengths that are large compared to the profile thickness can determine no more than these three integrals over the profile.

We shall list six dielectric function profiles and their invariants. Four of these have been studied already; the others (the linear and double exponential profiles) are useful simple models in the statistical mechanics and electrodynamics of interfaces.

In the following definitions z_1 is arbitrary, and $z_2 = z_1 + \Delta z$. For the first four profiles Δz is the total extent of the interface, while for the double exponential and hyperbolic tangent profiles Δz is a measure of the interface thickness. In the interval $z_1 \leqslant z \leqslant z_2$, the first four profiles take the form

$$\begin{aligned}
&\text{uniform layer: (or two-step)} && \varepsilon(z) = \varepsilon \quad \text{(a constant)} \\
&\text{linear:} && \varepsilon(z) = \varepsilon_1 + (\varepsilon_2 - \varepsilon_1)(z - z_1)/\Delta z \\
&\text{Rayleigh:} && \varepsilon(z) = [\varepsilon_1^{-1/2} + (\varepsilon_2^{-1/2} - \varepsilon_1^{-1/2})(z - z_1)/\Delta z]^{-2} \\
&\text{single exponential:} && \varepsilon(z) = \varepsilon_1 \exp[(z - z_1)\log(\varepsilon_2/\varepsilon_1)/\Delta z]
\end{aligned} \tag{68}$$

Table 3-1. First and second order integral invariants for six profiles. The first four interfaces have extent Δz, while for the double exponential and hyperbolic tangent profiles Δz is a measure of the interface thickness.

PROFILE	$\mathscr{I}_1/\Delta z$	$i_2/(\Delta z)^2$	$j_2/(\Delta z)^2$
uniform layer (two-step)	$(\varepsilon_1 - \varepsilon)(\varepsilon - \varepsilon_2)/\varepsilon$	$(\varepsilon_1 - \varepsilon)(\varepsilon - \varepsilon_2)$	$2(\varepsilon_1 - \varepsilon)(\varepsilon - \varepsilon_2)/\varepsilon$
linear	$\frac{1}{2}(\varepsilon_1 + \varepsilon_2) - \frac{\varepsilon_1\varepsilon_2}{\varepsilon_1 - \varepsilon_2} \log \frac{\varepsilon_1}{\varepsilon_2}$	$\frac{1}{12}(\varepsilon_1 - \varepsilon_2)^2$	$\frac{1}{2}(\varepsilon_1 + \varepsilon_2) - \frac{\varepsilon_1\varepsilon_2}{\varepsilon_1 - \varepsilon_2} \log \frac{\varepsilon_1}{\varepsilon_2}$
Rayleigh	$\frac{2}{3}(\sqrt{\varepsilon_1} - \sqrt{\varepsilon_2})^2$	$\frac{(\varepsilon_1 - \varepsilon_2)^2 \varepsilon_1\varepsilon_2}{(\sqrt{\varepsilon_1} - \sqrt{\varepsilon_2})^2} \left\{ \frac{\log \varepsilon_1/\varepsilon_2}{\varepsilon_1 - \varepsilon_2} - \frac{4}{(\sqrt{\varepsilon_1} + \sqrt{\varepsilon_2})^2} \right\}$	$\frac{2}{3}(\sqrt{\varepsilon_1} - \sqrt{\varepsilon_2})^2$
single exponential	$\varepsilon_1 + \varepsilon_2 - \frac{2(\varepsilon_1 - \varepsilon_2)}{\log \varepsilon_1/\varepsilon_2}$	$\left(\frac{\varepsilon_1 - \varepsilon_2}{\log \varepsilon_1/\varepsilon_2}\right)^2 - \varepsilon_1\varepsilon_2$	$\varepsilon_1 + \varepsilon_2 - \frac{2(\varepsilon_1 - \varepsilon_2)}{\log \varepsilon_1/\varepsilon_2}$
double exponential	$\varepsilon_1 \log \frac{\varepsilon_1 + \varepsilon_2}{2\varepsilon_2} + \varepsilon_2 \log \frac{\varepsilon_1 + \varepsilon_2}{2\varepsilon_1}$	$2(\varepsilon_1 - \varepsilon_2)^2$	$2(\varepsilon_1 - \varepsilon_2) \left\{ \log \frac{\varepsilon_1}{\varepsilon_2} + \int_{(\varepsilon_2 - \varepsilon_1)/2\varepsilon_1}^{(\varepsilon_1 - \varepsilon_2)/2\varepsilon_2} \frac{dx}{x} \log(1 + x) \right\}$
tanh	$(\varepsilon_1 - \varepsilon_2) \log \frac{\varepsilon_1}{\varepsilon_2}$	$\frac{\pi^2}{3}(\varepsilon_1 - \varepsilon_2)^2$	$2(\varepsilon_1 - \varepsilon_2)^2 \left\{ \frac{\pi^2}{6}\left(\frac{1}{\varepsilon_1} + \frac{1}{\varepsilon_2}\right) + \frac{\varepsilon_1 - \varepsilon_2}{\varepsilon_1\varepsilon_2} \int_0^1 \frac{dx}{1 + x} \times \left[\log\left(\frac{\varepsilon_1 x + \varepsilon_2}{\varepsilon_1 + \varepsilon_2 x}\right) + \frac{(\varepsilon_1^2 - \varepsilon_2^2) x \log x}{(\varepsilon_1 x + \varepsilon_2)(\varepsilon_1 + \varepsilon_2 x)} \right] \right\}$

The double exponential profile (illustrated in Figure 3-1) has two analytic forms:

$$\varepsilon(z) = \begin{cases} \varepsilon_1 + \frac{1}{2}(\varepsilon_2 - \varepsilon_1)\exp(z - z_1)/\Delta z & (z < z_1) \\ \varepsilon_2 + \frac{1}{2}(\varepsilon_1 - \varepsilon_2)\exp(z_1 - z)/\Delta z & (z > z_1) \end{cases} \tag{69}$$

The hyperbolic tangent profile is given by

$$\varepsilon(z) = \tfrac{1}{2}(\varepsilon_1 + \varepsilon_2) - \tfrac{1}{2}(\varepsilon_1 - \varepsilon_2)\tanh[(z - z_1)/2\Delta z]. \tag{70}$$

The evaluation of invariants is usually made simpler by choosing z_1 so as to make $\lambda_1 = 0$. For the symmetric profiles (linear, double-exponential, and hyperbolic tangent) the required positioning is at the origin ($z_1 = 0$). The invariants i_2 and j_2 then reduce to

$$i_2 = 2(\varepsilon_1 - \varepsilon_2)\int_{-\infty}^{\infty} dz\,[\varepsilon(z) - \varepsilon_0(z)]z \qquad (\lambda_1 = 0) \tag{71}$$

$$j_2 = 2(\varepsilon_1 - \varepsilon_2)\int_{-\infty}^{\infty} dz\,[\varepsilon(z) - \varepsilon_0(z)]\left\{\frac{z}{\varepsilon(z)} + \int_0^z \frac{d\zeta}{\varepsilon(\zeta)}\right\} \qquad (\lambda_1 = 0) \tag{72}$$

The integrations leading to the invariants are mostly elementary. Further details may be found in the references to Table 1 of Lekner (1984), and in Appendix C of that reference. The results are summarized in Table 3-1. Note that all the invariants are symmetric with respect to the interchange of ε_1 and ε_2, and that they are all non-negative when ε is positive and lies between ε_1 and ε_2 for all z. These properties are evident for $\mathscr{I}_1$ in the form

$$\mathscr{I}_1 = \int_{-\infty}^{\infty} dz\,\frac{(\varepsilon_1 - \varepsilon)(\varepsilon - \varepsilon_2)}{\varepsilon}. \tag{73}$$

For i_2 and j_2 they may be deduced from the definitions (33) and (39). For example, the integrands in (33) and (39) always keep the same sign when ε is positive and $\min(\varepsilon_1, \varepsilon_2) \leqslant \varepsilon \leqslant \max(\varepsilon_1, \varepsilon_2)$.

References

References quoted in the text

J. Lekner (1984) "Invariant formulation of the reflection of long waves by interfaces", Physica **128A**, 229–252.

J. Lekner (1982) "Second-order ellipsometric coefficients", Physica **113A**, 506–520.

The perturbation theory of Section 3-1 is based on

J. Lekner (1982) "Reflection of long waves by interfaces", Physica **112A**, 544–556.

The s wave Green's function for reflection at an inhomogeneity between like media (leading to (14)) was given by

P. M. Morse and H. Feshbach (1953) "Methods of theoretical physics", McGraw-Hill, p. 1071.

Perturbation theory for reflection of the s wave at an interface between unlike media has previously been developed by

D. G. Triezenberg (1973) "Capillary waves in a diffuse liquid-gas interface" (Ph.D thesis, University of Maryland, unpublished).

There have been several (mainly formal) derivations of expansions for reflection properties in powers of the interface thickness (Sections 3-2 and 3-4):

R. C. Maclaurin (1905) "Theory of the reflection of light near the polarising angle", Proc. Roy. Soc. **A76**, 49–65;

J. W. S. Rayleigh (1912) "On the propagation of waves through a stratified medium, with special reference to the question of reflection", Proc. Roy. Soc. **A86**, 207–266;

F. Abelès (1950) "Recherches sur la propagation des ondes électromagnétiques sinusoïdales dans les milieux stratifiés. Application aux couches minces", Annales de Physique **5**, 596–640;

P. G. Drazin (1963) "On one-dimensional propagation of long waves", Proc. Roy. Soc. **A273**, 400–411.

The formula (46) for the first order contribution to r_p/r_s is more than a century old. It is often attributed to Drude, but was, in essence, first obtained by L. Lorenz in 1860. Rayleigh (1912, quoted above) derives this result, and gives references to earlier work of Lorenz, Vam Ryn, Drude, Schott and Maclaurin.

4

Variational theory

In the previous chapter we have developed perturbation theories for dealing with reflection of electromagnetic or particle waves whose wavelength is long compared to the thickness of the interface. The variational theory of this chapter builds on these perturbation theories to provide formulae for the reflection amplitudes which have a greater range of validity, and which do not have the difficulties at grazing incidence such as the divergence of r^{pert} (as discussed in Section 3-5 in relation to equations (3.63) and (3.64)). The application of variational theory to the short wave case is also discussed.

4-1 A variational expression for the reflection amplitude

We shall first give a general formulation of variational theory for reflection of waves described by the equation

$$\frac{\mathrm{d}^2\psi}{\mathrm{d}z^2} + q^2\psi = 0, \qquad \mathrm{e}^{iq_1z} + r\,\mathrm{e}^{-iq_1z} \leftarrow \psi \rightarrow t\,\mathrm{e}^{iq_2z}, \tag{1}$$

and later specialize to the long wave, short wave, s and p cases. Suppose that a perturbation theory has been constructed, based on the solution ψ_0 of

$$\frac{\mathrm{d}^2\psi_0}{\mathrm{d}z^2} + q_0^2\psi_0 = 0, \qquad \mathrm{e}^{iq_1z} + r_0\,\mathrm{e}^{-iq_1z} \leftarrow \psi_0 \rightarrow t_0\,\mathrm{e}^{iq_2z}. \tag{2}$$

In particular, a Green's function $G(z, \zeta)$ is assumed known, satisfying

$$\frac{\partial^2 G}{\partial z^2} + q_0^2(z)G = \delta(z - \zeta), \tag{3}$$

and giving the correct asymptotic forms in the solution of the integral equation

$$\psi(z) = \psi_0(z) - \int_{-\infty}^{\infty} \mathrm{d}\zeta\ \Delta q^2(\zeta)G(z, \zeta)\psi(\zeta) \tag{4}$$

(here $\Delta q^2 = q^2 - q_0^2$). Such Green's functions are given in Sections 3-1 and 6-6 for the long and short wave cases. The results of perturbation theory follow by iteration of (4). Here we adapt Schwinger's variational method for the tangent of the phase shift produced by scattering off a central potential (Schwinger (1947); Blatt and Jackson (1949)) to the reflection problem. We rewrite (4) as

$$\psi(z) + \int_{-\infty}^{\infty} d\zeta \, \Delta q^2(\zeta) G(z, \zeta) \psi(\zeta) = \psi_0(z), \tag{5}$$

multiply by $\Delta q^2(z)\psi(z)$, and integrate over the whole range of z. The result is of the form

$$S = F \tag{6}$$

where S (of second degree in the unknown ψ) is given by

$$S = \int_{-\infty}^{\infty} dz \, \Delta q^2(z)\psi^2(z) + \int_{-\infty}^{\infty} dz \, \Delta q^2(z)\psi(z) \int_{-\infty}^{\infty} d\zeta \, \Delta q^2(\zeta)\psi(\zeta)G(z, \zeta), \tag{7}$$

and F (of first degree in ψ) is given by

$$F = \int_{-\infty}^{\infty} dz \, \Delta q^2(z)\psi(z)\psi_0(z). \tag{8}$$

From the comparison identity (2.6), which we can rewrite as

$$r = r_0 - \frac{1}{2iq_1} \int_{-\infty}^{\infty} dz \, \Delta q^2(z)\psi(z)\psi_0(z), \tag{9}$$

we see that, for the exact ψ,

$$F = 2iq_1(r_0 - r). \tag{10}$$

For the exact ψ we also have $S = F$. Consider now a shift to a neighbouring function (in the variational sense) $\psi + \delta\psi$, where ψ is the exact solution. The integrals F and S shift by

$$\delta F = \int_{-\infty}^{\infty} dz \, \delta\psi(z)\Delta q^2(z)\psi_0(z). \tag{11}$$

$$\delta S = 2\int_{-\infty}^{\infty} dz \, \delta\psi(z)\Delta q^2(z) \left\{\psi(z) + \int_{-\infty}^{\infty} d\zeta \, \Delta q^2(\zeta)G(z, \zeta)\psi(\zeta)\right\}. \tag{12}$$

From (5) the quantity in braces is equal to $\psi_0(z)$, and so $\delta S = 2\delta F$. But $S = F$, so $\delta S/S = 2\delta F/F$, or

$$\delta(F^2/S) = 0. \tag{13}$$

This is the variational principle: the correct ψ will extremize F^2/S. The extremal value of F^2/S approximates $F = 2iq_1(r_0 - r)$ and thus we have a variational estimate for the reflection amplitude:

$$r^{\text{var}} = r_0 - F^2/2iq_1 S. \tag{14}$$

In general one has a parametrized trial function $\psi^{\text{var}}(z)$, which when substituted for ψ gives the values F^{var} and S^{var}. The parameters which extremize $(F^{\text{var}})^2/S^{\text{var}}$ then give the best value (in the space spanned by the trial function) of the reflection amplitude. However, a useful variational estimate can be obtained without any variational parameters in ψ, provided ψ is well chosen. For example, we can take

$\psi^{\text{var}} = \psi_0$. Denote the corresponding values of F and S by F_0 and S_0, and the resulting variational estimate for r by $r_0 + r_1^{\text{var}}$. Then

$$r_1^{\text{var}} = -F_0^2/2iq_1 S_0. \tag{15}$$

From (7) and (8)

$$S_0 = F_0 + \int_{-\infty}^{\infty} \mathrm{d}z\, \Delta q^2(z)\psi_0(z) \int_{-\infty}^{\infty} \mathrm{d}\zeta\, \Delta q^2(\zeta)\psi_0(\zeta)G(z, \zeta); \tag{16}$$

while from (9) we see that

$$F_0 = -2iq_1 r_1 \tag{17}$$

where r_1 is the first order term in the perturbation series

$$r = r_0 + r_1 + r_2 + \dots . \tag{18}$$

Thus

$$r_1^{\text{var}} = r_1 F_0/S_0 = \frac{r_1}{1 + \dfrac{i}{2q_1 r_1}\displaystyle\int_{-\infty}^{\infty} \mathrm{d}z\, \Delta q^2(z)\psi_0(z) \int_{-\infty}^{\infty} \mathrm{d}\zeta\, \Delta q^2(\zeta)\psi_0(\zeta)G(z, \zeta)}. \tag{19}$$

We will apply this general expression to reflection of the electromagnetic s wave in the next two sections.

4-2 Variational estimate for r_s in the long wave case

The formulae of the previous section apply directly to the electromagnetic s wave, with $\psi = E$ and $q^2 = \varepsilon\omega^2/c^2 - K^2$. In the long wave case the appropriate starting point for perturbation theory is the step profile

$$q_0^2(z) = \tfrac{1}{2}(q_1^2 + q_2^2) - \tfrac{1}{2}(q_1^2 - q_2^2)\ \mathrm{sgn}\,(z), \tag{20}$$

and the corresponding solution

$$E_0(z) = \begin{cases} e^{iq_1 z} + r_0\, e^{-iq_1 z} & (z < 0) \\ t_0\, e^{iq_2 z} & (z > 0), \end{cases} \tag{21}$$

where $r_0 = (q_1 - q_2)/(q_1 + q_2)$ and $t_0 = 2q_1/(q_1 + q_2) = 1 + r_0$. The Green's function for this case was given in Section 3-1, equation (3.10). The first order term of perturbation theory is given by (3.13):

$$r_1 = \frac{i}{2q_1}\int_{-\infty}^{0} \mathrm{d}\zeta\, \Delta q^2(\zeta)\,(e^{iq_1\zeta} + r_0\, e^{-iq_1\zeta})^2$$

$$+ \frac{i}{q_1 + q_2}(1 + r_0)\int_0^{\infty} \mathrm{d}\zeta\, \Delta q^2(\zeta)\, e^{2iq_2\zeta}. \tag{22}$$

The variational estimate of r_s obtained from (19) on substituting (21), (22) and the Green's function of Section 3-1, has built into it two important general properties: it is correct to second order in the interfacial thickness/wavelength expansion, and it is correct at grazing incidence.

To verify the first statement, we express the expansion of r_s^{var} in terms the integrals

$$\lambda_n = \int_{-\infty}^{\infty} \mathrm{d}z\, (\varepsilon - \varepsilon_0) z^{n-1}, \tag{23}$$

defined in Section 3-2. We find

$$r_s^{\text{var}} = r_0 + \frac{2iq_1\omega^2/c^2}{(q_1 + q_2)^2}\left\{\lambda_1 + 2iq_2\lambda_2 + \frac{i\omega^2/c^2}{q_1 + q_2}\lambda_1^2 + \ldots\right\}. \tag{24}$$

This is correct to second order in the interface thickness, for all positioning of the reference profile $\varepsilon_0(z)$ (see equation (3.22)). In contrast, the expansion of r_1 is correct to this order only if the relative positioning of ε and ε_0 is such as to make λ_1 equal to zero. Such positioning is in general possible only when $\varepsilon_1 \neq \varepsilon_2$ and ε is real everywhere.

To verify that r_s^{var} correctly tends to -1 at grazing incidence (an exact general result, as shown in Section 2-3) we need to consider the $\varepsilon_1 \neq \varepsilon_2$ and $\varepsilon_1 = \varepsilon_2$ cases separately. When $\varepsilon_1 \neq \varepsilon_2$, r_1 and r_1^{var} become proportional to q_1 as $q_1 \to 0$, while $r_0 \to -1$ and so $r_s^{\text{var}} \to -1$. When $\varepsilon_1 = \varepsilon_2$, let q_0 be the common value of q_1 and q_2. As $q_0 \to 0$, $r_1 \to i(\omega^2/c^2)\lambda_1/2q_0$ and $S_0 - F_0 \to (\omega^4/c^4)\lambda_1^2/2iq_0$. From (19) we see that r_1^{var} correctly tends to -1 as q_0 tends to zero. This is in contrast to the *divergence* of the perturbation expression r_1 at grazing incidence, as can be seen by setting $q_1 = q_2 = q_0$ in (22):

$$r_{1s} = \frac{i}{2q_0}\int_{-\infty}^{\infty} \mathrm{d}\zeta\, \Delta q^2(\zeta)\, \mathrm{e}^{2iq_0\zeta}. \tag{25}$$

4-3 Exact, perturbation and variational results for the sech² profile

The comparative accuracy of the perturbation and variational expressions for the reflectivity is conveniently demonstrated by the sech² profile

$$\varepsilon(z) = \varepsilon_0 + \Delta\varepsilon\, \mathrm{sech}^2\,(z/a), \tag{26}$$

for which the exact, perturbation and variational results can all be found analytically for the s wave. The s wave equation reads

$$\frac{\mathrm{d}^2 E}{\mathrm{d}z^2} + \left[q_0^2 + \Delta\varepsilon \frac{\omega^2}{c^2}\mathrm{sech}^2\,(z/a)\right]E = 0, \tag{27}$$

where

$$q_0^2 = \varepsilon_0\frac{\omega^2}{c^2} - K^2 = \varepsilon_0\frac{\omega^2}{c^2}\cos^2\theta \tag{28}$$

gives the common value at $z = \pm\infty$ of the wavenumber component q_0 perpendicular to the interface, and θ is the common value of the angles of incidence and refraction.

A solution of (27) in terms of the hypergeometric function (defined and discussed in Section 2-5) can be found by changing to the variable $\xi = \tanh(z/a)$ (Epstein (1930), Budden (1961, 1985), Landau and Lifshitz (1965)). It is

$$E = (1 - \xi^2)^{-i\beta/2} F(-s - i\beta, 1 + s - i\beta; 1 - i\beta; \tfrac{1}{2}(1 - \xi)), \tag{29}$$

where $\beta = q_0 a$ and

$$s = \tfrac{1}{2}[-1 + (1 + 4\alpha)^{1/2}], \quad \alpha = \Delta\varepsilon(\omega a/c)^2. \tag{30}$$

As noted in Section 3-5, the problem is characterized by the two dimensionless parameters α (or s) and β. As $z \to \infty$, $1 - \xi \to 2e^{-2z/a}$, and E tends to $e^{iq_0 z}$ times a constant $(2^{-i\beta})$, as required for the reflection problem. The limiting form of E as $z \to -\infty$ (and $\xi \to -1$, $1 + \xi \to 2e^{2z/a}$) is found by using the relation between hypergeometric functions of ζ and $1 - \zeta$,

$$F(a, b; c; \zeta) = \frac{\Gamma(c)\Gamma(c - a - b)}{\Gamma(c - a)\Gamma(c - b)} F(a, b; 1 + a + b - c; 1 - \zeta) +$$

$$\frac{\Gamma(c)\Gamma(a + b - c)}{\Gamma(a)\Gamma(b)} (1 - \zeta)^{c-a-b} F(c - a, c - b; 1 + c - a - b; 1 - \zeta), \tag{31}$$

which holds for $|\arg(1 - \zeta)| < \pi$ (Oberhettinger (1964)). With $\zeta = \frac{1}{2}(1 - \xi)$ and $1 - \zeta = \frac{1}{2}(1 + \xi) \to e^{2z/a}$, this gives

$$2^{-i\beta} e^{-iq_0 z} \Gamma(1 - i\beta) \left\{ \frac{\Gamma(i\beta)}{\Gamma(1 + s)\Gamma(-s)} + \frac{\Gamma(-i\beta)}{\Gamma(-s - i\beta)\Gamma(1 + s - i\beta)} e^{2iq_0 z} \right\} \leftarrow E. \tag{32}$$

The reflection amplitude r_s is the ratio of the coefficient of $e^{-iq_0 z}$ to that of $e^{iq_0 z}$ and is therefore given by

$$r_s = \frac{\Gamma(i\beta)\Gamma(-s - i\beta)\Gamma(1 + s - i\beta)}{\Gamma(-i\beta)\Gamma(-s)\Gamma(1 + s)}$$

$$= \frac{\Gamma(i\beta)\Gamma(1 + s - i\beta) \sin \pi s}{\Gamma(-i\beta)\Gamma(1 + s + i\beta) \sin \pi(s + i\beta)} \tag{33}$$

(the second form is obtained by using $\Gamma(z)\Gamma(1 - z) = \pi/\sin \pi z$). Note that $r_s \to -1$ at grazing incidence (Section 2-3), since as $\beta = q_0 a$ tends to zero, $\Gamma(i\beta)/\Gamma(-i\beta) \to -1$ [see (2.86)].

The transmission amplitude is the ratio of the coefficients of $e^{iq_0 z}$ in the limiting forms as $z \to \pm\infty$:

$$t_s = \frac{\Gamma(1 + s - i\beta)\Gamma(-s - i\beta)}{\Gamma(1 - i\beta)\Gamma(i\beta)} = -\frac{\Gamma(1 + s - i\beta)\sin \pi i\beta}{\Gamma(1 + s + i\beta)\sin \pi(s + i\beta)}. \quad (34)$$

When s is real ($\alpha \geqslant -1/4$),

$$|r_s|^2 = \frac{\cos^2[\frac{1}{2}\pi(1 + 4\alpha)^{1/2}]}{\cos^2[\frac{1}{2}\pi(1 + 4\alpha)^{1/2}] + \sinh^2 \pi\beta}, \quad (35)$$

and $|t_s|^2 = 1 - |r_s|^2$, which is a special case of the conservation law $q_1(1 - |r|^2) = q_2|t|^2$ of Section 2-1. When $\alpha < -1/4$, s is complex:

$$s = -\tfrac{1}{2} + i\sigma, \qquad \sigma = \tfrac{1}{2}(4|\alpha| - 1)^{1/2}. \quad (36)$$

On setting $z = \frac{1}{2} + iy$ in $\Gamma(z)\Gamma(1 - z) = \pi/\sin \pi z$ we obtain

$$|\Gamma(\tfrac{1}{2} + iy)|^2 = \frac{\pi}{\cosh \pi y}; \quad (37)$$

using this in (33) and (34) gives

$$|r_s|^2 = \frac{\cosh^2 \pi\sigma}{\cosh^2 \pi\sigma + \sinh^2 \pi\beta}, \quad |t_s|^2 = \frac{\sinh^2 \pi\beta}{\cosh^2 \pi\sigma + \sinh^2 \pi\beta}. \quad (38)$$

For the comparison with perturbation and variational theories we restrict ourselves to $\Delta\varepsilon > 0$ (and thus $\alpha > 0$, s real). The reflectivity is then given by (35), with the corresponding perturbation and variational expressions found from (25) and (19):

$$|r_{1s}|^2 \doteq \left(\frac{\pi\alpha}{\sinh \pi\beta}\right)^2, \quad (39)$$

$$|r_s^{\mathrm{var}}|^2 = \frac{|r_{1s}|^2}{(1 + \alpha)^2 + (\alpha/\beta)^2}. \quad (40)$$

These expressions are compared in Figures 4-1 and 4-2.

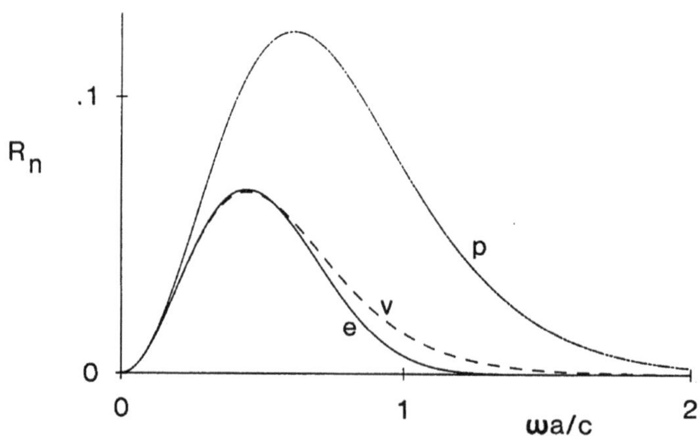

Figure 4-1. Normal incidence reflectivity for the $\mathrm{sech}^2(z/a)$ profile, as a function of the profile thickness. The curves are drawn for $\varepsilon_0 = 1$, $\Delta\varepsilon = 1$. The solid curve gives the exact reflectivity (e, equation (35)), the dashed curve the variational estimate (v, equation (40)) and the dot–dash curve the perturbation expression (p, equation (39)).

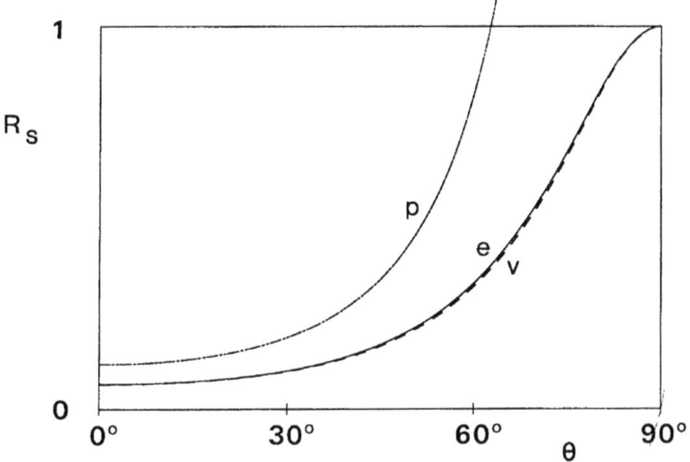

Figure 4-2. Reflectivity as a function of the angle of incidence for the $\text{sech}^2(z/a)$ profile. The exact, variational and perturbation results are denoted by e, v and p. The curves are drawn for $\varepsilon_0 = 1, \Delta\varepsilon = 1$, $(\omega/c)a = 1/2$.

4-4 Variational theory for the *p* wave

As in the *s* wave case, a variational estimate for the reflection amplitude builds on the corresponding perturbation theory. The long wave perturbation theory for the *p* wave was outlined in Section 3-4, and will be summarized here. The *p* wave has $\mathbf{B} = (0, \exp(iKx)B(z), 0)$ where K has the same meaning as for the *s* wave, and B satisfies

$$\frac{\mathrm{d}}{\mathrm{d}z}\left(\frac{1}{\varepsilon}\frac{\mathrm{d}B}{\mathrm{d}z}\right) + \left(\frac{\omega^2}{c^2} - \frac{K^2}{\varepsilon}\right)B = 0, \tag{41}$$

$$\mathrm{e}^{iq_1z} - r_p\,\mathrm{e}^{-iq_1z} \leftarrow B(z) \rightarrow \left(\frac{\varepsilon_2}{\varepsilon_1}\right)^{1/2} t_p\,\mathrm{e}^{iq_2z}. \tag{42}$$

The required Green's function satisfies

$$\frac{\partial}{\partial z}\left(\frac{1}{\varepsilon_0}\frac{\partial G}{\partial z}\right) + \left(\frac{\omega^2}{c^2} - \frac{K^2}{\varepsilon_0}\right)G = \delta(z - \zeta), \tag{43}$$

where $\varepsilon_0(z)$ is the step function profile, and $G(z, \zeta)$ was given in (3.48). B satisfies the integro-differential equation

$$B_0(z) = B(z) - \int_{-\infty}^{\infty} \mathrm{d}\zeta\;\Delta v(\zeta)\left\{K^2 B(\zeta)G(z, \zeta) + \frac{\mathrm{d}B}{\mathrm{d}\zeta}\frac{\partial G}{\partial \zeta}\right\}, \tag{44}$$

where $\Delta v = 1/\varepsilon - 1/\varepsilon_0$ and

$$B_0(z) = \begin{cases} e^{iq_1 z} - r_{p0}\, e^{-iq_1 z} & (z < 0) \\ \left(\dfrac{\varepsilon_2}{\varepsilon_1}\right)^{1/2} t_{p0}\, e^{iq_2 z} & (z > 0), \end{cases} \tag{45}$$

with

$$-r_{p0} = \frac{Q_1 - Q_2}{Q_1 + Q_2}, \qquad \left(\frac{\varepsilon_2}{\varepsilon_1}\right)^{1/2} t_{p0} = \frac{2Q_1}{Q_1 + Q_2}. \tag{46}$$

An exact expression for r_p is obtained from (44) by extracting the coefficient of $e^{-iq_1 z}$ as $z \to -\infty$. This is

$$\begin{aligned} r_p &= r_{p0} - \frac{1}{2iQ_1}\int_{-\infty}^{\infty} d\zeta\, \Delta v \left\{K^2 B B_0 + \frac{dB}{d\zeta}\frac{dB_0}{d\zeta}\right\} \\ &= r_{p0} - \frac{1}{2iQ_1}\int_{-\infty}^{\infty} d\zeta\, \{K^2 \Delta v B B_0 - \Delta\varepsilon C C_0\}, \end{aligned} \tag{47}$$

where $\Delta\varepsilon = \varepsilon - \varepsilon_0$, $C = dB/\varepsilon dz$ and $C_0 = dB_0/\varepsilon_0\, dz$. The first order perturbation theory expression for r_p is obtained by replacing B by B_0 and C by C_0 in (47). (This is equivalent to lowest order in Δv to replacing $dB/d\zeta$ by $dB_0/d\zeta$, but is preferable since C is continuous at a discontinuity in the dielectric function (as can be seen from (41)) while $dB/d\zeta$ is not, and since the resulting reflection amplitude gives the correct first order term in the interface thickness/wavelength expansion, to all orders in Δv). Thus in the expansion $r_p = r_{p0} + r_1 + r_2 + \ldots$,

$$r_1 = -\frac{1}{2iQ_1}\int_{-\infty}^{\infty} d\zeta \{K^2 \Delta v B_0^2 - \Delta\varepsilon C_0^2\}. \tag{48}$$

The variational expression for r_p is obtained by operating on (44) with

$$\int_{-\infty}^{\infty} dz \left\{\Delta v K^2 B - \frac{d}{dz}\left(\Delta v \frac{dB}{dz}\right)\right\}. \tag{49}$$

As in the s wave case, the resulting equation can again be put in the form $F = S$, where the term of the first degree in B is

$$F = \int_{-\infty}^{\infty} dz\, \Delta v \left\{K^2 B B_0 + \frac{dB}{dz}\frac{dB_0}{dz}\right\} = 2iQ_1(r_{p0} - r_p) \tag{50}$$

(the second equality follows from (47)). The term in second degree in the unknown B is

$$\begin{aligned} S = &\int_{-\infty}^{\infty} dz\, \Delta v K^2 B \left\{B - \int_{-\infty}^{\infty} d\zeta\, \Delta v \left[K^2 B G + \frac{dB}{d\zeta}\frac{\partial G}{\partial \zeta}\right]\right\} \\ &+ \int_{-\infty}^{\infty} dz\, \Delta v \frac{dB}{dz}\left\{\frac{dB}{dz} - \int_{-\infty}^{\infty} d\zeta\, \Delta v \left[K^2 B \frac{\partial G}{\partial z} + \frac{dB}{d\zeta}\frac{\partial^2 G}{\partial z \partial \zeta}\right]\right\}. \end{aligned} \tag{51}$$

As in the s wave case we find $\delta S = 2\delta F$ and hence the variational principle

$\delta(F^2/S) = 0$. Thus the device of operating on the integral equation with (49) produces a variational estimate

$$r_p^{\text{var}} = r_{p0} - \frac{F^2/S}{2iQ_1}, \tag{52}$$

in parallel with the corresponding expression for the s wave, equation (14).

The simplest trial function for $B(z)$ is $B_0(z)$. This gives the values F_0 and S_0 for F and S, where

$$F_0 = \int_{-\infty}^{\infty} dz\, \{\Delta v K^2 B_0^2 - \Delta\varepsilon C_0^2\} = -2iQ_1 r_1. \tag{53}$$

In the evaluation of S we must take care to include the $-\varepsilon_0(z)\delta(z - \zeta)$ singularity in $\partial^2 G/\partial z \partial\zeta$. We find, for general B,

$$\begin{aligned} S = {} & \int_{-\infty}^{\infty} dz\, \{\Delta v K^2 B^2 - \Delta\varepsilon C^2\} - K^4 \int_{-\infty}^{\infty} dz\, \Delta v B \int_{-\infty}^{\infty} d\zeta\, \Delta v B G \\ & + 2K^2 \int_{-\infty}^{\infty} dz\, \Delta v B \int_{-\infty}^{\infty} d\zeta\, \Delta\varepsilon C(\partial G/\varepsilon_0 \partial\zeta) \\ & - \int_{-\infty}^{\infty} dz\, \frac{\Delta\varepsilon}{\varepsilon_0} C \int_{-\infty}^{\infty} d\zeta\, \frac{\Delta\varepsilon}{\varepsilon_0} C \left(\frac{\partial^2 G}{\partial z \partial\zeta}\right)_r, \end{aligned} \tag{54}$$

where

$$\left(\frac{\partial^2 G}{\partial z \partial\zeta}\right)_r = \frac{\partial^2 G}{\partial z \partial\zeta} + \varepsilon_0(z)\delta(z - z) \tag{55}$$

is the regular part of $\partial^2 G/\partial z \partial\zeta$. The value of S_0 is found by replacing B with B_0 in (54), and the resulting variational estimate for the reflection amplitude has the form

$$r_p^{\text{var}} = r_{p0} + \frac{F_0}{S_0} r_1. \tag{56}$$

This expression gives a reflectivity which is correct to second order in the interface thickness/wavelength expansion, as may be shown by comparing the expansion of (56) with (3.50) (considerable reduction is required). Also built-in to the variational estimate (56) is the correct limiting value $r_p \to 1$ at grazing incidence (Section 2-3). The cases of equal and unequal ε_1 and ε_2 must be considered separately. When $\varepsilon_1 = \varepsilon_2$, $r_{p0} = 0$ and $(F_0/S_0)r_1 \to 1$ as $Q_1 \to 0$, Q_0 being the common value of Q_1 and Q_2. When $\varepsilon_1 \neq \varepsilon_2$, $(F_0/S_0)r_1 \to 0$ as $Q_1 \to 0$, and $r_p^{\text{var}} \to 1$ because $r_{p0} \to 1$.

4-5 Reflection by a non-uniform layer between like media

The variational formulae of Section 4-2 and 4-4 simplify considerably in the important special case of a reflecting layer between like media. As usual we set ε_0 equal to the common value of ε_1 and ε_2, and likewise for q_0; $\Delta\varepsilon$ stands for $\varepsilon - \varepsilon_0$, the deviation of the dielectric function from the ambient value. The variational theory based on long wave perturbation theory then gives r_s^{var} in terms of two integrals,

$$\lambda(k) = \int_{-\infty}^{\infty} \mathrm{d}z\, \mathrm{e}^{ikz}\, \Delta\varepsilon, \tag{57}$$

$$\sigma(k) = \int_{-\infty}^{\infty} \mathrm{d}z\, \Delta\varepsilon \left\{ \mathrm{e}^{ikz} \int_{-\infty}^{z} \mathrm{d}\zeta\, \Delta\varepsilon + \int_{z}^{\infty} \mathrm{d}\zeta\, \mathrm{e}^{ik\zeta}\, \Delta\varepsilon \right\}. \tag{58}$$

In terms of these integrals,

$$r_s^{\mathrm{var}} = \frac{-\dfrac{\omega^2/c^2}{2iq_0}\lambda(2q_0)}{1 + \dfrac{\omega^2/c^2}{2iq_0}\dfrac{\sigma(2q_0)}{\lambda(2q_0)}}. \tag{59}$$

The analogous result for the p wave is

$$r_p^{\mathrm{var}} = \frac{-[q_0^2\lambda(2q_0) - K^2\Lambda(2q_0)]/2iq_0}{1 + \dfrac{\{q_0^4\sigma(2q_0) - K^4\Sigma(2q_0) - 2q_0^2K^2\Gamma(2q_0)\}}{2iq_0[q_0^2\lambda(2q_0) - K^2\Lambda(2q_0)]}} \tag{60}$$

where

$$\Lambda(k) = \varepsilon_0 \int_{-\infty}^{\infty} \mathrm{d}z\, \mathrm{e}^{ikz} \frac{\Delta\varepsilon}{\varepsilon}, \tag{61}$$

$$\Sigma(k) = \varepsilon_0^2 \int_{-\infty}^{\infty} \mathrm{d}z \frac{\Delta\varepsilon}{\varepsilon} \left\{ \mathrm{e}^{ikz} \int_{-\infty}^{z} \mathrm{d}\zeta \frac{\Delta\varepsilon}{\varepsilon} + \int_{z}^{\infty} \mathrm{d}\zeta\, \mathrm{e}^{ik\zeta} \frac{\Delta\varepsilon}{\varepsilon} \right\}, \tag{62}$$

$$\Gamma(k) = \varepsilon_0 \int_{-\infty}^{\infty} \mathrm{d}z \frac{\Delta\varepsilon}{\varepsilon} \left\{ \mathrm{e}^{ikz} \int_{-\infty}^{z} \mathrm{d}\zeta\, \Delta\varepsilon - \int_{z}^{\infty} \mathrm{d}\zeta\, \mathrm{e}^{ik\zeta}\, \Delta\varepsilon \right\}. \tag{63}$$

(In both the σ and Σ expressions, the first and second terms are equal because of the z, ζ symmetry of the integrands). Of the five integrals, λ and Λ are Fourier transforms of $\Delta\varepsilon$ and $\varepsilon_0\Delta\varepsilon/\varepsilon$, respectively, and have the dimension of length. The integrals σ, Σ and Γ have dimensions of length squared. At grazing incidence, when $q_0 \to 0$, the results $r_s^{\mathrm{var}} \to -1$ and $r_p^{\mathrm{var}} \to 1$ follow from

$$\sigma(0) = \lambda^2(0), \qquad \Sigma(0) = \Lambda^2(0). \tag{64}$$

At normal incidence, when $K \to 0$ and $q_0 \to \sqrt{\varepsilon_0}\omega/c \equiv k_0$, both reflection amplitudes tend to

$$r_n^{\mathrm{var}} = \frac{-\dfrac{k_0}{2i\varepsilon_0}\lambda(2k_0)}{1 + \dfrac{k_0}{2i\varepsilon_0}\dfrac{\sigma(2k_0)}{\lambda(2k_0)}}. \tag{65}$$

From (60) we see that a layer between like media is transparent to the p wave (according to both the first order perturbation and variational theories) at an angle

$$\theta = \arctan\,\{\lambda(2q_0)/\Lambda(2q_0)\}^{1/2}. \tag{66}$$

This is an approximate extension of the rigorous result obtained in Section 3-5, that, to lowest order in the interface thickness, there is transparency at

$\theta_0 = \arctan(\lambda_1/\Lambda_1)^{1/2}$. Note that the ratio of $\lambda(2q_0)$ to $\Lambda(2q_0)$ is not (in general) real. Complete transparency at a certain angle is thus characteristic of thin films; as we saw in Section 3-5, it also characterizes uniform films of any thickness.

We shall compare the variational and perturbation theories with exact results for the uniform layer with dielectric constant ε for $z_1 \leqslant z \leqslant z_2 = z_1 + \Delta z$. In this case $\Delta\varepsilon$ and $\Delta\varepsilon/\varepsilon$ are both constant within the layer, and only the two integrals λ and σ are required for the reflection amplitudes, since

$$\Lambda(k) = \frac{\varepsilon_0}{\varepsilon}\lambda(k), \qquad \Sigma(k) = \left(\frac{\varepsilon_0}{\varepsilon}\right)^2\sigma(k), \qquad \Gamma(k) = 0. \tag{67}$$

The expression (60) then reduces to

$$r_p^{\text{var}} = \frac{\dfrac{-\omega/c}{2i\sqrt{\varepsilon_0}\cos\theta}\left[\cos^2\theta - \dfrac{\varepsilon_0}{\varepsilon}\sin^2\theta\right]\lambda(2q_0)}{1 + \dfrac{\omega/c}{2i\sqrt{\varepsilon_0}\cos\theta}\left[\cos^2\theta + \dfrac{\varepsilon_0}{\varepsilon}\sin^2\theta\right]\dfrac{\sigma(2q_0)}{\lambda(2q_0)}}, \tag{68}$$

correctly giving transparency at $\theta = \arctan(\varepsilon/\varepsilon_0)^{1/2}$.

For the integrals λ and σ we find

$$\lambda(2q_0) = \Delta\varepsilon\Delta z\, \mathrm{e}^{iq_0(z_1+z_2)}\, j_0(q_0\Delta z), \tag{69}$$

$$\sigma(2q_0) = (\Delta\varepsilon\Delta z)^2\, \mathrm{e}^{iq_0(z_1+z_2)}\, \{j_0(q_0\Delta z) + ij_1(q_0\Delta z)\}, \tag{70}$$

where $j_0(x) = x^{-1}\sin x$ and $j_1(x) = x^{-2}\sin x - x^{-1}\cos x$ are spherical Bessel functions.

In the figures below we compare the exact (e), perturbation (p) and variational (v) expressions for the reflectivity r_n at normal incidence as a function of the layer thickness (Figure 4-3), R_s and R_p as a function of the angle of incidence (Figures 4-4 and 4-5), and r_p/r_s in the complex plane as a function of the angle of incidence (Figure 4-6). The comparison is made for the values $\varepsilon_0 = 1$, $\varepsilon = 2$ (i.e., $\Delta\varepsilon = 1$).

4-6 The Hulthén–Kohn variational method applied to reflection

We have seen that the adaptation of Schwinger's variational technique in scattering theory to reflection has led, with the simplest trial function, to s and p reflection amplitudes which are correct to second order to the interface thickness, and are correct at grazing incidence. These desirable features have been obtained at the cost of some complexity, and we shall now show how the simpler method developed by Hulthén (1948) and Kohn (1948) for scattering problems may be applied to reflection.

We begin with the s wave, for which the exact field amplitude E satisfies

$$\frac{\mathrm{d}^2E}{\mathrm{d}z^2} + q^2E = 0, \qquad \mathrm{e}^{iq_1z} + r\,\mathrm{e}^{-iq_1z} \leftarrow E \rightarrow t\,\mathrm{e}^{iq_2z} \tag{71}$$

(we drop the subscript s except where needed to distinguish the s and p results).

Consider the functional

$$\Phi[E_t] = \int_{-\infty}^{\infty} dz\, E_t \left(\frac{d^2 E_t}{dz^2} + q^2 E_t \right) \tag{72}$$

of the trial function E_t, which we take to have the limiting forms

$$e^{iq_1 z} + r_t\, e^{-iq_1 z} \leftarrow E_t \rightarrow t_t\, e^{iq_2 z}. \tag{73}$$

The trial function differs from the exact E by δE: $E_t = E + \delta E$, with

$$e^{iq_1 z} + \delta r\, e^{-iq_1 z} \leftarrow \delta E \rightarrow \delta t\, e^{iq_2 z}, \tag{74}$$

where $\delta r = r_t - r$ and $\delta t = t_t - t$. We find

$$\delta\Phi = \Phi[E + \delta E] - \Phi[E] = \left[E \frac{d\delta E}{dz} - \delta E \frac{dE}{dz} \right]_{-\infty}^{\infty} + O(\delta E)^2. \tag{75}$$

(This result follows on two integrations-by-parts, and the use of the fact that, from (71), $\Phi[E] = 0$.) From (71) and (75) we obtain the result $\delta\Phi = 2iq_1\delta r$, which can be written in the form of a variational principle:

$$\delta(\Phi - 2iq_1 r) = 0. \tag{76}$$

In the application of this principle, we use a trial E_t, and the corresponding r_t and t_t, to evaluate $\Phi[E_t]$; then from (76)

$$\delta\Phi = \Phi[E_t] = 2iq_1(r_t - r) + O(\delta E)^2. \tag{77}$$

The variational estimate for the reflection amplitude is thus

$$r^{\text{var}} = r_t - \Phi[E_t]/2iq_1. \tag{78}$$

As an example, consider the simplest long-wave trial function $E_t = E_0$, the step profile solution given in (21). With $q^2 = q_0^2 + \Delta q^2$ as before,

$$\Phi[E_0] = \int_{-\infty}^{\infty} dz\, \Delta q^2 E_0^2, \tag{79}$$

which we recognise as the F_0 of the previous variational treatment. Thus the trial function E_0 leads to the first order perturbation result (compare (17))

$$r_s^{\text{var}} = r_{s0} - F_0/2iq_1 = r_{s0} + r_1. \tag{80}$$

Similarly, in the short wave case the trial function $E_t = (q_1/q)^{1/2}\, e^{i\phi}$ produces the perturbation result (6.55):

$$r_s^{\text{var}} = \frac{1}{4i} \int_{-\infty}^{\infty} dz \frac{d}{dz} \left(\frac{dq/dz}{q^{3/2}} \right) q^{-1/2}\, e^{2i\phi}. \tag{81}$$

The corresponding results for the p wave are not as satisfactory: if one defines the functional

$$\Phi[B_t] = \int_{-\infty}^{\infty} dz\, B_t \left\{ \frac{d}{dz} \left(\frac{1}{\varepsilon} \frac{dB_t}{dz} \right) + \left(\frac{\omega^2}{c^2} - \frac{K^2}{\varepsilon} \right) B_t \right\} \tag{82}$$

of the trial function B_t, the variational principle takes the form

$$\delta(\Phi + 2iQ_1 r_p) = 0. \tag{83}$$

For the zeroth-order trial function B_0 defined in (45),

$$\begin{aligned} r_p^{\text{var}} &= r_{p0} + \Phi[B_0]/2iQ_1 \\ &= r_{p0} - \frac{1}{2iQ_1}\int_{-\infty}^{\infty} dz\, \Delta v\, \{K^2 B_0 + (dB_0/dz)^2\}. \end{aligned} \tag{84}$$

This agrees with the perturbation result (48) only to lowest order in $\Delta v = 1/\varepsilon - 1/\varepsilon_0$. In consequence, (84) does not give the correct result to first order in the interface thickness (given by (3.44)), and does not agree with r_s^{var} at normal incidence.

The adaptation of the Hulthén–Kohn variational method to reflection problems is thus seen to give results which are inferior, for the simplest trial functions, to that obtained from adapting the Schwinger method. However, the greater simplicity of the Hulthén–Kohn method makes possible the use of more sophisticated trial functions (Joachain (1975), Chapter 10).

4-7 Variational estimates in the short wave case

We consider the s wave first. The variational theory is built on the perturbation theory of Section 6-5. The appropriate Green's function is (6.69):

$$2i\{q(z)q(\zeta)\}^{1/2} G(z, \zeta) = \begin{cases} \exp i(\phi(\zeta) - \phi(z)) & z < \zeta \\ \exp i(\phi(z) - \phi(\zeta)) & z > \zeta, \end{cases} \tag{85}$$

where ϕ is the phase integral

$$\phi(z) = \int^z d\zeta\, q(\zeta) \tag{86}$$

which is discussed in Section 6-2. The simplest variational expression uses

$$\psi_1^+ = \left(\frac{q_1}{q}\right)^{1/2} e^{i\phi} \tag{87}$$

as the trial function. Setting $\psi_0 = \psi_1^+$ in (19) we obtain the variational estimate

$$r_s^{\text{var}} = r_s^{(1)} \Big/ \left\{1 + \frac{i}{2r_s^{(1)}} \int_{-\infty}^{\infty} dz\, q^{-1/2} \Delta q^2\, e^{i\phi} \int_{-\infty}^{\infty} d\zeta\, q^{-1/2} \Delta q^2\, e^{i\phi}\, G(z, \zeta)\right\}, \tag{88}$$

where the numerator

$$r_s^{(1)} = \frac{1}{4i}\int_{-\infty}^{\infty} dz\, q^{-1/2} \frac{d}{dz}\left(\frac{dq/dz}{q^{3/2}}\right) e^{2i\phi} \tag{89}$$

is the short wave perturbation result. On using (85) and the expression (6.73), namely

$$\Delta q^2 = -\tfrac{1}{2}q^{1/2}\frac{\mathrm{d}}{\mathrm{d}z}\left(\frac{\mathrm{d}q/\mathrm{d}z}{q^{3/2}}\right),$$

the double integral in (88) reduces to

$$\frac{1}{4i}\int_{-\infty}^{\infty}\mathrm{d}z\left(\frac{\mathrm{d}\gamma}{\mathrm{d}z}+\tfrac{1}{2}q\gamma^2\right)\mathrm{e}^{2i\phi}\int_{-\infty}^{z}\mathrm{d}\zeta\left(\frac{\mathrm{d}\gamma}{\mathrm{d}\zeta}+\tfrac{1}{2}q\gamma^2\right). \tag{90}$$

Here, as in Chapter 6, γ is the dimensionless function $\mathrm{d}q/q^2\,\mathrm{d}z$, and we have used the fact that

$$q^{-1/2}\frac{\mathrm{d}}{\mathrm{d}z}\left(q^{-3/2}\frac{\mathrm{d}q}{\mathrm{d}z}\right) = \frac{\mathrm{d}\gamma}{\mathrm{d}z}+\tfrac{1}{2}q\gamma^2. \tag{91}$$

The evaluation of (90) involves a triple integration (unless the phase integral ϕ can be evaluated analytically), making it more difficult to apply than the long wave variational expression.

A variational theory for r_p in the short wave case may be derived along the same lines, since $b = (\varepsilon_1/\varepsilon)^{1/2}B$ satisfies an equation of the same form as E. The results are however so complex that they are unlikely to have practical value.

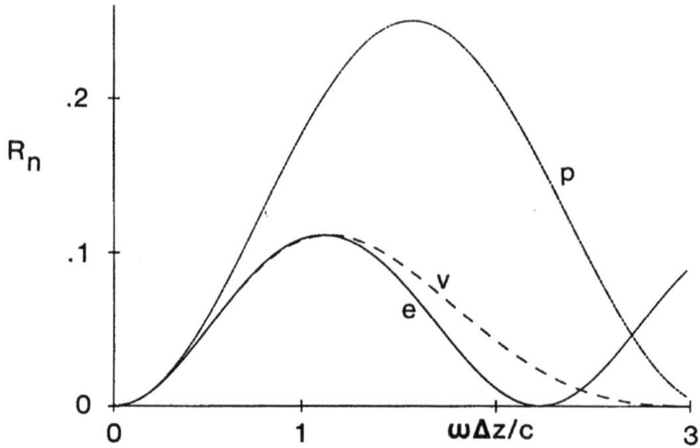

Figure 4-3. Reflectivity at normal incidence as a function of the layer thickness Δz. The exact reflectivity (e) is the solid curve, the perturbation result (p) is the dash–dot–dot curve, and the variational result (v) is the dashed curve. In this and the following figures, $\varepsilon_0 = 1$ and $\varepsilon = 2$.

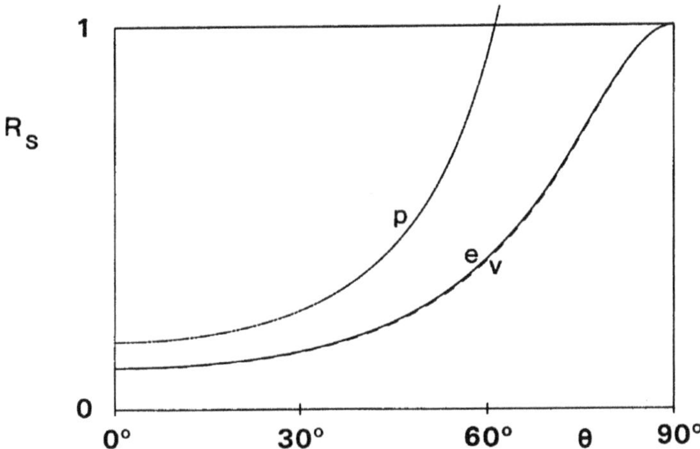

Figure 4-4. Reflectivity for the s wave as a function of the angle of incidence, at $(\omega/c)\Delta z = 1$. The exact, perturbation, and variational results are denoted by e(——), p(–··) and v(– – –)

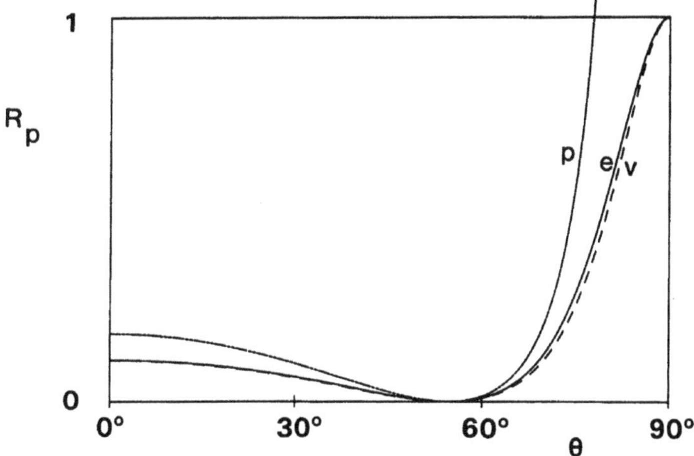

Figure 4-5. Reflectivity for the p wave as a function of the angle of incidence, at $(\omega/c)\Delta z = 1$. The exact, perturbation and variational reflectivities are all zero at $\theta = \arctan\sqrt{2} \simeq 54.7°$.

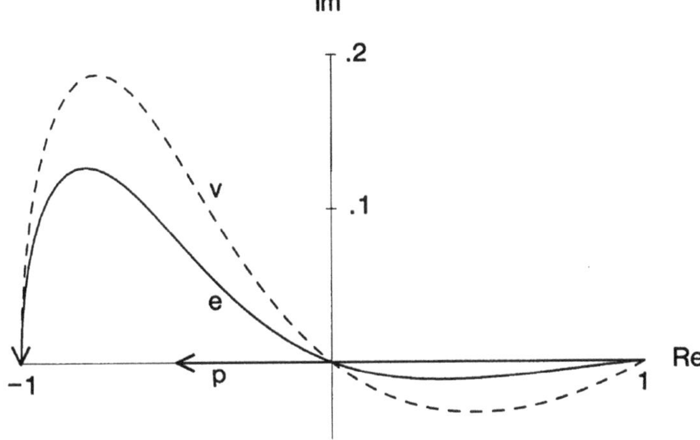

Figure 4-6. The ratio r_p/r_s in the complex plane, for $(\omega/c)\Delta z = 1$. The exact (e) and variational (v) trajectories are shown by solid and dashed lines; the perturbation trajectory lies along the real axis between $+1$ and $-1/2$. All three trajectories start at $+1$ at normal incidence and pass through the origin at $\theta = \arctan(\varepsilon/\varepsilon_0)^{1/2} \simeq 54.7°$. Only the perturbation trajectory does not end at -1 at grazing incidence.

References

References quoted in text

J. Schwinger (1947) "A variational principle for scattering problems", Phys. Rev. **72**, 742.

J. M. Blatt and J. D. Jackson (1949) "On the interpretation of neutron–proton scattering data by the Schwinger variational method", Phys. Rev. **26**, 18–28.

L. Huthén (1948) "On the Sturm–Liouville problem connected with a continuous spectrum", Arkiv. Mat. Astr. Fys. **35A**, paper 25, 14pp.

P. S. Epstein (1930) "Reflection of waves in an inhomogeneous absorbing medium", Proc. Nat. Acad. Sci. **16**, 627–637.

K. G. Budden (1961) "Radio waves in the ionosphere", Cambridge, Section 17.16.

K. G. Budden (1985) "The propagation of radio waves", Cambridge, Section 15.17.

L. D. Landau and E. M. Lifshitz (1965) "Quantum mechanics", Pergamon, p. 79.

F. Oberhettinger (1964) "Hypergeometric functions", Chapter 15 of Handbook of Mathematical Functions, edited by M. Abramowitz and I. A. Stegun, NBS Applied Mathematics Series No. 55.

L. Hulthén (1948) "On the Sturm–Liouville problem connected with a continuous spectrum", Arkiv. Mat. Astr. Fys. **35A**, paper 25, 14pp.

W. Kohn (1948) "Variational methods in nuclear collision problems", Phys. Rev. **74**, 1763–1772.

C. J. Joachain (1975) "Quantum collision theory", North Holland.

A general discussion of variational methods may be found in

P. M. Morse and H. Feshbach (1953) "Methods of theoretical physics", McGraw-Hill, Section 9.4.

This chapter is based on three papers by the author:

J. Lekner (1985) "Variational theory of reflection", Aust. J. Phys. **38**, 113–123.

J. Lekner (1986) "Reflection of light by a non-uniform film between like media", J. Opt. Soc. Amer., **A3**, 9–15.

J. Lekner (1986) "Variational theory of the reflection of light by interfaces", J. Opt. Soc. Amer., **A3**, 16–21.

5

Equations for the reflection amplitudes

For some purposes, both analytical and numerical, it is useful to transform the linear second order differential equation for the wave amplitude into a non-linear first order Riccati type differential equation for a quantity related to the reflection amplitude. The advantage lies in dealing directly with the quantity one wants to calculate. A disadvantage is the non-linearity of the resulting equations.

Early applications of this approach to the calculation of reflection amplitudes were by Walker and Wax (1946), Kofink (1947), Brekhovskikh (1949, 1980) and Schelkunoff (1951). This was preceded by development of related techniques in scattering theory, beginning with Morse and Allis (1933), and fully covered in Calogero's book (1967). Analogous methods were used by Courant and Hilbert (1924/1953) to obtain the asymptotic forms of Bessel functions.

5-1 A first order non-linear equation for an *s* wave reflection coefficient

We first rewrite the second order differential equation for the electromagnetic *s* wave (and, equivalently, for particle waves) as a pair of coupled first order equations: the equation

$$E'' + q^2 E = 0 \tag{1}$$

is equivalent to the pair

$$E' = D, \qquad D' = -q^2 E \tag{2}$$

(primes denote differentiation with respect to z). In turn, new functions F and G are introduced, defined by (2) and the equations

$$E = F + G, \qquad D = iq(F - G). \tag{3}$$

We shall see shortly that F and G have the character of incident and reflected waves, tending to a constant times $e^{\pm iq_1 z}$ as $z \to -\infty$. The ratio of these functions thus tends to r_s times $e^{-2iq_1 z}$ as $z \to -\infty$, since r_s is defined as the ratio of the coefficient of $e^{-iq_1 z}$ to that of $e^{+iq_1 z}$ as $z \to -\infty$.

On substituting (3) into (2) we obtain a pair of coupled first order equations for F and G, which may be solved for F' and G' to give

$$F' = iqF - \frac{q'}{2q}(F - G), \tag{4}$$

$$G' = -iqG + \frac{q'}{2q}(F - G). \tag{5}$$

(When q is constant we see that F and G are proportional to $e^{\pm iqz}$.) We now multiply (4) by G, (5) by F, subtract, and divide the result by F^2, obtaining an equation for $\varrho = G/F$:

$$\varrho' + 2iq\varrho - \frac{q'}{2q}(1 - \varrho) = 0. \tag{6}$$

The limiting forms of ϱ as $z \to \pm\infty$ are

$$r_s\, e^{-2iq_1 z} \leftarrow \varrho \to 0. \tag{7}$$

The absolute square of ϱ at $z = -\infty$ gives the reflectivity:

$$|\varrho(-\infty)|^2 = |r_s|^2 = R_s. \tag{8}$$

When q is real everywhere, the equation for the complex conjugate of ϱ is

$$\varrho^{*\prime} - 2iq\varrho^* - \frac{q'}{2q}(1 - \varrho^{*2}) = 0. \tag{9}$$

From (6) and (9) we may obtain an equation for the reflectivity function $R = \varrho\varrho^*$:

$$R' = \frac{q'}{q}\,\mathrm{Re}\,(\varrho)(1 - R). \tag{10}$$

The boundary conditions on $R(z)$ are

$$R(\infty) = 0, \qquad R(-\infty) = R_s. \tag{11}$$

On dividing both sides of (10) by $1 - R$, and integrating from $-\infty$ to $+\infty$, we obtain

$$\log(1 - R_s) = \int_{-\infty}^{\infty} dz\, \frac{q'}{q}\,\mathrm{Re}\,(\varrho). \tag{12}$$

The right-hand side is real, and thus $R_s \leqslant 1$ (cf Section 2-2). More generally, (10) divided by $1 - R$ can be integrated from z_0 to infinity, giving

$$\log(1 - R(z_0)) = \int_{z_0}^{\infty} dz\, \frac{q'}{q}\,\mathrm{Re}\,(\varrho), \tag{13}$$

so that $R(z)$ is less than unity everywhere. This is in accord with the physical interpretation of $|\varrho(z_0)|^2 = R(z_0)$, which is that of the reflectivity of a profile truncated at z_0, as shown in Figure 5-1.

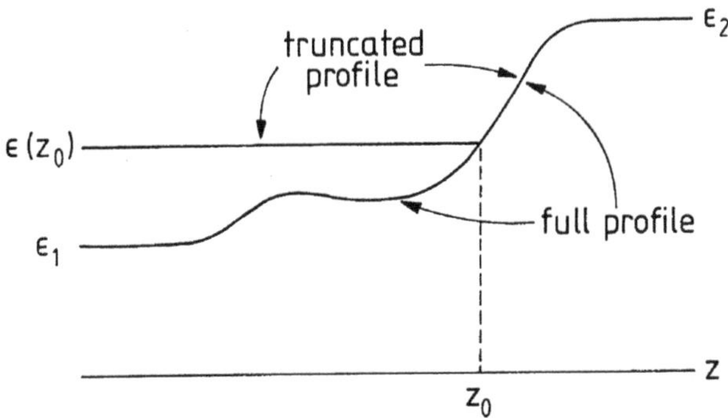

Figure 5-1. The functions $\varrho(z_0)$ and $R(z_0)$ correspond to a profile truncated at z_0, that is having $\varepsilon = \varepsilon(z)$ for $z \geqslant z_0$, and $\varepsilon = \varepsilon(z_0)$ for $z \leqslant z_0$. As z_0 increases the reflectivity $R(z_0)$ changes from the value R_s for the full profile to zero. (The decrease need not be monotonic.)

5-2 An example: reflection by the linear profile

We will illustrate the variation of R with z_0 for the linear profile

$$\varepsilon(z) = \begin{cases} \varepsilon_1 & z \leqslant z_1 \\ \varepsilon_1 + \dfrac{\Delta\varepsilon}{\Delta z}(z - z_1) & z_1 \leqslant z \leqslant z_2 \\ \varepsilon_2 & z \geqslant z_2 \end{cases} \tag{14}$$

where $\Delta\varepsilon = \varepsilon_2 - \varepsilon_1$, $\Delta z = z_2 - z_1$. At the same time we will give some of the properties of the solutions of (1) when ε is linear in z (the Airy functions), which will be useful in other applications. Within (z_1, z_2) ε is linear in z; as an intermediate step we introduce ε as the independent variable. Equation (1) transforms to

$$\frac{d^2E}{d\varepsilon^2} + \left(\frac{\Delta z}{\Delta\varepsilon}\frac{\omega}{c}\right)^2\left[\varepsilon - \left(\frac{cK}{\omega}\right)^2\right]E = 0. \tag{15}$$

We now transform to the variable $x = s\varepsilon$, where

$$s = \left(\left|\frac{\Delta z}{\Delta\varepsilon}\right|\frac{\omega}{c}\right)^{2/3}; \tag{16}$$

the equation becomes

$$\frac{d^2E}{dx^2} + \left[x - s\left(\frac{cK}{\omega}\right)^2\right]E = 0. \tag{17}$$

This equation has the general solution

$$E = \alpha Ai\left[s\left(\frac{cK}{\omega}\right)^2 - x\right] + \beta Bi\left[s\left(\frac{cK}{\omega}\right)^2 - x\right], \tag{18}$$

where Ai and Bi are Airy functions, the solutions of

$$\frac{d^2E}{d\zeta^2} - \zeta E = 0. \tag{19}$$

This is known as the Airy (or sometimes, the Stokes) differential equation (Heading 1962, Appendix A.3; Olver 1964, Section 10.4; Miller 1971). It has two power series solutions which are convergent for all ζ:

$$f(\zeta) = 1 + \zeta^3/3! + 1.4\zeta^6/6! + 1.4.7\zeta^9/9! + \ldots \tag{20}$$

$$g(\zeta) = \zeta + 2\zeta^4/4! + 2.5\zeta^7/7! + 2.5.8\zeta^{10}/10! + \ldots \tag{21}$$

The standard pair of independent solutions are

$$Ai(\zeta) = c_1 f(\zeta) - c_2 g(\zeta), \qquad Bi(\zeta) = \sqrt{3}[c_1 f(\zeta) + c_2 g(\zeta)], \tag{22}$$

where

$$\begin{aligned} c_1 &= Ai(0) = Bi(0)/\sqrt{3} = (3^{2/3}\Gamma(2/3))^{-1} = 0.355028\ldots \\ c_2 &= -Ai'(0) = Bi'(0)/\sqrt{3} = (3^{1/3}\Gamma(1/3))^{-1} = 0.258819\ldots. \end{aligned} \tag{23}$$

From the solution (18) and the general formula (2.25) we can immediately obtain the s wave reflection amplitude, care being taken to convert between derivatives with respect to z and x via

$$\frac{dE}{dz} = \frac{\Delta\varepsilon}{\Delta z}\frac{dE}{d\varepsilon} = \frac{\Delta\varepsilon}{\Delta z} s \frac{dE}{dx}. \tag{24}$$

The result is

$$r_s = e^{2iq_1z_1} \frac{q_1q_2(A_1B_2 - B_1A_2) + iq_1(A_1B_2' - B_1A_2') + iq_2(A_1'B_2 - B_1'A_2) - (A_1'B_2' - B_1'A_2')}{q_1q_2(A_1B_2 - B_1A_2) + iq_1(A_1B_2' - B_1A_2') - iq_2(A_1'B_2 - B_1'A_2) + (A_1'B_2' - B_1'A_2')}, \tag{25}$$

where

$$\begin{aligned} A_1 &= Ai\left[s\left(\frac{cK}{\omega}\right)^2 - s\varepsilon_1\right] = Ai\left[-s\left(\frac{cq_1}{\omega}\right)^2\right], \\ A_2 &= Ai\left[s\left(\frac{cK}{\omega}\right)^2 - s\varepsilon_2\right] = Ai\left[-s\left(\frac{cq_2}{\omega}\right)^2\right], \end{aligned} \tag{26}$$

with similar definition of B_1 and B_2 in terms of Bi, and

$$A_1' = \frac{\Delta\varepsilon}{\Delta z} s\, Ai'\left[-s\left(\frac{cq_1}{\omega}\right)^2\right], \quad \text{etc.} \tag{27}$$

Here we are interested in the variation of $|\varrho(z)|^2$ with z as z varies between z_1 and z_2. This may be obtained from the above by treating ε_1 as a variable. More instructive in the present context is a calculation of $\varrho = G/F$, where the functions F and G are found in terms of the known E: from (3)

$$F = \tfrac{1}{2}(E + \mathrm{d}E/iq\,\mathrm{d}z), \qquad G = \tfrac{1}{2}(E - \mathrm{d}E/iq\,\mathrm{d}z). \tag{28}$$

The condition $G = 0$ at $z = z_2$ (or $\varepsilon = \varepsilon_2$) determines the ratio of the coefficients α and β in (18):

$$\frac{\alpha}{\beta} = -\frac{B_2 - B_2'/iq_2}{A_2 - A_2'/iq_2}. \tag{29}$$

The function $\varrho(z)$ and the reflectivity function $R = \varrho\varrho^*$ are then obtained from

$$\varrho = \frac{G}{F} = \frac{iqE - \mathrm{d}E/\mathrm{d}z}{iqE + \mathrm{d}E/\mathrm{d}z}. \tag{30}$$

The results are equivalent to (25), with z_1, ε_1 being replaced by z, ε. Some reflectivity function curves are shown in Figure 5-2.

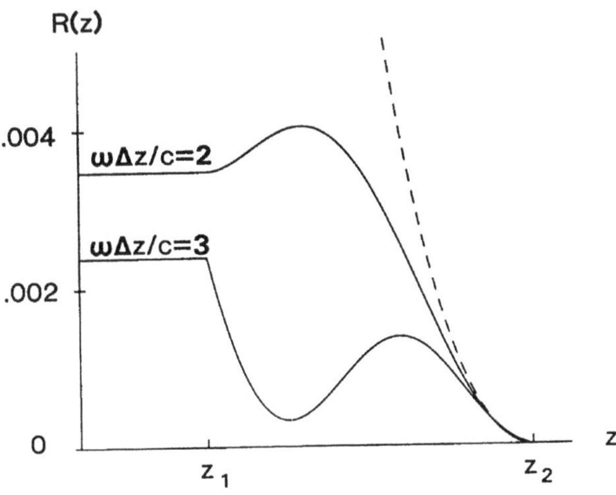

Figure 5-2. Reflectivity function for the linear profile of extent $\Delta z = z_2 - z_1$, with $\varepsilon_1 = 1$ and $\varepsilon_2 = (3/2)^2$. The curves are drawn for $(\omega/c)\Delta z = 2$ and 3, at normal incidence. Also shown (dashed line) is $[(\sqrt{\varepsilon} - \sqrt{\varepsilon_2})/(\sqrt{\varepsilon} + \sqrt{\varepsilon_2})]^2$, the reflectivity of a step from (ε to ε_2), which all such curves approach as $z \to z_2$.

5-3 Differential equation for a *p* wave reflection coefficient

The second order equation for the p wave,

$$\frac{\mathrm{d}}{\mathrm{d}z}\left(\frac{1}{\varepsilon}\frac{\mathrm{d}B}{\mathrm{d}z}\right) + \left(\frac{\omega^2}{c^2} - \frac{K^2}{\varepsilon}\right) B = 0, \tag{31}$$

maybe written as a pair of coupled first order equations,

$$\frac{1}{\varepsilon}\frac{\mathrm{d}B}{\mathrm{d}z} = C, \qquad \frac{\mathrm{d}C}{\mathrm{d}z} = -\frac{q^2}{\varepsilon}B. \tag{32}$$

We now write

$$B = F + G, \qquad C = \frac{iq}{\varepsilon}(F - G). \tag{33}$$

On eliminating B and C, we find the companion relations to (4) and (5):

$$F' = iqF - \frac{Q'}{2Q}(F - G), \tag{34}$$

$$G' = -iqG + \frac{Q'}{2Q}(F - G), \tag{35}$$

where $Q = q/\varepsilon$. Thus F and G again have the character of incident and reflected waves as $z \to -\infty$, being proportional to $e^{\pm iq_1 z}$. The ratio $\varrho = G/F$ now has the limiting forms

$$-r_p\, e^{-2iq_1 z} \leftarrow \varrho \to 0, \tag{36}$$

and satisfies the nonlinear first order equation

$$\varrho' + 2iq\varrho - \frac{Q'}{2Q}(1 - \varrho^2) = 0. \tag{37}$$

When q is real everywhere, the p reflectivity function $R = \varrho\varrho^*$ satisfies

$$R' = \frac{Q'}{Q}\,\mathrm{Re}\,(\varrho)(1 - R), \tag{38}$$

with the boundary conditions

$$R(\infty) = 0, \qquad R(-\infty) = R_p. \tag{39}$$

The fact that $R_p \leqslant 1$ follows from integrating (31) over the whole range of z:

$$\log\,(1 - R_p) = \int_{-\infty}^{\infty} \mathrm{d}z\, \frac{Q'}{Q}\,\mathrm{Re}\,(\varrho). \tag{40}$$

This physically necessary upper bound of unity can be much improved, as we shall see in the next section.

5-4 Upper bounds on R_s and R_p

It is intuitively plausible that the s wave will reflect less from a monotonically increasing or decreasing profile than from a step profile with the same values of ε_1 and ε_2. This is in accord with the long wave result of Chapter 3,

$$R_s = R_{s0} - \frac{4q_1 q_2 \omega^4/c^4}{(q_1 + q_2)^4}\, i_2 + \ldots, \tag{41}$$

where the invariant i_2 was shown to be non-negative if $\varepsilon(z)$ lies between ε_1 and ε_2. We will show now that a monotonic profile cannot reflect more of the s wave than the corresponding step profile, at any angle of incidence:

$$R_s \leqslant R_{s0} = \left(\frac{q_1 - q_2}{q_1 + q_2}\right)^2. \tag{42}$$

To prove this result we write $\varrho = |\varrho| \, e^{i\theta}$ in (6), and obtain a pair of coupled equations for the modulus $|\varrho|$ and the phase θ by separating the real and imaginary parts:

$$|\varrho|' - \frac{q'}{2q}(1 - |\varrho|^2)\cos\theta = 0, \tag{43}$$

$$\theta' + 2q + \frac{q'}{2q}(|\varrho|^{-1} + |\varrho|)\sin\theta = 0. \tag{44}$$

We can rewrite (43) as

$$|\varrho|'\left\{\frac{1}{1 - |\varrho|} + \frac{1}{1 + |\varrho|}\right\} = \frac{q'}{q}\cos\theta, \tag{45}$$

and integrating from $-\infty$ to ∞ and using (8), we obtain

$$\log\frac{1 + |r_s|}{1 - |r_s|} = -\int_{-\infty}^{\infty} dz \, \frac{dq}{q dz}\cos\theta. \tag{46}$$

This holds for any profile. Suppose now that $\varepsilon(z)$ increases monotonically from ε_1 to ε_2; the normal component of the wavevector then increases monotonically from q_1 to q_2, and the right-hand side of (46) has the upper bound $\log(q_2/q_1)$. Thus

$$\frac{1 + |r_s|}{1 - |r_s|} \leqslant \frac{q_2}{q_1}, \qquad |r_s| \leqslant \frac{q_2 - q_1}{q_2 + q_1}, \tag{47}$$

and (42) follows. The same bound on R_s holds for monotonic decrease from ε_1 to ε_2.

The corresponding result for the p wave reflectivity cannot be true without restriction, since we know that the reflectivity due to a sharp interface is zero at the Brewster angle, where a general interface does not show perfect transparency. Nevertheless a useful result can be obtained from the p wave equation corresponding to (43), namely

$$|\varrho|' = \frac{Q'}{2Q}(1 - |\varrho|^2)\cos\theta. \tag{48}$$

On integrating this as before we find

$$\log\frac{1 + |r_p|}{1 - |r_p|} = -\int_{-\infty}^{\infty} dz \, \frac{dQ}{Q dz}\cos\theta. \tag{49}$$

An upper bound of $\log[\max(Q_1, Q_2)/\min(Q_1, Q_2)]$ again follows, *provided $Q(z)$* is monotonic. Thus

$$R_p \leqslant \left(\frac{Q_1 - Q_2}{Q_1 + Q_2}\right)^2, \quad \text{if } Q \text{ monotonic}. \tag{50}$$

Suppose $\varepsilon(z)$ is monotonic. Under what circumstances is then Q monotonic also? We have

$$Q^2 = q^2/\varepsilon^2 = \frac{1}{\varepsilon}\frac{\omega^2}{c^2} - \frac{K^2}{\varepsilon^2}, \tag{51}$$

and so

$$\frac{dQ^2}{dz} = -\frac{d\varepsilon/dz}{\varepsilon^3}\left[\varepsilon\frac{\omega^2}{c^2} - 2K^2\right]. \tag{52}$$

Thus if $d\varepsilon/dz$ does not change sign, Q^2 will increase or decrease monotonically provided $\varepsilon - 2\varepsilon_1 \sin^2\theta_1$ does not change sign. This will be true if $2\varepsilon_1 \sin^2\theta_1 \leqslant \varepsilon_1$, i.e., $\theta_1 \leqslant 45°$, and also if $2\varepsilon_1 \sin^2\theta_1 \geqslant \varepsilon_2$ (we have assumed $\varepsilon_1 \leqslant \varepsilon \leqslant \varepsilon_2$). Thus $R_p \leqslant R_{p0}$ is guaranteed for angles of incidence in the ranges

$$\theta_1 \leqslant \pi/4, \qquad \theta_1 \geqslant \arcsin\,(\varepsilon_2/2\varepsilon_1)^{1/2}, \tag{53}$$

Note that the Brewster angle $\theta_B = \arctan\,(\varepsilon_2/\varepsilon_1)^{1/2}$ lies between these two limits. In the opposite case, when $\varepsilon_1 \geqslant \varepsilon \geqslant \varepsilon_2$, $R_p \leqslant R_{p0}$ is guaranteed in the ranges

$$\theta_1 \leqslant \arcsin\left(\frac{\varepsilon_2}{2\varepsilon_1}\right)^{1/2}, \qquad \theta_1 \geqslant \pi/4. \tag{54}$$

The Brewster angle again lies between these two limits. Figure 5-3 illustrates the reflectivity ratios R_s/R_{s0} and R_p/R_{p0} for the uniform layer for which r_p/r_s was displayed in Figure 2-8.

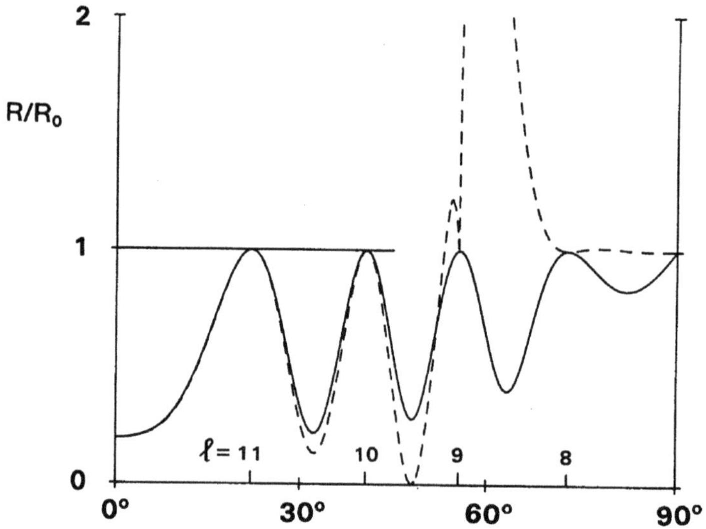

Figure 5-3. The ratios R_s/R_{s0} (——) and R_p/R_{p0} (- - -) for a uniform layer, with $\varepsilon_1 = 1$, $\varepsilon = (4/3)^2$, $\varepsilon_2 = (3/2)^2$, $(\omega/c)\Delta z = 27$, representing a layer of water (about four wavelengths thick) on glass. These parameters are the same as in Figure 2-8. Unity is the upper bound for R_s/R_{s0} at all angles of incidence, and for R_p/R_{p0} for $\theta_1 \leqslant 45°$. The value 1 is attained by both the s and p reflectivity ratios when $q\Delta z = l\pi$ (l an integer), that is at angles of incidence $\theta_1 = \arcsin\,[\{\varepsilon - (l\pi/(\omega/c)\Delta z)^2\}/\varepsilon_1]^{1/2}$. In this case the values $l = 11$, 10, 9 and 8 give the angles indicated.

5-5 Long wave approximations

Systematic approximations based on the non-linear equation for ϱ have been developed by Brekhovskikh (1949, 1980), and will be outlined here. We have seen that the physical meaning of $\varrho(z)\,\mathrm{e}^{2iq(z)z}$ is that of the reflection amplitude of a profile truncated at z (Figure 5-1). In the long wave limit this would be

$$r(z) \simeq \mathrm{e}^{2iq(z)z}\,\frac{q(z) - q_2}{q(z) + q_2}, \qquad \varrho(z) = \frac{q(z) - q_2}{q(z) + q_2}. \tag{55}$$

Brekhovskikh writes the exact $\varrho(z)$ in terms of two functions $u(z)$, $v(z)$, in analogy with (55):

$$\varrho = \frac{qu - q_2 v}{qu + q_2 v}. \tag{56}$$

Then from (6) and (56) it follows that

$$\frac{u'}{u} - \frac{v'}{v} = iq\left\{\frac{q_2 v}{qu} - \frac{qu}{q_2 v}\right\}. \tag{57}$$

This equation is satisfied by

$$u' = iq_2 v, \qquad v' = iq^2 u/q_2. \tag{58}$$

As $z \to z_2$, $\varrho(z) \to 0$, it being assumed that $\varepsilon = \varepsilon_2$ for $z \geqslant z_2$. Thus the boundary conditions on u and v may be taken as

$$u(z_2) = 1 = v(z_2) \tag{59}$$

(any nonzero constant other than unity would do as well). The equations (58) and (59) are equivalent to

$$u(z) = 1 - iq_2 \int_z^{z_2} \mathrm{d}\zeta\, v(\zeta), \quad v(z) = 1 - \frac{i}{q_2}\int_z^{z_2} \mathrm{d}\zeta\, q^2(\zeta)u(\zeta). \tag{60}$$

These coupled integral equations may be iterated to give $u = \Sigma\, u_n$ and $v = \Sigma\, v_n$, starting with $u_0 = 1 = v_0$. The first iterates are

$$u_1(z) = -iq_2 \int_z^{z_2} \mathrm{d}\zeta = -iq_2(z_2 - z), \tag{61}$$

$$v_1(z) = -\frac{i}{q_2}\int_z^{z_2} \mathrm{d}\zeta\, q^2(\zeta). \tag{62}$$

The nth order is, for $n \geqslant 1$,

$$u_n(z) = -iq_2 \int_z^{z_2} \mathrm{d}\zeta\, v_{n-1}(\zeta), \quad v_n(z) = -\frac{i}{q_2}\int_z^{z_2} \mathrm{d}\zeta\, q^2(\zeta)u_{n-1}(\zeta). \tag{63}$$

This iteration gives a series in powers of interface thickness/wavelength, and should thus duplicate the long wave results of Chapter 3. We will verify this to second order, for a profile of extent $\Delta z = z_2 - z_1$ (it is assumed that $\varepsilon = \varepsilon_1$ for $z \leqslant z_1$). For such a profile, the exact ϱ is given by

$$\varrho(z_1) = \frac{q_1 u(z_1) - q_2 v(z_1)}{q_1 u(z_1) + q_2 v(z_1)}. \tag{64}$$

To second order we have

$$u(z_1) = 1 - iq_2 \Delta z - \int_{z_1}^{z_2} \mathrm{d}z\,(z - z_1) q^2(z) + \dots, \tag{65}$$

$$v(z_1) = 1 - \frac{i}{q_2} \int_{z_1}^{z_2} \mathrm{d}z\, q^2(z) - \int_{z_1}^{z_2} \mathrm{d}z\,(z_2 - z) q^2(z) + \dots. \tag{66}$$

After some simplification, the reflectivity $R_s = |\varrho(z_1)|^2$ reduces to

$$R_s = R_{s0} + \frac{4q_1 q_2 \omega^4/c^4}{(q_1 + q_2)^4} [(\varepsilon_2 - \varepsilon_1) \int_{z_1}^{z_2} \mathrm{d}z\,(2z - z_1 - z_2)\varepsilon$$

$$+ \int_{z_1}^{z_2} \mathrm{d}z\,(\varepsilon - \varepsilon_1) \int_{z_1}^{z_2} \mathrm{d}\zeta\,(\varepsilon - \varepsilon_2)] + \dots. \tag{67}$$

The quantity within the square bracket in (67) is independent of the angle of incidence, as it should be, since we know from Chapter 3 that the universal form for the s wave reflectivity is

$$R_s = R_{s0} - \frac{4q_1 q_2 \omega^4/c^4}{(q_1 + q_2)^4} i_2 + \dots. \tag{68}$$

Here i_2 is the second order invariant of Section 3-3:

$$i_2 = 2(\varepsilon_1 - \varepsilon_2) \int_{-\infty}^{\infty} \mathrm{d}z\,(\varepsilon - \varepsilon_0) z - \left\{ \int_{-\infty}^{\infty} \mathrm{d}z\,(\varepsilon - \varepsilon_0) \right\}^2, \tag{69}$$

where ε_0 is the step function: $\varepsilon_0 = \varepsilon_1$ for $z < 0$, $\varepsilon_0 = \varepsilon_2$ for $z > 0$. The integrands in (67) do not go to zero at the end-points z_1 and z_2, and thus (67) appears to have no meaning as it stands for profiles which attain the limiting values ε_1 and ε_2 at infinity. To convert (67) to a universally applicable form we replace ε by $\varepsilon - \varepsilon_0 + \varepsilon_0$ in the integrands, and use the results

$$(\varepsilon_2 - \varepsilon_1) \int_{z_1}^{z_2} \mathrm{d}z\,(2z - z_1 - z_2)\varepsilon_0 + \int_{z_1}^{z_2} \mathrm{d}z\,(\varepsilon_0 - \varepsilon_1) \int_{z_1}^{z_2} \mathrm{d}z\,(\varepsilon_0 - \varepsilon_2) = 0 \tag{70}$$

$$\int_{z_1}^{z_2} \mathrm{d}z\,(2\varepsilon_0 - \varepsilon_1 - \varepsilon_2) = (\varepsilon_2 - \varepsilon_1)(z_1 + z_2). \tag{71}$$

The negative of the quantity within the square brackets of (67) then reduces to i_2 as given by (69), and we have regained the second order s wave result of Chapter 3 in its invariant form.

The p wave results are obtained similarly, but are much more complex. Both s and p thickness/wavelength expansions will be discussed again briefly in the chapter on matrix methods.

5-6 Differential equations for the reflection amplitudes

In the preceding sections we have derived and used generalized Riccati equations for the quantity $\varrho = G/F$, which takes the values $\pm r(\varepsilon(z), \varepsilon_2)\, e^{-2iqz}$ at z for the s and p waves. Thus as $z \to -\infty$, ϱ tends to $\pm r(\varepsilon_1, \varepsilon_2)$ times an oscillatory function of unit modulus. As we shall see here, it is sometimes advantageous to work directly with the reflection amplitude itself. This is particularly so in the short wave limit, which is discussed briefly here and forms the subject of the next chapter.

A non-linear first order equation for the s wave reflection amplitude may be obtained as follows: we set

$$E = f\,e^{i\phi} + g\,e^{-i\phi}, \qquad D = iq(f\,e^{i\phi} - g\,e^{-i\phi}), \tag{72}$$

where f and g are functions determined from (2) and (72), and ϕ is the phase integral (discussed in detail in Chapter 6),

$$\phi(z) = \int^z d\zeta\, q(\zeta). \tag{73}$$

In this instance it is convenient to choose the normally unspecified lower limit of integration in (73) so as to make $\phi(z) \to q_1 z$ as $z \to -\infty$. For example, if $\varepsilon(z) = \varepsilon_1$ for $z \leqslant z_1$, one can locate the origin at z_1 and set $\phi = \int_0^z d\zeta\, q(\zeta)$. We shall see shortly that f and g tend to constants as z tends to minus infinity; since r_s is defined as the ratio of the coefficient of $e^{-iq_1 z}$ to that of $e^{iq_1 z}$, for this choice for ϕ the ratio $r = g/f$ tends to r_s as $z \to -\infty$.

On eliminating E and D from (2) and (72), using $\phi' = q$, we find

$$f' + \frac{q'}{2q}(f - g\,e^{-2i\phi}) = 0, \tag{74}$$

$$g' + \frac{q'}{2q}(g - f\,e^{2i\phi}) = 0. \tag{75}$$

Thus f and g are changing only where ε (and thus q) is changing, verifying that f and g tend to constants at $\pm\infty$. An equation for $r = g/f$ is obtained by multiplying (74) by g, (75) by f, subtracting, and dividing the result by f^2. It is

$$r' = \frac{q'}{2q}(e^{2i\phi} - r^2\,e^{-2i\phi}), \tag{76}$$

with limiting values of $r(z)$ as $z \to \pm\infty$ given by

$$r_s \leftarrow r \rightarrow 0. \tag{77}$$

We note that, in contrast to the equation for ϱ, variation of r occurs only where the dielectric function is varying.

The analogous equation for the p wave reflection amplitude is obtained by setting

$$B = f\,e^{i\phi} + g\,e^{-i\phi}, \qquad C = iQ(f\,e^{i\phi} - g\,e^{-i\phi}). \tag{78}$$

The result of eliminating B and C from (32) and (78) is

$$f' + \frac{Q'}{2Q}(f - g\,\mathrm{e}^{-2i\phi}) = 0, \tag{79}$$

$$g' + \frac{Q'}{2Q}(g - f\,\mathrm{e}^{2i\phi}) = 0. \tag{80}$$

Thus the p wave reflection amplitude $r = -g/f$ satisfies

$$-r = \frac{Q'}{2Q}(\mathrm{e}^{2i\phi} - r^2\,\mathrm{e}^{-2i\phi}), \tag{81}$$

with limiting values

$$r_p \leftarrow r \rightarrow 0. \tag{82}$$

At normal incidence $q'/q = \varepsilon'/2\varepsilon = -Q'/Q$, and so the equations for the s and p wave amplitudes are the same, as they must be.

On integrating (76) and (81) from $-\infty$ to $+\infty$ we find

$$r_s = -\int_{-\infty}^{\infty} \mathrm{d}z\,\frac{q'}{2q}(\mathrm{e}^{2i\phi} - r^2\,\mathrm{e}^{-2i\phi}), \tag{83}$$

$$r_p = \int_{-\infty}^{\infty} \mathrm{d}z\,\frac{Q'}{2Q}(\mathrm{e}^{2i\phi} - r^2\,\mathrm{e}^{-2i\phi}). \tag{84}$$

These exact relationships lead naturally to the approximations of the next section.

5-7 Weak reflection: the Rayleigh approximation

We have seen in Section 5-1 that the meaning of $|\varrho(z_0)|^2$ is that of the reflectivity of a profile truncated at z_0 (i.e., one which has $\varepsilon = \varepsilon(z_0)$ for $z \leqslant z_0$). The quantity $|r(z_0)|^2$ has the same meaning, and does not exceed unity (compare (13)). When the reflection is weak, as for a monotonic profile in the long wave limit with $|q_1 - q_2| \ll q_1 + q_2$, or for a smooth profile in the short wave limit (the absence of total internal reflection or regions of negative q^2 is assumed in both cases), it is reasonable to assume that $|r^2(z)| \ll 1$ everywhere. Good approximations for r_s and r_p are then obtained by neglecting the terms in r^2 in (83) and (84):

$$r_s \simeq r_s^R = -\int_{-\infty}^{\infty} \mathrm{d}z\,\frac{q'}{2q}\,\mathrm{e}^{2i\phi}, \tag{85}$$

$$r_p \simeq r_p^R = \int_{-\infty}^{\infty} \mathrm{d}z\,\frac{Q'}{2Q}\,\mathrm{e}^{2i\phi}. \tag{86}$$

We have called these Rayleigh approximations, r_s^R, r_p^R, since they were first derived by Rayleigh (1912). They could also be called the weak reflection approximations, or associated with the names of Brekhovskikh (1949) or Bremmer (1951), who independently derived closely related approximations.

The physical basis of (for example) equation (85) can be seen by considering the profile as a series of small steps. As z changes by δz, the dielectric function changes by $\delta\varepsilon$ and the normal component of the wavenumber by δq. The contribution to the total reflection amplitude from this change is $\delta r = -(\delta q/2q)\,e^{2i\phi}$, assuming that the reflection at all preceding steps is weak enough to be ignored. The contribution written down above follows from the single-step formula (1.15), namely

$$\delta r = \frac{q-(q-\delta q)}{q+(q+\delta q)}\,e^{2iqz},$$

with ϕ (the accumulated phase at z) replacing qz. Adding up the contributions δr gives, in the limit of a large number of small steps, the result (85).

The weak reflection approximations lead to

$$r_s \simeq \frac{1}{2}\log\frac{q_1}{q_2}, \qquad r_p \simeq \frac{1}{2}\log\frac{Q_2}{Q_1} \tag{87}$$

in the long wave limit, for a profile located near the origin. These expressions are good representations of the exact limiting values (see Section 2-2)

$$r_{s0} = \frac{q_1-q_2}{q_1+q_2}, \qquad r_{p0} = \frac{Q_2-Q_1}{Q_2+Q_1} \tag{88}$$

provided these are small compared to unity; that is, provided $\delta = q_1/q_2 - 1$ and $\Delta = Q_2/Q_1 - 1$ are small. More precisely, the s wave reflection amplitudes (87) and (88) agree to second order in δ, both having the leading terms $\delta/2 - \delta^2/4 + \ldots$. The Rayleigh expressions fail completely in the long wave limit when δ or Δ are not small: for sufficiently large or small values of the ratios q_1/q_2 and Q_1/Q_2 the expressions (87) will give reflectivities greater than 1. Since $|\frac{1}{2}\log x|$ is no smaller than $|(x-1)/(x+1)|$ for $x > 0$, the expressions (87) give reflectivities which are never less than the Fresnel values.

On comparing the exact expression (83) with the Rayleigh approximation (85), we see that

$$r_s = r_s^R + \int_{-\infty}^{\infty} dz\,\frac{q'}{2q}\,r^2\,e^{-2i\phi} \equiv r_s^R + \Delta r_s. \tag{89}$$

Since $|r| \leqslant 1$ everywhere,

$$|\Delta r_s| \leqslant \int_{-\infty}^{\infty} dz\left|\frac{q'}{2q}\right|. \tag{90}$$

When $\varepsilon(z)$ is monotonic,

$$|\Delta r_s| \leqslant \frac{1}{2}\left|\log\frac{q_2}{q_1}\right|. \tag{91}$$

Similar results follow for the p wave, with Q replacing q in (90) and (91), the latter holding only if Q is monotonic. Thus simple bounds may be put on the error in the Rayleigh approximation. An example of the accuracy of r^R is given in the next section.

5-8 Iteration of the integral equation for r

The differential equation (76), together with the condition that $r \to 0$ as $z \to \infty$, may be integrated from z to ∞ to give

$$r(z) = -\int_z^\infty d\zeta \frac{q'}{2q}(e^{2i\phi} - r^2 e^{-2i\phi}). \tag{92}$$

Iteration of this non-linear integral equation gives successive approximations for $r(z)$ and thus for $r_s = r(-\infty)$. If we label these functions $r^{(l)}(z)$, then

$$r^{(l+1)}(z) = -\int_z^\infty d\zeta \frac{q'}{2q}(e^{2i\phi} - [r^{(l)}(\zeta)]^2 e^{-2i\phi}). \tag{93}$$

The natural starting point for this iteration is $r^{(0)}(\zeta) = 0$, giving

$$r^{(1)}(z) = -\int_z^\infty d\zeta \frac{q'}{2q} e^{-2i\phi} \tag{94}$$

and $r_s^{(1)} = r_s^R$, the Rayleigh approximation.

As an example, we apply this method to the Rayleigh profile of Section 2-5, for which the reciprocal of the refractive index is linear in z in the interval (z_1, z_2):

$$\varepsilon^{-1/2}(z) = \eta(z) = \eta_1 + (z - z_1)\frac{\Delta\eta}{\Delta z}, \tag{95}$$

where $\eta_1 = \varepsilon_1^{-1/2}$, $\eta_2 = \varepsilon_2^{-1/2}$, $\Delta\eta = \eta_2 - \eta_1$, $\Delta z = z_2 - z_1$. At normal incidence the phase integral is given by

$$\phi_n(z) = \frac{\omega}{c}\int_{z_1}^z dz\, \eta^{-1}(z) = \frac{\omega}{c}\frac{\Delta z}{\Delta\eta}\log \eta/\eta_1, \tag{96}$$

and $q'/q = \frac{1}{2}\varepsilon'/\varepsilon = -\eta'/\eta = -(\Delta\eta/\Delta z)/\eta$. Thus

$$r^{(1)}(z) = \frac{1}{2}\int_\eta^{\eta_2} d\eta\, \eta^{-1}\left(\frac{\eta}{\eta_1}\right)^{2i\alpha}, \tag{97}$$

where α stands for the dimensionless parameter $(\omega/c)(\Delta z/\Delta\eta)$. The Rayleigh approximation is

$$r_n^R = r^{(1)}(z_1) = \frac{1}{4i\alpha}[(\eta_2/\eta_1)^{2i\alpha} - 1] = \left(\frac{\eta_2}{\eta_1}\right)^{i\alpha}\frac{\sin[\alpha \log(\eta_2/\eta_1)]}{2\alpha}, \tag{98}$$

and gives the reflectivity

$$R_n^R = \left(\frac{\sin\left[\frac{1}{2}\alpha \log\frac{\varepsilon_2}{\varepsilon_1}\right]}{2\alpha}\right)^2. \tag{99}$$

The exact reflectivity at normal incidence is given by the two formulae (2.107) and (2.108) according as α^2 is smaller or greater than 1/4. The exact and Rayleigh approximation reflectivities are compared in Figure 5-4:

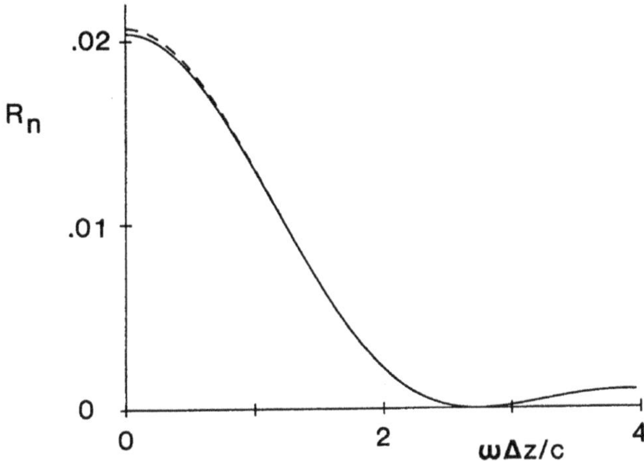

Figure 5-4. Exact (——) and weak reflection approximation (– – –) reflectivities for the Rayleigh profile. The normal incidence reflectivities are shown, for $\varepsilon_1 = 1$, $\varepsilon_2 = (4/3)^2$. The exact and approximate curves have their first zero at $(\omega/c)\Delta z \simeq 2.733$ and 2.730; successive exact and approximate zeros come progressively closer.

References

L. R. Walker and N. Wax (1946) "Non-uniform transmission lines and reflection coefficients", J. Applied Phys. **17**, 1043–1045.

W. Kofink (1947) "Reflexion elektromagnetischer Wellen an einer inhomogenen Schicht", Annalen der Physik **1**, 119–132.

L. M. Brekhovskikh (1949) "Reflection of plane waves from layered inhomogeneous media", Zhurnal Technicheskoi Fiziki **19**, 1126–1135.

L. M. Brekhovskikh (1980) "Waves in layered media", Academic Press, second edition.

S. A. Schelkunoff (1951) "Remarks concerning wave propagation in stratified media", Commun. Pure and App. Math. **4**, 117–128.

P. M. Morse and W. P. Allis (1933) "The effect of exchange on the scattering of electrons", Phys. Rev. **44**, 269–276.

F. Calogero (1967) "Variable phase approach to potential scattering", Academic Press.

R. Courant and D. Hilbert (1953) "Methods of mathematical physics", Interscience Publishers, Volume 1, p. 322.

J. Heading (1962) "An introduction to phase integral methods", Methuen.

F. W. J. Olver (1964) "Bessel functions of fractional order" (Chapter 10 of "Handbook of mathematical functions", edited by M. Abramowitz and I. A. Stegun, N.B.S. Applied Math. Series **55**).

J. C. P. Miller (1971) "The Airy integral", British Association Mathematical Tables, Part-volume B, Cambridge.

J. W. S. Rayleigh (1912) "On the propagation of waves through a stratified medium, with special reference to the question of reflection", Proc. Roy. Soc. **86**, 207–266.

H. Bremmer (1951) "The W.K.B. approximation as the first term of a geometric-optical series", Commun. Pure and App. Math. *4*, 105–115.

6

Reflection of short waves

Consider electromagnetic waves of angular frequency ω, incident on a planar inhomogeneity of thickness (or characterizing length) Δz. When $(\omega/c)\Delta z \gg 1$, and at normal incidence, there are many wavelengths within the inhomogeneity and (for smooth profiles) the change in the dielectric function within a wavelength is small. This is known as the short wave limit. We shall see that at general angle of incidence the short wave limiting forms are attained when $q\Delta z$ is large. Since $q_1 \to 0$ as $\theta_1 \to \pi/2$, the short wave approximations fail at grazing incidence. Special techniques are also needed when $q^2(z)$ passes through zero, and when $\varepsilon(z)$ has discontinuities in gradient.

6-1 Short wave limiting forms for some solvable profiles

It will be useful to look at some profiles for which the reflection amplitude is known analytically, both to orient ourselves and to have examples for comparison with the approximate expressions to be derived. We shall see that there is no *universal* form for the short wave expressions, in contrast to the long wave case of Chapter 3, where we showed that (for instance) the s wave reflectivity takes the form

$$R_s = R_{s0} - \frac{4q_1 q_2 \omega^4/c^4}{(q_1 + q_2)^4} i_2 + \dots . \tag{1}$$

There is greater variety and complexity in the short wave case, because short waves are sensitive to details in the dielectric function profile, while long waves are not.

Hyperbolic tangent profile: from (2.88) and (2.89) we have

$$r_s = \exp\left[2i \sum_{n=1}^{\infty} \arctan\left(\frac{2y_1}{n} \frac{y_1^2 - y_2^2}{n^2 + 3y_1^2 + y_2^2}\right)\right] \frac{\sinh \pi(y_1 - y_2)}{\sinh \pi(y_1 - y_2)}, \tag{2}$$

where $y_1 = q_1 \Delta z$ and $y_2 = q_2 \Delta z$. Suppose $\varepsilon_1 < \varepsilon_2$. Then $q_1^2 < q_2^2$, there is no total internal reflection, and (for large $y_2 - y_1$)

$$R_s = e^{-4\pi q_1 \Delta z} + \dots . \tag{3}$$

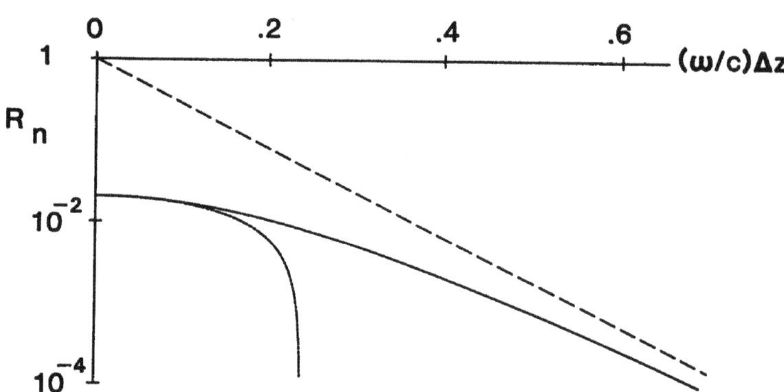

Figure 6-1. Normal incidence reflectivity for the hyperbolic tangent profile. The solid, dashed and dot–dashed curves are respectively the exact, short wave and long wave forms. The curves are drawn for $\varepsilon_1 = 1$, $\varepsilon_2 = (4/3)^2$.

This short wave limiting form, the long wave limiting form

$$R_s = R_{s0}\left(1 - \frac{4\pi^2}{3} q_1 q_2 (\Delta z)^2 + \ldots\right), \tag{4}$$

(obtained from (1) and the value $i_2 = (\pi^2/3)(\varepsilon_1 - \varepsilon_2)^2(\Delta z)^2$ from Table 3-1), and the exact reflectivity are compared in Figure 6-1.

When $\varepsilon_1 > \varepsilon_2$, $q_1^2 > q_2^2$ and for $\theta_1 > \theta_c = \arcsin(\varepsilon_2/\varepsilon_1)^{1/2}$, $q_2 = i|q_2|$, with total internal reflection. The leading terms of (2) for large $y_1 = q_1 \Delta z$ are thus unity for $\theta_1 > \theta_c$ and

$$R_s = e^{-4\pi q_2 \Delta z} + \ldots \qquad (\theta_1 < \theta_c). \tag{5}$$

Rayleigh profile: from (2.108) the reflectivity at normal incidence is

$$R_n = \frac{\sin^2[\frac{1}{2}|\nu| \log(\varepsilon_1/\varepsilon_2)]}{4|\nu|^2 + \sin^2[\frac{1}{2}|\nu| \log(\varepsilon_1/\varepsilon_2)]}, \tag{6}$$

where

$$|\nu| = \left[\left(\frac{\omega}{c}\frac{\Delta z}{\Delta\eta}\right)^{1/2} - \frac{1}{4}\right]^{1/2}, \qquad \Delta\eta = \varepsilon_2^{-1/2} - \varepsilon_1^{-1/2} \tag{7}$$

(these forms apply for $((\omega/c)\Delta z)^2 > (\Delta\eta/2)^2$, which is the appropriate case in the short wave limit). For this profile the $(\omega/c)\Delta z \gg 1$ limiting form shows oscillatory decay:

$$R_n = \left\{\frac{\sin\left(\frac{1}{2}\frac{\omega}{c}\frac{\Delta z}{\Delta\eta}\log\frac{\varepsilon_1}{\varepsilon_2}\right)}{2\frac{\omega}{c}\frac{\Delta z}{\Delta\eta}}\right\}^2. \tag{8}$$

Note the power law decrease with increasing $(\omega/c)\Delta z$, in contrast to the exponential decay for the smoother hyperbolic tangent profile.

Exponential profile: the reflectivity at normal incidence is given by (2.102):

$$R_n = \frac{A_0^2 + B_0^2 + C_0^2 + D_0^2 - 8/\pi^2 u_1 u_2}{A_0^2 + B_0^2 + C_0^2 + D_0^2 + 8/\pi^2 u_1 u_2}. \tag{9}$$

when $(\omega/c)\Delta z \gg 1$, the variable $u = 2\sqrt{\varepsilon}(\omega/c)\Delta z/\log(\varepsilon_2/\varepsilon_1)$ is large, and we can use Hankel's asymptotic expansions (Olver (1964), Section 9.2)

$$\begin{aligned} J_0(u) &= \left(\frac{2}{\pi u}\right)^{1/2} (\alpha \cos w - \beta \sin w), \\ Y_0(u) &= \left(\frac{2}{\pi u}\right)^{1/2} (\alpha \sin w + \beta \cos w), \\ J_0'(u) &= \left(\frac{2}{\pi u}\right)^{1/2} (-\gamma \sin w - \delta \cos w), \\ Y_0'(u) &= \left(\frac{2}{\pi u}\right)^{1/2} (\gamma \cos w - \delta \sin w), \end{aligned} \tag{10}$$

where $w = u - \pi/4$. The functions α, β, γ and δ have asymptotic expansions which are conveniently expressed in terms of $v = (8u)^{-1}$:

$$\begin{aligned} \alpha &\sim 1 - \tfrac{9}{2}v^2 + \dots, & \beta &\sim -v + \dots, \\ \gamma &\sim 1 + \tfrac{15}{2}v^2 + \dots, & \delta &\sim 3v + \dots, \end{aligned} \tag{11}$$

The cross products A_0 to D_0 are given by

$$\begin{aligned} A_0 &= \frac{2}{\pi(u_1 u_2)^{1/2}} \{(\alpha_1\alpha_2 + \beta_1\beta_2)\sin(u_2 - u_1) + (\alpha_1\beta_2 - \beta_1\alpha_2)\cos(u_2 - u_1)\}, \\ B_0 &= \frac{2}{\pi(u_1 u_2)^{1/2}} \{(\alpha_1\gamma_2 + \beta_1\delta_2)\cos(u_2 - u_1) + (\beta_1\gamma_2 - \alpha_1\delta_2)\sin(u_2 - u_1)\}, \\ C_0 &= \frac{2}{\pi(u_1 u_2)^{1/2}} \{-(\gamma_1\alpha_2 + \delta_1\beta_2)\cos(u_2 - u_1) - (\delta_1\alpha_2 - \gamma_1\beta_2)\sin(u_2 - u_1)\}, \\ D_0 &= \frac{2}{\pi(u_1 u_2)^{1/2}} \{\gamma_1\gamma_2 + \delta_1\delta_2)\sin(u_2 - u_1) + (\gamma_1\delta_2 - \delta_1\gamma_2)\cos(u_2 - u_1)\}. \end{aligned} \tag{12}$$

The leading terms of the asymptotic expansions give

$$A_0^2 + B_0^2 + C_0^2 + D_0^2 \sim \frac{8}{\pi^2 u_1 u_2}\{1 + 16[v_1^2 + v_2^2 - 2v_1 v_2 \cos 2(u_2 - u_1)]\}. \tag{13}$$

The short wave limiting form of R_n is thus

$$\begin{aligned} R_n &\sim 4[v_1^2 + v_2^2 - 2v_1 v_2 \cos 2(u_2 - u_1)] \\ &= \frac{1}{16}\left[\frac{1}{u_1^2} + \frac{1}{u_2^2} - \frac{2}{u_1 u_2}\cos 2(u_2 - u_1)\right]. \end{aligned} \tag{14}$$

Like the Rayleigh profile expression (8), this decreases as the inverse square of $(\omega/c)\Delta z$. The period of the oscillatory term (that is, the change in $(\omega/c)\Delta z$ during which the oscillatory term goes through one cycle) is, for large $(\omega/c)\Delta z$,

$$P_e = \frac{\frac{\pi}{2}\log\frac{\varepsilon_2}{\varepsilon_1}}{\sqrt{\varepsilon_2} - \sqrt{\varepsilon_1}}, \tag{15}$$

while the Rayleigh profile reflectivity has the period

$$P_R = \frac{2\pi(\sqrt{\varepsilon_2} - \sqrt{\varepsilon_1})}{(\varepsilon_1\varepsilon_2)^{1/2}\log(\varepsilon_2/\varepsilon_1)}. \tag{16}$$

These expressions look dissimilar, but give similar values provided ε_1 and ε_2 are not too different. For example, when $\varepsilon_1 = 1$ and $\varepsilon_2 = (4/3)^2$, $P_e \simeq 2.71$ and $P_R \simeq 2.73$. For small $|\varepsilon_1 - \varepsilon_2|$ the two profiles are both approximately linear in z, and $P_e \simeq P_R \simeq \pi/(\varepsilon_1\varepsilon_2)^{1/4}$.

The short and long wave limiting forms for the exponential profile are compared with the exact reflectivity in Figure 6-2.

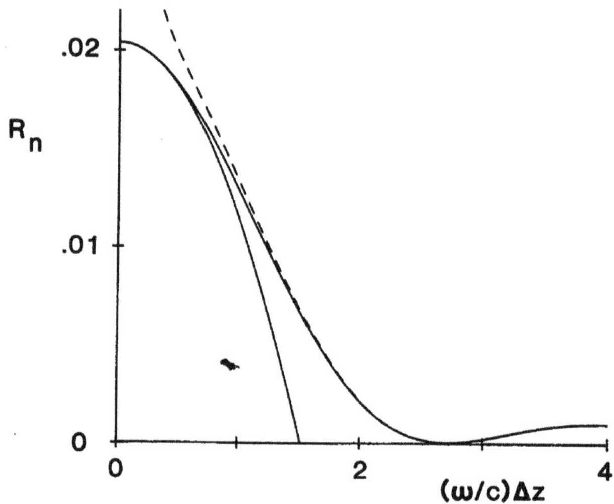

Figure 6-2. Normal incidence reflectivity for the exponential profile. The solid curve is the exact reflectivity derived in Section 2-5, the dot–dash curve is the long wave approximation (3.51), and the dashed curve is the short wave approximation (14). The ε_1, ε_2 values are 1 and $(4/3)^2$.

The three examples discussed in this section are sufficient to show the greater variety and complexity in the short wave limiting forms compared to the long wave case, where the reflectivity is always the Fresnel value plus a term proportional to the square of the interfacial thickness. The long wave limit is simpler because the difference between the actual profile and a step is sensed only in an averaged or cumulative way (through integrals over the difference between the two profiles). When the wavelength is short compared to the profile extent, finer details are sensed. For example, discontinuities in slope in the Rayleigh and exponential profiles give a sinusoidal times inverse square dependence of the reflectivity on the parameter $(\omega/c)\Delta z$, in contrast to the exponential decrease with $(\omega/c)\Delta z$ for the smooth tanh profile.

6-2 Approximate waveforms

We can consider together the electromagnetic s and p waves and particle waves, as in Section 2-1. The z variation in all three cases is given by solutions of equations of the form

$$\frac{d^2\psi}{dz^2} + q^2\psi = 0. \tag{17}$$

We assume here that $q^2(z)$ is everywhere positive; solutions in the neighbourhood of zeros of q^2 will be discussed in Sections 6-7 and 6-8.

When q is constant the propagating solutions are $e^{\pm iqz}$. When q varies we may expect solutions of the form $A(z)\, e^{\pm i\phi(z)}$, where the *phase integral* ϕ is given by

$$\phi(z) = \int^z d\zeta\, q(\zeta). \tag{18}$$

The phase integral gives the accumulated phase at z, being the sum of phase differences equal to $(2\pi/\lambda) \times$ (path difference), where for motion in the z direction described by (17) the effective local wavelength λ is $2\pi/q$. Thus $q(\zeta)\, d\zeta$ is the increase in the phase on going from ζ to $\zeta + d\zeta$, and (18) gives the accumulated phase at z.

The x dependence for planar stratified media is contained in the factor e^{iKx}. Thus the total phase is

$$\Phi(z, x) = \phi(z) + Kx. \tag{19}$$

Wave functions whose dominant z and x dependence is contained in the phase factor $\exp i\Phi$ correspond to geometrical optics rays, or to semiclassical particle trajectories. For, consider the wavefronts of $\exp i\Phi$. These are surfaces of equal phase, $\Phi(z, x) =$ constant. By definition, the rays are normal to the wavefronts, and will thus have the direction of

$$\nabla\Phi = \left(K, 0, \frac{d\phi}{dz}\right) = (K, 0, q(z)). \tag{20}$$

Thus the function $\exp i\Phi$ can be interpreted as a system of rays propagating in the z, x plane, with x and z components of the wavevector being K and $q(z)$. This is precisely the geometrical optics or semiclassical particle picture of propagation in a medium which is stratified with z.

A sequence of approximations for the solution of (17) may be obtained by setting $\psi = \exp \int^z d\zeta\, \chi(\zeta)$. Then χ satisfies an equation of the generalized Riccati type:

$$\frac{d\chi}{dz} + \chi^2 + q^2 = 0. \tag{21}$$

By the short wave limit we mean that for an inhomogeneous region of extent Δz, the normal component of the wavevector q is large compared to $(\Delta z)^{-1}$. The physical meaning of $q\Delta z \gg 1$ is that the inhomogeneity extends over many wavelengths. In the zeroth approximation one neglects $d\chi/dz$ in (21), since (if χ varies smoothly) this is of order $(q\Delta z)^{-1}$ smaller in magnitude than the other two terms. Thus $\chi_0^{\pm} = \pm iq$, and

$$\psi_0^{\pm}(z) = e^{\pm i\phi(z)}. \tag{22}$$

In the next approximation we set

$$\begin{aligned} \chi_1^{\pm} &= \left(-q^2 - \frac{d\chi_0^{\pm}}{dz}\right)^{1/2} = \pm iq\left(1 \pm \frac{idq/dz}{q^2}\right)^{1/2} \\ &\simeq \pm iq\left(1 \pm \frac{idq/dz}{2q^2}\right), \end{aligned} \tag{23}$$

on the assumption that the dimensionless quantity $\gamma = q^{-2}\, dq/dz$ is small. The corresponding wavefunctions $\psi_1 = \exp \int^z d\zeta\chi_1$ are thus

$$\psi_1^{+}(z) = \left(\frac{q_1}{q(z)}\right)^{1/2} e^{i\phi(z)}, \quad \psi_1^{-}(z) = \left(\frac{q_2}{q(z)}\right)^{1/2} e^{-i\phi(z)}. \tag{24}$$

(The square roots of q_1 and q_2 are inserted for later convenience.) It is instructive to compare the differential equations satisfied by the approximate waveforms $\psi_0^{\pm}$ and $\psi_1^{\pm}$ with the original wave equation (17). These are

$$\frac{d^2\psi_0^{\pm}}{dz^2} + \left(q^2 \mp \frac{idq}{dz}\right)\psi_0^{\pm} = 0, \tag{25}$$

and

$$\frac{d^2\psi_1^{\pm}}{dz^2} + \left[q^2 + \frac{d^2q/dz^2}{2q} - \frac{3}{4}\left(\frac{dq/dz}{q}\right)^2\right]\psi_1^{\pm} = 0. \tag{26}$$

Note that ψ_0^{+} and ψ_0^{-} satisfy different equations.

The approximate solutions $\psi_1^{\pm}$ go back to Liouville (1837) and Green (1837), with later contributions by Rayleigh (1912) and Gans (1915). A historical survey can be found in Heading (1962). The initials WKB or JWKB are often associated with $\psi_1^{\pm}$, the initials standing for Jeffreys (1924), Wentzel (1926), Kramers (1926), and Brillouin (1926). However, as Olver (1974) remarks, the contribution of these authors was not the construction of the approximation (which was already known), but the determination of connection formulae for linking exponential and oscillatory approximations across a zero of q^2 on the real axis. The latter problem is discussed in Sections 6-7 and 6-8.

Exact solutions of (17) satisfy the flux conservation condition, the physical basis of which (conservation of energy in the electromagnetic case, and of probability density current in the particle case) was discussed in Section 2-1. The mathematical statement follows directly from (17) and its complex conjugate, which for real q^2 is

$$\frac{d^2\psi^*}{dz^2} + q^2\psi^* = 0. \tag{27}$$

We multiply (20) by ψ^* and (28) by ψ, and subtract. The result is

$$\frac{d}{dz}\left(\psi^* \frac{d\psi}{dz} - \psi \frac{d\psi^*}{dz}\right) = 0, \tag{28}$$

so that

$$\psi^* \frac{d\psi}{dz} - \psi \frac{d\psi^*}{dz} = 2i \operatorname{Im}\left(\psi^* \frac{d\psi}{dz}\right)$$

is a constant. For reflection problems the asymptotic forms of ψ are

$$e^{iq_1 z} + r\, e^{-iq_1 z} \leftarrow \psi \rightarrow t\, e^{iq_2 z}, \tag{29}$$

and the constancy of Im $(\psi^* d\psi/dz)$ implies the flux conservation law

$$q_1(1 - |r|^2) = q_2|t|^2. \tag{30}$$

Any approximation to ψ which satisfies an equation of the form $(d^2\psi/dz^2) + \tilde{q}^2\psi = 0$ with real $\tilde{q}^2$ (as do $\psi_1^\pm$) will have Im $(\psi^* d\psi/dz)$ = constant, and thus conserve flux. The zeroth approximations $\psi_0^\pm$ on the other hand have

$$\operatorname{Im}(\psi_0^{\pm *} d\psi_0^\pm/dz) = \pm q(z), \tag{31}$$

which varies through the interface. This variation is associated with the absorption-type term $i dq/dz$ in (25). Nevertheless, we will find it instructive to examine the zeroth approximations obtained with $\psi_0^\pm$ for the reflectivity in the short wave limit.

6-3 Profiles of finite extent with discontinuities in slope at the endpoints

Consider profiles for which q varies continuously in the interval $[z_1, z_2]$, and takes the values q_1 for $z \leqslant z_1$ and q_2 for $z \geqslant z_2$; and let $F(z)$ and $G(z)$ be two linearly independent solutions of (17) within $[z_1, z_2]$. We showed in Section 2-2 that the exact reflection amplitude is given by

$$r = e^{2iq_1 z_1} \times \frac{q_1 q_2(F_1 G_2 - G_1 F_2) + iq_1(F_1 G_2' - G_1 F_2') + iq_2(F_1' G_2 - G_1' F_2) - (F_1' G_2' - G_1' F_2')}{q_1 q_2(F_1 G_2 - G_1 F_2) + iq_1(F_1 G_2' - G_1 F_2') - iq_2(F_1' G_2 - G_1' F_2) + (F_1' G_2' - G_1' F_2')}. \tag{32}$$

In this Section we consider the approximations $r^{(0)}$ and $r^{(1)}$ to r, obtained by setting F, G equal to ψ_0^+, ψ_0^- or ψ_1^+, ψ_1^-, respectively. We assume $q^2(z)$ to be continuous at z_1 and z_2; this excludes (for example) the uniform layer or two-step profile.

Consider $r^{(0)}$ first, obtained by approximating F by ψ_0^+ and G by ψ_0^+. On comparing (25) with (17), we see that such an approximation can be good in general only if $\gamma = q^{-2} dq/dz$ is small within the interface. If $q^2(z)$ varies smoothly within $[z_1, z_1 + \Delta z]$, γ is of order $(q\Delta z)^{-1}$. Thus when $q\Delta z$ is large, and the profile is smooth, $|r - r^{(0)}|$ is expected to be small. We have $F = e^{i\phi}$, $F' = iq\, e^{i\phi}$, $G = e^{-i\phi}$, $G' = -iq\, e^{-i\phi}$. Let $\Delta\phi = \phi(z_2) - \phi(z_1) = \int_{z_1}^{z_2} dz\, q(z)$ be the phase difference betweeen z_2 and z_1. We find

$$\begin{aligned} F_1G_2 - G_1F_2 &= -2i \sin \Delta\phi \\ F_1G_2' - G_1F_2' &= -2iq_2 \cos \Delta\phi \\ F_1'G_2 - G_1'F_2 &= 2iq_1 \cos \Delta\phi \\ F_1'G_2' - G_1'F_2' &= -2iq_1q_2 \sin \Delta\phi. \end{aligned} \tag{33}$$

On substitution of (33) into (32) to find $r^{(0)}$ we see that the zeroth order reflectivity is *zero*. This is the correct short wave limiting value in the absence of turning points (values of z where $q^2(z) = 0$, at which a classical particle or a ray described by geometrical optics would turn back).

The next approximation is obtained by substituting $F = \psi_1^+$ and $G = \psi_1^-$ into (32). Using $F' = q(i - \gamma/2)F$ and $G' = -q(i + \gamma/2)G$, where γ stands for the dimensionless quantity $\mathrm{d}q/q^2\,\mathrm{d}z$, we find

$$\begin{aligned} F_1G_2 - G_1F_2 &= -2i \sin \Delta\phi \\ F_1G_2' - G_1F_2' &= iq_2(-2 \cos \Delta\phi + \gamma_2 \sin \Delta\phi) \\ F_1'G_2 - G_1'F_2 &= iq_1(2 \cos \Delta\phi + \gamma_1 \sin \Delta\phi) \\ F_1'G_2' - G_1'F_2' &= iq_1q_2[-2 \sin \Delta\phi + (\gamma_1 - \gamma_2) \cos \Delta\phi - \tfrac{1}{2}\gamma_1\gamma_2 \sin \Delta\phi]. \end{aligned} \tag{34}$$

On keeping first order terms in the small quantity γ, we find

$$r^{(1)} = \frac{e^{2iq_1z_1 + i\Delta\phi}}{4i}\{(\gamma_1 - \gamma_2)\cos \Delta\phi - i(\gamma_1 + \gamma_2)\sin \Delta\phi\}. \tag{35}$$

The corresponding reflectivity is

$$R^{(1)} = |r^{(1)}|^2 = \tfrac{1}{16}\{\gamma_1^2 + \gamma_2^2 - 2\gamma_1\gamma_2 \cos 2\Delta\phi\}. \tag{36}$$

We can compare this formula with the short wave limiting forms obtained for the Rayleigh and exponential profiles in Section 6-1. These results were given for normal incidence, for which $q = \sqrt{\varepsilon}\omega/c$, and

$$\gamma(z) = q^{-2}\,\mathrm{d}q/\mathrm{d}z = (c/\omega)\varepsilon^{-1}\,\mathrm{d}\sqrt{\varepsilon}/\mathrm{d}z. \tag{37}$$

For the Rayleigh profile defined by (2.103), γ is independent of z at normal incidence:

$$\gamma = \frac{-\Delta\eta}{(\omega/c)\Delta z}, \qquad \Delta\eta = \varepsilon_2^{-1/2} - \varepsilon_1^{-1/2}, \tag{38}$$

and

$$\Delta\phi = \frac{1}{2}\frac{\omega}{c}\frac{\Delta z}{\Delta\eta}\log\frac{\varepsilon_1}{\varepsilon_2}. \tag{39}$$

Thus (36) is in accord with (8).

For the exponential profile given by (2.93),

$$\gamma = \frac{\log(\varepsilon_2/\varepsilon_1)}{2(\omega/c)\Delta z\sqrt{\varepsilon}} \equiv \frac{1}{u}, \tag{40}$$

and

$$\Delta\phi = \frac{2(\omega/c)\Delta z}{\log(\varepsilon_2/\varepsilon_1)}(\sqrt{\varepsilon_2} - \sqrt{\varepsilon_1}) \equiv u_2 - u_1. \tag{41}$$

Again the expression (36), derived by approximating the solutions by $\psi_{\mathrm{I}}^{\pm}$, is in agreement with the exact limiting form (14).

It is interesting that the factor $q^{-1/2}$, which transforms the approximate waveforms $\psi_0^{\pm}$ into the better approximations $\psi_{\mathrm{I}}^{\pm}$, takes us from zero reflectivity to the useful expression (36). This can be better understood by looking at the total wavefunctions $\alpha F + \beta G$ within the inhomogeneous region, with F, G respectively given by ψ_0^+, ψ_0^- and ψ_{I}^+, ψ_{I}^-. With ψ given by

$$\psi(z) = \begin{cases} e^{iq_1 z} + r\,e^{-iq_1 z} & (z < z_1) \\ \alpha F(z) + \beta G(z) & (z_1 \leqslant z \leqslant z_2) \\ t\,e^{iq_2 z} & (z > z_2), \end{cases} \tag{42}$$

the continuity of ψ and $d\psi/dz$ at z_1 and z_2 gives us four linear equations in the four unknowns α, β, r, t. In Section 2-2 we found r and t (equations 2.25 and 2.26). The corresponding formulae for α and β are

$$\alpha = 2iq_1\,e^{iq_1 z_1}(G_2' - iq_2 G_2)/D, \qquad \beta = -2iq_1\,e^{iq_1 z_1}(F_2' - iq_2 F_2)/D, \tag{43}$$

where D is the denominator of (32). We see at once that $\beta = 0$ for $F = \psi_0^+ = e^{i\phi}$, since $F' = iqF$. Thus there is no backward propagating wave, consistent with zero reflection. The values of α, β when $F = \psi_0^+$ and $G = \psi_0^-$ are

$$\alpha_0 = e^{i(q_1 z_1 - \phi_1)}, \qquad \beta_0 = 0. \tag{44}$$

The corresponding expressions obtained with $F = \psi_{\mathrm{I}}^-$ and $G = \psi_{\mathrm{I}}^-$ are, to first order in the small quantity γ,

$$\alpha_1 = \left(1 - \frac{i\gamma_2}{4}\right) e^{i(q_1 z_1 - \phi_1)},$$
$$\beta_1 = \left(\frac{q_1}{q_2}\right)^{1/2} \frac{i\gamma_2}{4}\, e^{i(q_1 z_1 - \phi_1 + 2\phi_2)}. \tag{45}$$

The coefficient of the backward propagating wave is now nonzero, and the coefficient of the forward propagating wave has a first order correction term in it.

The transmission amplitude for a profile of finite extent may be found from (2.26) and (34). The result is

$$t^{(1)} = \frac{4(q_1/q_2)^{1/2}\,e^{i(q_1 z_1 - q_2 z_2)}}{4\cos\Delta\phi + (\gamma_1 - \gamma_2)\sin\Delta\phi - i(4\sin\Delta\phi - (\gamma_1 - \gamma_2)\cos\Delta\phi + \frac{1}{2}\gamma_1\gamma_2\sin\Delta\phi)}. \tag{46}$$

Thus, to second order in γ_1 and γ_2,

$$|t^{(1)}|^2 = \frac{q_1}{q_2}[1 - \tfrac{1}{16}(\gamma_1^2 + \gamma_2^2 - 2\gamma_1\gamma_2\cos 2\Delta\phi)]. \tag{47}$$

On comparing with (36), we see that the conservation law $q_1(1 + |r|^2) = q_2|t|^2$ is obeyed to this order in γ.

6-4 Reflection amplitude estimates from a comparison identity

Let ψ be the solution of

$$\frac{d^2\psi}{dz^2} + q^2\psi = 0, \qquad e^{iq_1 z} + r\,e^{-iq_1 z} \leftarrow \psi \rightarrow t\,e^{iq_2 z}, \tag{48}$$

and $\tilde{\psi}$ the solution of

$$\frac{d^2\tilde{\psi}}{dz^2} + \tilde{q}^2\tilde{\psi} = 0, \qquad e^{iq_1 z} + \tilde{r}\,e^{-iq_1 z} \leftarrow \tilde{\psi} \rightarrow \tilde{t}\,e^{iq_2 z}. \tag{49}$$

The functions $q(z)$ and $\tilde{q}(z)$ have the same asymptotic values q_1 and q_2 at $\pm\infty$. We showed in Section 2-1 that r and $\tilde{r}$ are related by

$$r = \tilde{r} - \frac{1}{2iq_1}\int_{-\infty}^{\infty} dz\,(q^2 - \tilde{q}^2)\psi\tilde{\psi}. \tag{50}$$

We will use this identity to obtain approximations for r in the short wave limit. Consider first the result which comes from substituting $\tilde{\psi} = \psi_0^+$ into (50), where $\psi_0^+ = e^{i\phi}$ is the zeroth approximation to ψ in the short wave limit. Since ψ_0^+ satisfies (25), $\tilde{q}^2 = q^2 - i\,dq/dz$; also $\tilde{r} = 0$ since there is no backward propagating component in ψ_0^+, as can be verified by examining its limiting form as $z \rightarrow -\infty$. Thus we have the identity

$$r = -\frac{1}{2q_1}\int_{-\infty}^{\infty} dz\,\frac{dq}{dz}\,\psi\psi_0^+. \tag{51}$$

This is an exact relation, for all $q(z)$ and the corresponding ψ and ψ_0^+. Contributions to r come from regions where q is changing; since $q^2(z) = \varepsilon(z)\omega^2/c^2 - K^2$, these are the regions where the dielectric function is changing.

We saw in Section 6-2 that ψ_0^+ is a fair approximation to ψ provided the dimensionless quantity $\gamma = q^{-2}\,dq/dz$ is small. Assuming this is so, the expression

$$r^{(0)} = -\frac{1}{2q_1}\int_{-\infty}^{\infty} dz\,\frac{dq}{dz}\,(\psi_0^+)^2 = -\frac{1}{2q_1}\int_{q_1}^{q_2} dq\,e^{2i\phi} \tag{52}$$

should be a fair approximation to r. We will not discuss the properties of $r^{(0)}$, since there is better theoretical basis for the similar approximation $r^{(1)}$, which is obtained from substituting $\tilde{\psi} = \psi_1^+ = (q_1/q)^{1/2}\,e^{i\phi}$ into the comparison identity (50). (The factor $\sqrt{q_1}$ in the expression for ψ_1^+ gives the correct coefficient of $e^{iq_1 z}$ in $\tilde{\psi}$ as $z \rightarrow -\infty$.) Since ψ_1^+ satisfies (26),

$$\tilde{q}^2 = q^2 + \frac{1}{2q}\frac{d^2q}{dz^2} - \frac{3}{4q^2}\left(\frac{dq}{dz}\right)^2 = q^2 + \tfrac{1}{2}q^{1/2}\frac{d}{dz}\left(\frac{dq/dz}{q^{3/2}}\right), \tag{53}$$

and we obtain the identity

$$r = \frac{1}{4i\sqrt{q_1}} \int_{-\infty}^{\infty} \mathrm{d}z \frac{\mathrm{d}}{\mathrm{d}z}\left(\frac{\mathrm{d}q/\mathrm{d}z}{q^{3/2}}\right) \psi \,\mathrm{e}^{i\phi}. \tag{54}$$

On approximating ψ by ψ_1^+ we have another estimate for r, which can be put into several equivalent forms by integration by parts and a change of variable from z to ϕ:

$$\begin{aligned} r^{(1)} &= \frac{1}{4i} \int_{-\infty}^{\infty} \mathrm{d}z \frac{\mathrm{d}}{\mathrm{d}z}\left(\frac{\mathrm{d}q/\mathrm{d}z}{q^{3/2}}\right) q^{-1/2}\,\mathrm{e}^{2i\phi} \\ &= -\frac{1}{2}\int_{-\infty}^{\infty} \mathrm{d}z \frac{\mathrm{d}q}{q\mathrm{d}z}\,\mathrm{e}^{2i\phi} + \frac{1}{8i}\int_{-\infty}^{\infty} \mathrm{d}z \left(\frac{\mathrm{d}q/\mathrm{d}z}{q^{3/2}}\right)^2 \mathrm{e}^{2i\phi} \\ &= -\frac{1}{2}\int_{-\infty}^{\infty} \mathrm{d}\phi \left(\frac{\mathrm{d}q}{q\mathrm{d}\phi}\right) \mathrm{e}^{2i\phi} + \frac{1}{8i}\int_{-\infty}^{\infty} \mathrm{d}\phi \left(\frac{\mathrm{d}q}{q\mathrm{d}\phi}\right)^2 \mathrm{e}^{2i\phi}. \end{aligned} \tag{55}$$

We note that the part equal to $-\frac{1}{2}\int \mathrm{d}q\, q^{-1}\,\mathrm{e}^{2i\phi}$ is the Rayleigh (or weak reflection) approximation of Section 5-7, r^R. This is the dominant part of $r^{(1)}$: in the last line of (55) the dimensionless quantity $\mathrm{d}q/q\mathrm{d}\phi$ is precisely the function $\gamma = \mathrm{d}q/q^2\,\mathrm{d}z$ which has to be small for ψ_0 and ψ_1 to approximate ψ well. The last term in (55) could thus be omitted in the calculation of the reflection amplitude to lowest order in this quantity. However, the expression (55) came directly from the identity (54), and will be seen to follow from first order perturbation theory in the next section, so there are reasons both for retaining and for dropping the last term in (55).

In the short wave limit the factor $\mathrm{e}^{2i\phi}$ oscillates rapidly with z, and a smooth variation of q with z will lead to exponentially small values of the integral. If however q has discontinuities in any of its derivatives, the main contributions to the integral come from the neighbourhood of the discontinuities. We will give examples of these possibilities.

Consider the case of a medium which is nonuniform only within the interval $[z_1, z_2]$, and which has discontinuities in the slope or higher derivatives of ε only at the end points z_1 and z_2. The exponential and Rayleigh profiles are in this category. In the short wave limit the factor $\mathrm{e}^{2i\phi}$ oscillates rapidly with z, averaging to an exponentially small value any slowly varying contribution in the integrals defining $r^{(1)}$. There will be a contribution in (55) arising from discontinuities in slope: we write (55) in the form

$$r^{(1)} = \frac{1}{4i}\int_{-\infty}^{\infty} \mathrm{d}z \left\{\frac{\mathrm{d}^2 q/\mathrm{d}z^2}{q^2} - \frac{3}{2}\frac{(\mathrm{d}q/\mathrm{d}z)^2}{q^3}\right\} \mathrm{e}^{2i\phi}, \tag{56}$$

and note that for a profile in which the dimensionless function $\gamma = \mathrm{d}q/q^2\,\mathrm{d}z$ changes from 0 to γ_1 at z_1, and from γ_2 to 0 at z_2, the second derivative term in (56) has the delta function singularities $\gamma_1\delta(z - z_1) - \gamma_2\delta(z - z_2)$. Thus (56) takes the value

$$r^{(1)} = \frac{1}{4i}\{\gamma_1 e^{2i\phi_1} - \gamma_2 e^{2i\phi_2}\}$$
$$= \frac{e^{i(\phi_1+\phi_2)}}{4i}\{\gamma_1 e^{-i\Delta\phi} - \gamma_2 e^{i\Delta\phi}\}, \tag{57}$$

plus exponentially small terms from the smooth part of the profile. This expression gives the same reflectivity as the theory of Section 6-3 (equation 36). It is in agreement with (35) in phase as well as in absolute magnitude if $\phi_1 = \phi(z_1) = q_1 z_1$. This condition is in fact a requirement arising in the setting up of the comparison identity (54), in which the asymptotic form $e^{iq_1 z}$ was assumed for ψ_1^+ in the limit as $z \to -\infty$. This requirement fixes the phase function to be

$$\phi(z) = q_1 z_1 + \int_{z_1}^{z} d\zeta\, q(\zeta). \tag{58}$$

In the case of a smooth profile, with all of its derivatives continuous everywhere, the approximation $r^{(1)}$ of (55), and its dominant part, the Rayleigh approximation

$$r^R = -\int_{-\infty}^{\infty} dz \frac{dq/dz}{2q} e^{2i\phi}, \tag{59}$$

work well down to surprisingly small values of $(\omega/c)\Delta z$. However, these approximations give (in general) only the correct exponent in the short wave limiting form, and not the correct prefactor. We shall illustrate with the hyperbolic tangent profile, for which the phase integral may be evaluated analytically. We have

$$q^2(z) = \tfrac{1}{2}(q_1^2 + q_2^2) + \tfrac{1}{2}(q_2^2 - q_1^2) \tanh z/2\Delta z. \tag{60}$$

In the phase integral

$$\phi(z) = \int_0^z d\zeta\, q(\zeta) \tag{61}$$

we transform first to the variable $x = \tanh(\zeta/2\Delta z)$, and then to the variable $y = q\Delta z$. We find

$$\phi(z) = \int_0^{\tanh z/2\Delta z} dx \left(\frac{1}{1-x} + \frac{1}{1+x}\right) y(x)$$
$$= 2\int_{\bar{y}}^{q\Delta z} dy\, y^2 \left\{\frac{1}{y_2^2 - y^2} + \frac{1}{y^2 - y_1^2}\right\} \tag{62}$$
$$= f(y) - f(\bar{y}),$$

where $\bar{y}^2 = \frac{1}{2}(y_1^2 + y_2^2)$ and

$$f(y) = y_2 \log\frac{y_2 + y}{y_2 - y} - y_1 \log\frac{y + y_1}{y - y_1} \tag{63}$$

(this form applies for $y_1 < y_2$, which is the case when $\varepsilon_1 < \varepsilon_2$). The Rayleigh approximation for the reflection amplitude thus becomes

$$r^R = -\tfrac{1}{2}\,\mathrm{e}^{-2i f(\bar{y})} \int_{y_1}^{y_2} \frac{\mathrm{d}y}{y} \left(\frac{y_2 + y}{y_2 - y}\right)^{2iy_2} \left(\frac{y - y_1}{y + y_1}\right)^{2iy_1}. \tag{64}$$

A related form, more convenient for numerical calculation, is

$$r^R = -\tfrac{1}{8}(y_2^2 - y_1^2)\,\mathrm{e}^{-2i f(\bar{y})} \int_{-\infty}^{\infty} \mathrm{d}x\,(y \cosh x)^{-2}\,\mathrm{e}^{2i f(y)}, \tag{65}$$

where $x = z/2\Delta z$. The Rayleigh approximation is compared with the exact result in Figure 6-3.

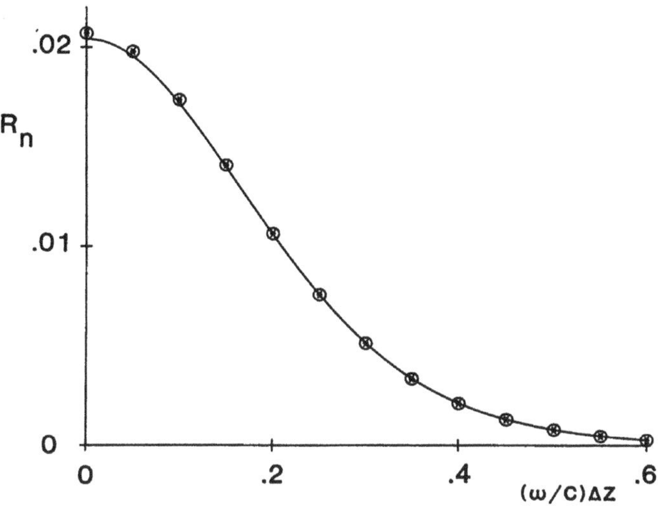

Figure 6-3. Reflectivity at normal incidence for the tanh profile. The curve is the exact reflectivity from (2), the points are the Rayleigh approximation reflectivities from (65). The values $\varepsilon_1 = 1$, $\varepsilon_2 = (4/3)^2$ are used, as in Figure 6-1 (where the long and short wave limiting forms are shown).

We see that the Rayleigh approximation gives good agreement with the exact reflectivity over the entire range of wavelengths. One might expect that at short wavelengths the agreement becomes perfect. This is not so: the asymptotic value has been given by Berry and Mount (1972), and they find that the reflectivity resulting from this approximation is not expression (3), but $\pi^2/9 \simeq 1.0966$ times that value. The exponential is given correctly, but not the prefactor. The theoretical reason for such discrepancies was shown by Pokrovskii, Savvinykh and Ulinich (1958) to lie in the nature of the perturbation series of which r^R is the first term. This is called the Bremmer series; a related approach is the Brekhovskikh iteration method of Section 5-8. These will be discussed further in the next Section. Here we note only that the discrepancy is usually of little practical importance. For example, for the hyperbolic tangent profile the correct asymptotic form $\exp(-4\pi q_1 \Delta z)$ is better than the Rayleigh approximation only beyond $(\omega/c)\Delta z \simeq 2$, where the reflectivity is so small (about 10^{-11}) as to be extremely difficult to measure.

6-5 Perturbation theory for short waves

As in Section 3-1 we wish to express ψ, the solution of

$$\frac{\mathrm{d}^2\psi}{\mathrm{d}z^2} + q^2\psi = 0, \qquad \mathrm{e}^{iq_1 z} + r\,\mathrm{e}^{iq_1 z} \leftarrow \psi \rightarrow t\,\mathrm{e}^{iq_2 z}, \tag{66}$$

in terms of a known function $\tilde{\psi}$, the solution of

$$\frac{d^2\tilde{\psi}}{dz^2} + \tilde{q}^2\tilde{\psi} = 0, \qquad e^{iq_1 z} + \tilde{r}\, e^{iq_1 z} \leftarrow \tilde{\psi} \rightarrow \tilde{t}\, e^{iq_2 z}. \tag{67}$$

To do this we need a Green's function $G(z, \zeta)$ satisfying the equation

$$\frac{\partial^2 G}{\partial z^2} + \tilde{q}^2 G = \delta(z - \zeta), \tag{68}$$

and giving the appropriate boundary conditions for ψ.

In Section 3-1 we used $e^{\pm iq_0 z}$ to construct G, with $q_0(z) = q_1$ for $z < 0$ and q_2 for $z > 0$. These functions are appropriate for the long wave case. Here we use $\psi_1^+ = (q_1/q)^{1/2}\, e^{i\phi}$, $\psi_1^- = (q_2/q)^{1/2}\, e^{-i\phi}$ to construct G, these being wavefunctions which approximate ψ in the short wave case. The required Green's function is

$$G(z, \zeta) = \begin{cases} \dfrac{1}{2i(q_1 q_2)^{1/2}}\, \psi_1^-(z)\psi_1^+(\zeta) & z < \zeta \\[2ex] \dfrac{1}{2i(q_1 q_2)^{1/2}}\, \psi_1^+(z)\psi_1^-(\zeta) & z > \zeta. \end{cases} \tag{69}$$

This G satisfies (68) when $z \neq \zeta$, with

$$\tilde{q}^2 = q^2 + \frac{d^2q/dz^2}{2q} - \frac{3}{4}\left(\frac{dq}{q dz}\right)^2 = q^2 + \tfrac{1}{2}q^{1/2}\frac{d}{dz}\left(\frac{dq/dz}{q^{3/2}}\right), \tag{70}$$

because of (26). The derivative of G is given by (again using $\gamma = q^{-2}\, dq/dz$)

$$\frac{\partial G}{\partial z} = \begin{cases} -q(z)\,(i + \gamma(z)/2)G(z, \zeta) & z < \zeta \\ q(z)\,(i - \gamma(z)/2)G(z, \zeta) & z > \zeta. \end{cases} \tag{71}$$

When $z = \zeta$, G takes the value $1/2iq(z)$. Thus $\partial G/\partial z$ has the required unit discontinuity at $z = \zeta$, leading to the delta function on the right-hand side of (68). The integral equation satisfied by ψ appropriate to the reflection problem is

$$\psi(z) = \psi_1^+(z) - \int_{-\infty}^{\infty} d\zeta\; G(z, \zeta)\Delta q^2(\zeta)\psi(\zeta), \tag{72}$$

where

$$\Delta q^2 = q^2 - \tilde{q}^2 = -\tfrac{1}{2}q^{1/2}\frac{d}{dz}\left(\frac{dq/dz}{q^{3/2}}\right). \tag{73}$$

As $z \rightarrow -\infty$ this gives the asymptotic form (on choosing the lower limit of integration in ϕ to make $\phi \rightarrow q_1 z$ as $z \rightarrow -\infty$)

$$\begin{aligned} \psi(z) &\rightarrow e^{iq_1 z} - \frac{1}{2i(q_1 q_2)^{1/2}}\int_{-\infty}^{\infty} d\zeta\; \psi_1^-(z)\psi_1^+(\zeta)\Delta q^2(\zeta)\psi(\zeta) \\ &= e^{iq_1 z} + e^{-iq_1 z}\frac{1}{4iq_1^{1/2}}\int_{-\infty}^{\infty} d\zeta\, \frac{d}{d\zeta}\left(\frac{dq/d\zeta}{q^{3/2}}\right) e^{i\phi(\zeta)}\, \psi(\zeta). \end{aligned} \tag{74}$$

The coefficient of $e^{-iq_1 z}$ is the exact reflection amplitude r; the expression above is equivalent to (54), obtained from the comparison identity (50). The first order perturbation result is obtained by setting $\psi = \psi_1^+$ on the right hand sides of (72) and (74), and reproduces $r^{(1)}$ as given by (55). Higher order approximations are obtained by iteration of the integral equation, as in the long wave case of Chapter 3.

The integral equation (72) and the resulting perturbation series are closely related to the coupled equations derived by Bremmer (1951), and the resulting *Bremmer series*. This in turn has an intimate connection with the Brekhovskikh series of Section 5-8, as we shall show by deriving the Bremmer equations from the results of Section 5-6. In (5.74) and (5.75) we put $f = q^{-1/2}F, g = q^{-1/2}G$, to obtain the coupled, first order, linear differential equations

$$F' = \frac{q'}{2q} G\,e^{-2i\phi}, \tag{75}$$

$$G' = \frac{q'}{2q} F\,e^{2i\phi}. \tag{76}$$

These have the same form as the Bremmer equations (Bremmer (1951), equation 15); the relation of F and G to the Bremmer functions $u_\uparrow$ and $u_\downarrow$ is

$$F = q^{1/2}\,e^{-i\phi}\,u_\uparrow, \qquad G = q^{1/2}\,e^{i\phi}\,u_\downarrow. \tag{77}$$

For waves incident from $z = -\infty$ we have $G(+\infty) = 0$. If further we set $F(-\infty) = 1$, then $G(-\infty) = r_s$, and integrating the equation for G' from $-\infty$ to $+\infty$ gives

$$r_s = -\int_{-\infty}^{\infty} dz\,\frac{q'}{2q} F\,e^{2i\phi}. \tag{78}$$

When we approximate F by unity (weak reflection) we regain the Rayleigh expression for r_s.

The Bremmer series, obtained by iteration of the pair (75, 76), has been investigated by Bellman and Kalaba (1959) and Atkinson (1960), the latter showing that the series converges if

$$\int_{-\infty}^{\infty} dz \left|\frac{dq/dz}{q}\right| < \pi. \tag{79}$$

When $\varepsilon(z)$ is monotonically increasing, so is $q(z)$, and (79) may be written as

$$\int_{q_1}^{q_2} \frac{dq}{q} = \log\frac{q_2}{q_1} < \pi. \tag{80}$$

At normal incidence this condition reads $\varepsilon_2/\varepsilon_1 < e^{2\pi} \simeq 535$, a rather weak constraint. The condition for convergence progressively tightens as the angle of incidence is increased: taking the example of the air-water interface at optical frequencies ($\varepsilon_1 = 1$, $\varepsilon_2 \simeq (4/3)^2$), we find that the inequality (80) is satisfied by a factor of more than three hundred at normal incidence, but is violated at $\theta_1 \simeq 87.8°$, that is at about 2.2° from grazing incidence. The failure of the short wave approximation near grazing incidence was implied in Section 6-2, where we noted that the function $\gamma = q^{-2}\,dq/dz$ had to remain small. Near grazing incidence, or near the critical angle (where respectively q_1 and q_2 tend to zero) γ becomes large within the interface, and the short wave perturbation theories fail.

6-6 Short wave results for r_p and r_p/r_s

We shall first summarize the results obtained for the s wave, rewriting the results in electromagnetic notation. The electric field is $(0, E(z)\,\mathrm{e}^{i(Kx-\omega t)}, 0)$, with

$$\frac{\mathrm{d}^2 E}{\mathrm{d}z^2} + q^2 E = 0, \qquad \mathrm{e}^{iq_1 z} + r_s\,\mathrm{e}^{-iq_1 z} \leftarrow E \rightarrow t_s\,\mathrm{e}^{iq_2 z}. \tag{81}$$

Approximate solutions of (81) are

$$\psi_1^+ = \left(\frac{q_1}{q}\right)^{1/2} \mathrm{e}^{i\phi}, \qquad \psi_1^- = \left(\frac{q_2}{q}\right)^{1/2} \mathrm{e}^{-i\phi}, \tag{82}$$

which in fact satisfy

$$\frac{\mathrm{d}^2\psi_1^\pm}{\mathrm{d}z^2} + \left[q^2 + \frac{1}{2q}\frac{\mathrm{d}^2 q}{\mathrm{d}z^2} - \frac{3}{4}\left(\frac{\mathrm{d}q}{q\mathrm{d}z}\right)^2\right]\psi_1^\pm = 0. \tag{83}$$

The comparison identity

$$r_s = \frac{1}{4iq_1^{1/2}}\int_{-\infty}^{\infty} \mathrm{d}z\,\frac{\mathrm{d}}{\mathrm{d}z}\left(\frac{\mathrm{d}q/\mathrm{d}z}{q^{3/2}}\right) E\,\mathrm{e}^{i\phi}, \tag{84}$$

or the perturbation theory of the previous section, lead to the approximation

$$r_s^{(1)} = \frac{1}{4i}\int_{-\infty}^{\infty} \mathrm{d}z\,\frac{\mathrm{d}}{\mathrm{d}z}\left(\frac{\mathrm{d}q/\mathrm{d}z}{q^{3/2}}\right) q^{-1/2}\,\mathrm{e}^{2i\phi}. \tag{85}$$

For the interfaces which extend from z_1 to z_2, and have discontinuities in the slope of ε at the end points but are otherwise smooth, (85) leads to

$$r_s^{(1)} = \frac{\mathrm{e}^{i(\phi_1+\phi_2)}}{4i}\{\gamma_1\,\mathrm{e}^{-i\Delta\phi} - \gamma_2\,\mathrm{e}^{i\Delta\phi}\} + \ldots \tag{86}$$

(exponentially small terms from the smooth part of the interface being omitted).

We now wish to derive corresponding results for the p wave, for which

$$\mathbf{B} = \left[0, \left(\frac{\varepsilon}{\varepsilon_1}\right)^{1/2} b\,\mathrm{e}^{i(Kx-\omega t)}, 0\right],$$

with

$$\frac{\mathrm{d}^2 b}{\mathrm{d}z^2} + q_b^2 b = 0, \qquad \mathrm{e}^{iq_1 z} - r_p\,\mathrm{e}^{-iq_1 z} \leftarrow b \rightarrow t_p\,\mathrm{e}^{iq_2 z}, \tag{87}$$

and

$$\begin{aligned} q_b^2 &= q^2 + \tfrac{1}{2}\varepsilon^{1/2}\frac{\mathrm{d}}{\mathrm{d}z}\left(\frac{\mathrm{d}\varepsilon/\mathrm{d}z}{\varepsilon^{3/2}}\right) = q^2 - \varepsilon^{1/2}\frac{\mathrm{d}^2\varepsilon^{-1/2}}{\mathrm{d}z^2} \\ &= q^2 + \frac{1}{2\varepsilon}\frac{\mathrm{d}^2\varepsilon}{\mathrm{d}z^2} - \frac{3}{4}\left(\frac{1}{\varepsilon}\frac{\mathrm{d}\varepsilon}{\mathrm{d}z}\right)^2 \end{aligned} \tag{88}$$

(compare equations (1.22) or (2.3)). For smooth profiles the difference $q_b^2 - q^2$ is of order $(\Delta z)^{-2}$, and thus smaller by the factor $[(\omega/c)\Delta z]^{-2}$ in comparison with q^2. It is therefore negligible in the short wave case, except at grazing incidence or near turning points. For profiles which have discontinuities in their first derivative (typically at the end points z_1 and z_2), the second derivative has delta functions contributions at such points.

We first derive a comparison identity linking b and ψ_1^+, which satisfies

$$\frac{d^2\psi_1^+}{dz^2} + \left[q^2 + \frac{1}{2q}\frac{d^2q}{dz^2} - \frac{3}{4}\left(\frac{dq}{q dz}\right)^2\right]\psi_1^+ = 0,$$
$$e^{iq_1 z} \leftarrow \psi_1^+ \rightarrow \left(\frac{q_1}{q_2}\right)^{1/2} e^{iq_2 z}. \tag{89}$$

The expression in brackets will again be shortened to $\tilde{q}^2$. The identity resulting from multiplying (87) by ψ_1^+, (89) by b, subtracting, and integrating the result from $-\infty$ to $+\infty$ is

$$r_p = \frac{1}{2iq_1}\int_{-\infty}^{\infty} dz\,(q_b^2 - \tilde{q}^2)\psi_1^+ b. \tag{90}$$

At normal incidence $q \rightarrow k = \sqrt{\varepsilon}\,\omega/c$,

$$\tilde{q}^2 - q^2 \rightarrow \frac{1}{4}\frac{\varepsilon''}{\varepsilon} - \frac{5}{15}\left(\frac{\varepsilon'}{\varepsilon}\right)^2, \tag{91}$$

and

$$q_b^2 - \tilde{q}^2 \rightarrow \frac{1}{4}\frac{\varepsilon''}{\varepsilon} - \frac{7}{16}\left(\frac{\varepsilon'}{\varepsilon}\right)^2. \tag{92}$$

Thus (90) gives

$$r_n = \frac{1}{2ik_1}\int_{-\infty}^{\infty} dz\left[\frac{1}{4}\frac{\varepsilon''}{\varepsilon} - \frac{7}{16}\left(\frac{\varepsilon'}{\varepsilon}\right)^2\right]\psi_1^+ b, \tag{93}$$

which is to be compared with the corresponding expression obtained from (85):

$$r_n = \frac{1}{2ik_1}\int_{-\infty}^{\infty} dz\left[\frac{1}{4}\frac{\varepsilon''}{\varepsilon} - \frac{5}{16}\left(\frac{\varepsilon'}{\varepsilon}\right)^2\right]\psi_1^+ E. \tag{94}$$

These are both identities, but give slightly different values for r_n when b and E are both approximated by ψ_1^+. (There is no difference in the leading term for profiles which have discontinuities in the first derivative, because this comes from the $\varepsilon''/\varepsilon$ term.)

At general angle of incidence, (90) reads

$$r_p = \frac{1}{4iq_1}\int_{-\infty}^{\infty} dz\left[\frac{\varepsilon''}{\varepsilon} - \frac{3}{2}\left(\frac{\varepsilon'}{\varepsilon}\right)^2 - \frac{q''}{q} + \frac{3}{2}\left(\frac{q'}{q}\right)^2\right]\psi_1^+ b, \tag{95}$$

and when b is approximated by ψ_1^+, it leads to

$$r_p^{(1)} = \frac{1}{4i}\int_{-\infty}^{\infty} dz\left[\frac{\varepsilon''}{\varepsilon} - \frac{3}{2}\left(\frac{\varepsilon'}{\varepsilon}\right)^2 - \frac{q''}{q} + \frac{3}{2}\left(\frac{q'}{q}\right)^2\right]q^{-1}\,e^{2i\phi}. \tag{96}$$

For profiles of finite extent which have discontinuities in the slope of ε at the endpoints z_1 and z_2, the dominant short wave contribution comes from these discontinuities. Let (as in the s wave case) the quantity $\gamma = dq/q^2\,dz$ change from 0 to γ_1 at z_1, and from γ_2 to 0 at z_2. The resulting contribution to the integrand of (96) is

$$\gamma_1 \cos 2\theta_1\, \delta(z - z_1) - \gamma_2 \cos 2\theta_2\, \delta(z - z_2) \tag{97}$$

(at z_1 the q''/q^2 term has delta function strength γ_1, and the $\varepsilon''/\varepsilon q$ term has delta function strength $2\gamma_1 \cos^2\theta_1$). If the profile is smooth everywhere except at z_1 and z_2, the leading term in the short wave limit is

$$r_p^{(1)} = \frac{1}{4i}\{\gamma_1 \cos 2\theta_1\, e^{2i\phi_1} - \gamma_2 \cos 2\theta_2\, e^{2i\phi_2}\} + \ldots \tag{98}$$

(... denotes exponentially small term terms). This may be rewritten as

$$r_p^{(1)} = \frac{e^{i(\phi_1+\phi_2)}}{4i}\{\gamma_1 \cos 2\theta_1\, e^{-i\Delta\phi} - \gamma_2 \cos 2\theta_2\, e^{i\Delta\phi}\} \tag{99}$$

(where $\Delta\phi = \phi_2 - \phi_1$). At normal incidence this is in agreement with $r_s^{(1)}$ as given by (86), and the ratio $r_p^{(1)}/r_s^{(1)}$ correctly takes the value $+1$. At grazing incidence the approximations $r_p^{(1)}$ and $r_s^{(1)}$ both fail, since the assumption that $\gamma = q^{-2}\,dq/dz$ is small compared to unity cannot hold as $q_1 = \sqrt{\varepsilon_1}(\omega/c)\cos\theta_1$ tends to zero. Neither does $r_p^{(1)}$ tend to 1 nor does $r_s^{(1)}$ tend to -1 (the correct limiting values at grazing incidence, as shown in Section 2-3), but $r_p^{(1)}/r_s^{(1)}$ does tend to the correct limiting value of -1, since the γ_1 terms dominate and $\cos 2\theta_1$ tends to -1.

The s and p reflectivities in the short wave limit are

$$R_s^{(1)} = |r_s^{(1)}|^2 = \tfrac{1}{16}\{\gamma_1^2 + \gamma_2^2 - 2\gamma_1\gamma_2 \cos 2\Delta\phi\}, \tag{100}$$

$$R_p^{(1)} = |r_p^{(1)}|^2 = \tfrac{1}{16}\{\gamma_1^2 \cos^2 2\theta_1 + \gamma_2^2 \cos^2 2\theta_2 - 2\gamma_1\gamma_2 \cos 2\theta_1 \cos 2\theta_2 \cos 2\Delta\phi\}. \tag{101}$$

We will compare these formulae with the exact results for the single exponential profile, for which (compare Sections 2-5 and 6-1)

$$\gamma = \tfrac{1}{2}q^{-3}\frac{dq^2}{dz} = \frac{\varepsilon\omega^2/c^2}{2aq^3}, \tag{102}$$

$$\Delta\phi = 2a\left\{q_2 - q_1 - K\left[\arctan\frac{q_2}{K} - \arctan\frac{q_1}{K}\right]\right\}, \tag{103}$$

where $a = \Delta z/\log(\varepsilon_2/\varepsilon_1)$. The reflectivities as a function of angle of incidence are shown in Figures 6-4 and 6-5.

We see from these figures that the short wave approximations work well at normal incidence with the rather small value $(\omega/c)\Delta z = 2$, but at this value their accuracy is poor near the Brewster angle and beyond. Thus care must be taken in the ellipsometric application of the formulae (86) and (99) in the intermediate region when the interfacial thickness is of the same order of magnitude as the

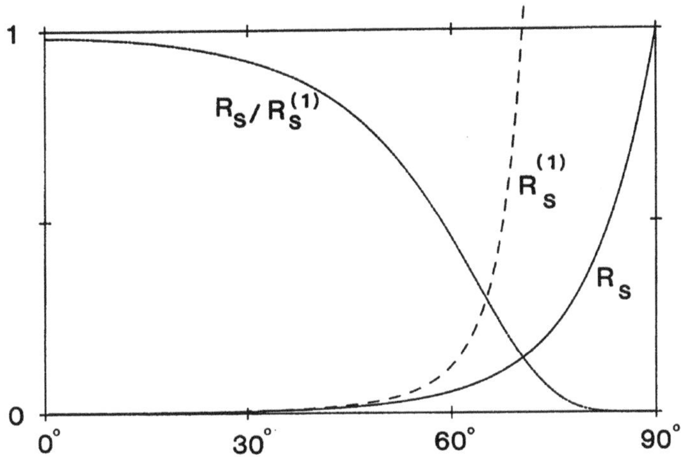

Figure 6-4. Angular dependence of the s wave reflectivity for the exponential profile, at $(\omega/c)\Delta z = 2$ and with $\varepsilon_1 = 1$, $\varepsilon_2 = (4/3)^2$. The solid curve is the exact reflectivity obtained from (2.99); the dashed curve is the short wave approximation (100). The ratio $R_s/R_s^{(1)}$ is also shown; $R_s^{(1)}$ is about 2% too large at normal incidence, and about a factor of 2 too large at 60°.

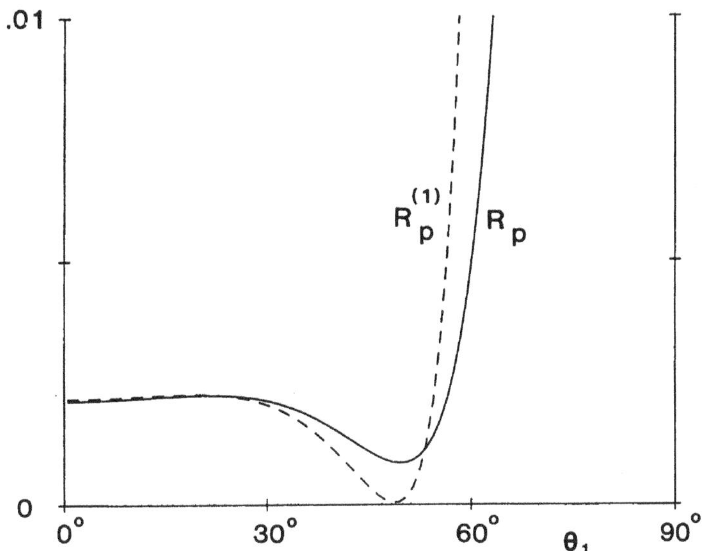

Figure 6-5. Angular dependence of the p wave reflectivity for the exponential profile, at $(\omega/c)\Delta z = 2$ and with $\varepsilon_1 = 1$, $\varepsilon_2 = (4/3)^2$. The solid curve is the exact reflectivity obtained from (2.98); the dashed curve is the short wave approximation (101). Note the vertical scale is enlarged one hundred times relative to Figure 6-4. The minima of R_p and $R_p^{(1)}$ are at 49.7° and 48.9°; the zero-thickness Brewster angle is arctan (4/3) $\simeq$ 53.1°.

wavelength ($(\omega/c)\Delta z = 2$ corresponds to a wavelength about three times the interfacial thickness, giving the normal incidence values $\gamma_1 \simeq 1/7$ and $\gamma_2 \simeq 1/10$ for the profile used in Figure 6-4 and 6-5).

From (86) and (99) we find that the real and imaginary parts of the ellipsometric ratio r_p/r_s have the short wave limiting forms

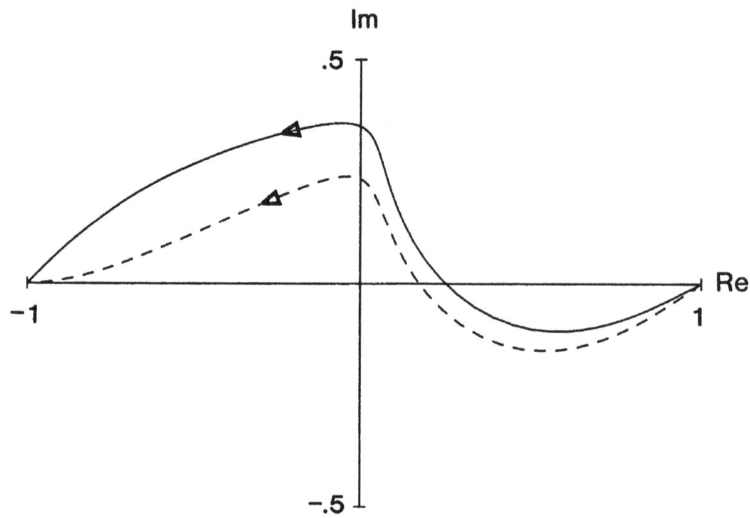

Figure 6-6. Exact (———) and approximate (- - - -) trajectories of r_p/r_s in the complex plane, as a function of the angle of incidence. The curves are drawn from the exponential profile, with $\varepsilon_1 = 1$, $\varepsilon_2 = (4/3)^2$, $(\omega/c)\Delta z = 3$. The arrows indicate direction of increasing angle of incidence; the arrowheads are located at the zero thickness Brewster angle $\theta_B = \arctan(4/3) \simeq 53.13°$.

$$\mathrm{Re}\left(\frac{r_p}{r_s}\right)^{(1)} = \frac{\gamma_1 \cos 2\theta_1 + \gamma_2 \cos 2\theta_2 - \gamma_1\gamma_2(\cos 2\theta_1 + \cos 2\theta_2)\cos 2\Delta\phi}{\gamma_1^2 + \gamma_2^2 - 2\gamma_1\gamma_2 \cos 2\Delta\phi}, \tag{104}$$

$$\mathrm{Im}\left(\frac{r_p}{r_s}\right)^{(1)} = \frac{\gamma_1\gamma_2(\cos 2\theta_1 - \cos 2\theta_2)\sin 2\Delta\phi}{\gamma_1^2 + \gamma_2^2 - 2\gamma_1\gamma_2 \cos 2\Delta\phi}. \tag{105}$$

The trajectory of r_p/r_s in the complex plane, as a function of the angle of incidence, is shown in Figure 6-6 for the exponential profile.

We see that while the short wave approximations for the reflectivities work well down to $(\omega/c)\Delta z = 2$ at normal incidence, the agreement is poor at intermediate angles even at $(\omega/c)\Delta z = 3$.

So far in this Section we have given results based on the approximations $r_s^{(1)}$ and $r_p^{(1)}$, which may be written in the form

$$r_s^{(1)} = \frac{1}{4i}\int_{-\infty}^{\infty} dz \left[\frac{q''}{q} - \frac{3}{2}\left(\frac{q'}{q}\right)^2\right] q^{-1}\, e^{2i\phi}, \tag{106}$$

$$r_p^{(1)} = \frac{1}{4i}\int_{-\infty}^{\infty} dz \left[\frac{\varepsilon''}{\varepsilon} - \frac{3}{2}\left(\frac{\varepsilon'}{\varepsilon}\right)^2 - \frac{q''}{q} + \frac{3}{2}\left(\frac{q'}{q}\right)^2\right] q^{-1}\, e^{2i\phi}. \tag{107}$$

We will compare these with the Rayleigh approximations of Section 5-7:

$$r_s^R = -\int_{-\infty}^{\infty} dz\, \frac{q'}{2q}\, e^{2i\phi}, \tag{108}$$

$$r_p^R = \int_{-\infty}^{\infty} dz\, \frac{Q'}{2Q}\, e^{2i\phi}. \tag{109}$$

By changing the variable of integration temporarily to ϕ, integrating by parts, and changing back, these may be written in the form

$$r_s^R = \frac{1}{4i}\int_{-\infty}^{\infty} dz \left[\frac{q''}{q} - 2\left(\frac{q'}{q}\right)^2\right] q^{-1}\, e^{2i\phi}, \tag{110}$$

$$r_p^R = \frac{1}{4i}\int_{-\infty}^{-\infty} dz \left[\frac{\varepsilon''}{\varepsilon} - \left(\frac{\varepsilon'}{\varepsilon}\right)^2 - \frac{q'\varepsilon'}{q\varepsilon} - \frac{q''}{q} + 2\left(\frac{q'}{q}\right)^2\right] q^{-1}\, e^{2i\phi}. \tag{111}$$

When written in this form the difference between the two approximate sets is seen to lie in the coefficients of the square of the first derivative, but not in the second derivative. Thus for profiles with a discontinuity in the first derivative of ε, and consequently a delta function contribution to ε'' or q'', there is no difference in the short wave limiting forms.

At normal incidence the two Rayleigh forms agree, giving

$$r_n^R = \frac{1}{8i}\int_{-\infty}^{\infty} dz \left[\frac{\varepsilon''}{\varepsilon} - \frac{3}{2}\left(\frac{\varepsilon'}{\varepsilon}\right)^2\right] k^{-1}\, e^{2i\phi}, \tag{112}$$

while there is a difference between the expressions (106) and (107) when $q \to \sqrt{\varepsilon}\,\omega/c = k$, the two averaging to (112):

$$r_s^{(1)} \to \frac{1}{8i}\int_{-\infty}^{\infty} dz \left[\frac{\varepsilon''}{\varepsilon} - \frac{5}{4}\left(\frac{\varepsilon'}{\varepsilon}\right)^2\right] k^{-1}\, e^{2i\phi}, \tag{113}$$

$$r_s^{(1)} \to \frac{1}{8i}\int_{-\infty}^{\infty} dz \left[\frac{\varepsilon''}{\varepsilon} - \frac{7}{4}\left(\frac{\varepsilon'}{\varepsilon}\right)^2\right] k^{-1}\, e^{2i\phi}. \tag{114}$$

6-7 A single turning point: total reflection

The preceding sections have dealt with the case $\varepsilon_1 < \varepsilon_2$ (or $V_1 > V_2$ in the quantum particle case), where $q^2(z) = \varepsilon(z)\omega^2/c^2 - K^2$ is positive everywhere. We now examine the opposite case where $\varepsilon_1 > \varepsilon_2$ (or $V_1 < V_2$), of which examples are: light incident on an interface from the optically denser medium, radio waves incident on an ionospheric layer, or particles moving up a potential gradient. This was illustrated in Figures 1-8 and 1-9. For $\theta_1 > \theta_c = \arcsin\,(\varepsilon_2/\varepsilon_1)^{1/2}$ there will be total internal reflection, since then $q_2^2 < 0$, and the wave deep inside medium 2 decays exponentially as $\exp\,(-|q_2|z)$. In geometrical optics and classical particle physics the reflection occurs at the point z_0 defined by $q^2(z_0) = 0$. This is called the *turning point*: a classical particle turns back at z_0, being unable to penetrate into a region where the kinetic energy of the motion in the z direction would become negative. Waves do penetrate beyond this point, but decay exponentially for $z > z_0$, and there is no propagating wave at infinity.

For a given profile, the location of the turning point is a function of the angle of incidence. The location is given by $q^2(z_0) = \varepsilon(z_0)\omega^2/c^2 - K^2 = 0$, and since $K^2 = \varepsilon_1(\omega^2/c^2)\sin^2\theta_1$, z_0 is determined by

$$\varepsilon(z_0) = \varepsilon_1 \sin^2\theta_1. \tag{115}$$

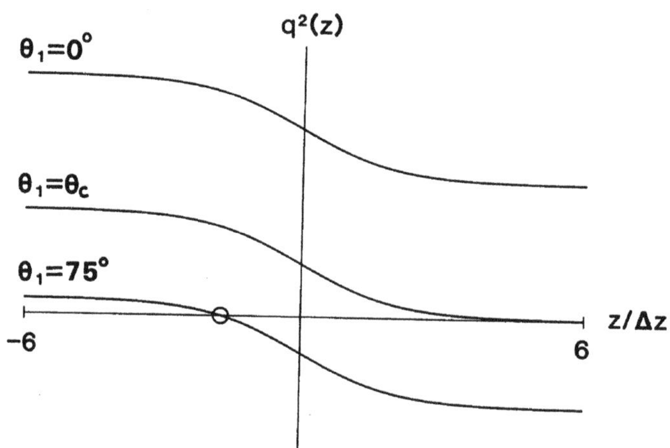

Figure 6-7. Variation of $q^2(z)$ with angle of incidence. The curves are drawn for the hyperbolic tangent profile, with $\varepsilon_1 = (4/3)^2$ and $\varepsilon_2 = 1$, representing the water–air interface at optical frequencies. The turning point for $\theta_1 = 75°$ is circled.

For example, for the hyperbolic tangent profile, for which

$$\varepsilon(z) = \frac{\varepsilon_1 + \varepsilon_2 \, e^{z/\Delta z}}{1 + e^{z/\Delta z}}, \tag{116}$$

$$z_0(\theta_1) = \Delta z \log\left(\frac{\cos^2\theta_1}{\sin^2\theta_1 - \varepsilon_2/\varepsilon_1}\right). \tag{117}$$

This varies from $+\infty$ at θ_c to $-\infty$ at grazing incidence. Three curves of q^2 versus z for this profile are shown in Figure 6-7.

A turning point, and the consequent total reflection, may be present even at normal incidence if the dielectric function passes through zero. An example is provided by the dielectric function of an electron plasma, approximating the electron gas in metals or electrons in the ionosphere. If electron collisions and the consequent damping are neglected, this takes the form (see for example Budden 1961, 1985 or Kittel 1966)

$$\varepsilon(z, \omega) = 1 - \frac{\omega_p^2(z)}{\omega^2}, \tag{118}$$

where ω_p is the plasma angular frequency, and is a function of z through its proportionality to the square root of the electron density. For this simplified dielectric function, there is a turning point at normal incidence at z_0 given by $\omega_p(z_0) = \omega$, and at a general angle of incidence at $z_0(\omega, \theta_1)$ given by

$$\omega_p(z_0) = \omega \cos\theta_1 \tag{119}$$

(the value $\varepsilon_1 = 1$ is assumed in (118)). This model will be considered again, with dissipation included, in Chapter 8. An analogous case of total reflection at normal incidence occurs for particles when their energy is less than V_2, their potential energy in the second medium.

When total reflection occurs we know that $|r|^2 = 1$ (in the absence of dissipation) so there is little point in calculating the magnitude of r. But there is information

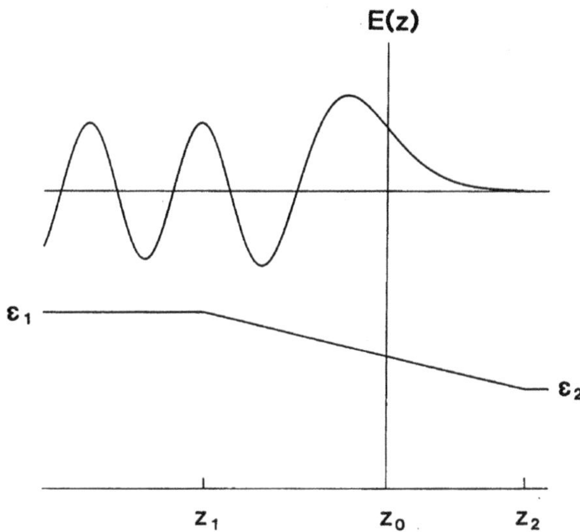

Figure 6-8. Wavefunction $E(z)$ (upper diagram) for total internal reflection off the linear dielectric function profile (lower diagram). The parameters used are $\varepsilon_1 = (4/3)^2$, $\varepsilon_2 = 1$, $\theta_1 = 60°$, $(\omega/c)\Delta z = 27$. The real and imaginary parts of $E(z)$ are proportional to each other when there is total reflection: Im (E)/Re (E) = tan $\delta_s/2$, when $r_s = \mathrm{e}^{i\delta_s}$. Only the real part is shown in this figure.

in the phase of r: it gives for example the location of the ellipsometric ratio r_p/r_s on the unit circle, and determines the time of arrival and shape of reflected pulses. A simple argument shows that the phase of r_s is always a bit less than $2(\phi_0 - \phi_-)$, where $\phi(z) = \int^z \mathrm{d}\zeta q(\zeta)$ is the phase integral, and takes the limiting form $q_1 z + \phi_-$ as $z \to -\infty$, and ϕ_0 is the value of ϕ at z_0. The argument is based on the behaviour of the wave function near a turning point, illustrated in Figure 6-8 for the linear profile considered in Section 5-2.

The fact that the phase of r_s is a bit less than $2(\phi_0 - \phi_-)$ follows from the shape of the wavefunction near the turning point, where it changes from oscillatory behaviour to monotonic decay, with a consequent extremum at $z_0 - \delta z$.

The zeroth approximation for the wave is, for $z < z_0$,

$$E \simeq \psi_0 = \mathrm{e}^{i(\phi - \phi_-)} + r_s \mathrm{e}^{-i(\phi - \phi_-)}. \tag{120}$$

This expression fails near z_0, but in the short wave limit its region of validity approaches it. Since E has an extremum at $z_0 - \delta z$, we obtain an estimate of r_s by setting $\mathrm{d}\psi_0/\mathrm{d}z = 0$ at this point, where ϕ takes the value $\phi_0 - \delta\phi$. This gives $r_s \simeq \mathrm{e}^{2i(\phi_0 - \delta\phi - \phi_-)}$. For profiles which are smooth enough near the turning point to be approximated by a linear variation in this region, it turns out that $\delta\phi$ takes the *universal value* $\pi/4$ in the short wave limit:

$$r_s \simeq \mathrm{e}^{2i(\phi_0 - \phi_- - \pi/4)}. \tag{121}$$

Equation (121) is derived by constructing an accurate solution in the neighbourhood of the turning point, on the assumption that $q^2(z) = \varepsilon(z)\omega^2/c^2 - K^2$ is approximately linear near its zero z_0, with

$$q^2(z) \simeq (z - z_0)\frac{\omega^2}{c^2}\left(\frac{\mathrm{d}\varepsilon}{\mathrm{d}z}\right)_0. \tag{122}$$

For waves incident from the left and totally reflected, q^2 and ε are decreasing functions of z, and the derivative of ε at z_0 is negative. We set

$$\zeta = (z - z_0)\left[\frac{\omega^2}{c^2}\left(-\frac{\mathrm{d}\varepsilon}{\mathrm{d}z}\right)_0\right]^{1/3}, \tag{123}$$

and the wave equation $\mathrm{d}^2E/\mathrm{d}z^2 + q^2E = 0$ transforms to Airy's equation

$$\frac{\mathrm{d}^2E}{\mathrm{d}\zeta^2} - \zeta E = 0. \tag{124}$$

The solutions $Ai(\zeta)$ and $Bi(\zeta)$ of (124) were discussed in Section 5-2. Here we need only the asymptotic forms for large $|\zeta|$:

$$\begin{aligned} Ai(\zeta) &\sim \tfrac{1}{2}\pi^{-1/2}\zeta^{-1/4}\exp\left[-\tfrac{2}{3}\zeta^{3/2}\right] && (|\arg\zeta| < \pi)\\ Ai(-\zeta) &\sim \pi^{-1/2}\zeta^{-1/4}\sin\left[\tfrac{2}{3}\zeta^{3/2} + \pi/4\right] && (|\arg\zeta| < 2\pi/3)\\ Bi(\zeta) &\sim \pi^{-1/2}\zeta^{-1/4}\exp\left[\tfrac{2}{3}\zeta^{3/2}\right] && (|\arg\zeta| < \pi/3)\\ Bi(-\zeta) &\sim \pi^{-1/2}\zeta^{-1/4}\cos\left[\tfrac{2}{3}\zeta^{3/2} + \pi/4\right] && (|\arg\zeta| < 2\pi/3). \end{aligned} \tag{125}$$

The main features of the above (for real ζ) can be obtained from the approximate waveforms $\psi_{\mathrm{I}}^{\pm}$ of Section 6-2 (for example when $\zeta > 0$ these are $\zeta^{-1/4}\exp(\mp\tfrac{2}{3}\zeta^{3/2})$). For more details see Heading (1962, Appendix A3), Olver (1964, 1974) and Budden (1985, Chapter 8).

We are now able to complete the derivation of (121) by matching the approximate solution $q^{-1/2}\psi_0$ (ψ_0 given by (120)), which breaks down in the neighbourhood of z_0, to the asymptotic form of the solution $Ai(\zeta)$ which is accurate near z_0. (The coefficient of $Bi(\zeta)$ goes to zero in the short wave limit, since $Bi(\zeta)$ diverges for large positive ζ). For large negative ζ, $Ai(\zeta)$ is (from (125)) proportional to $(-\zeta)^{-1/4}\sin[\tfrac{2}{3}(-\zeta)^{3/2} + \pi/4]$. The quantity $\tfrac{2}{3}(-\zeta)^{3/2}$ is, from (122), (123) and the definition of the phase integral, equal to $\phi_0 - \phi$. Also $(-\zeta)^{-1/4}$ is proportional to $q^{-1/2}$. Thus $\sin(\phi_0 - \phi + \pi/4)$ must be proportional to ψ_0 as given by (120)), which proportionality holds when r_s is given by (121). The result (121) is, in essence, due to Hartree (1931).

The derivation has assumed an overlap of regions of validity of the Airy function solution, and of the ψ_0 form. The Airy function is the solution of the equation (124), which is itself an approximation, valid when $(z - z_0)(\mathrm{d}\varepsilon/\mathrm{d}z)_0$ is small. On the other hand, the variable ζ must be large to ensure the accuracy of the asymptotic forms assumed in the derivation. As for the short wave form $q^{-1/2}\psi_0$, we know this to fail when $\gamma = q^{-2}\,\mathrm{d}q/\mathrm{d}z = \tfrac{1}{2}q^{-3}(\omega^2/c^2)\,\mathrm{d}\varepsilon/\mathrm{d}z$ is not small. We note from (122) that the quantity $|\gamma\zeta^{3/2}|$ is of order unity, and thus overlap of the regions of validity of the two forms will exist only if $[(\omega/c)/|\mathrm{d}\varepsilon/\mathrm{d}z|_0]^{2/3}$ is substantially larger than the value of $|\zeta|$ for which the asymptotic forms become useful. If this is (say) 4, we need $(\omega/c)/|\mathrm{d}\varepsilon/\mathrm{d}z|_0$ substantially larger than 8 as a necessary condition for the validity of (121).

For the hyperbolic tangent profile

$$\varepsilon(z) = \tfrac{1}{2}(\varepsilon_1 + \varepsilon_2) - \tfrac{1}{2}(\varepsilon_1 - \varepsilon_2)\tanh(z/2\Delta z) \tag{126}$$

the phase integral may be evaluated exactly. As in Section 6-4 we use the variables $x = \tanh(z/2\Delta z)$ and then $y = q\Delta z$. In the present case of total reflection, with $q_2 = i|q_2|$, it is convenient to set the lower limit in the phase integral equal to z_0, the turning point:

$$\phi(z) = \int_{z_0}^{z} d\zeta\, q(\zeta), \tag{127}$$

making $\phi_0 = 0$. We need ϕ for $z \leqslant z_0$; this is given by

$$\phi(y) = 2|y_2| \arctan\left(\frac{y}{|y_2|}\right) - y_1 \log \frac{y_1 + y}{y_1 - y}. \tag{128}$$

To evaluate ϕ_-, defined by $\phi \to q_1 z + \phi_-$ as $z \to -\infty$, we set $y = y_1 - \delta y$ in (128), using

$$\delta y \to \frac{y_1^2 + |y_2|^2}{2y_1} e^{z/\Delta z}.$$

This gives ϕ_-:

$$\phi_- = 2|y_2| \arctan\left(\frac{y_1}{|y_2|}\right) - y_1 \log \frac{4y_1^2}{y_1^2 + |y_2|^2}. \tag{129}$$

The approximate phase is $2(\phi_0 - \phi_- - \pi/4)$. At grazing incidence, $q_1 = \sqrt{\varepsilon_1}(\omega/c)\cos\theta_1$ tends to zero, so ϕ_- tends to zero, and (121) gives the incorrect limiting value $r_s \to -i$ (we saw in Section 2-3 that the exact r_s always tends to -1 at grazing incidence). This is due to the breakdown of the short wave approximations at grazing incidence, where even if $(\omega/c)\Delta z$ is large, $q_1\Delta z$ eventually tends to zero. The approximate phase is compared in Figure 6-9 with the exact phase, calculated from (10.44).

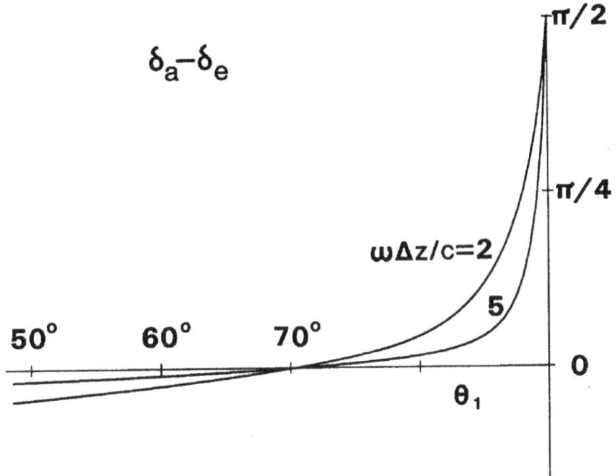

Figure 6-9. The difference between the approximate and exact phases of the reflection amplitude for the hyperbolic tangent profile, with $\varepsilon_1 = (4/3)^2$, $\varepsilon_2 = 1$, for $(\omega/c)\Delta z = 2$ and 5. The phase difference is plotted as a function of the angle of incidence for $\theta_1 > \theta_c \simeq 48.6°$, when total internal reflection occurs.

The treatment of the p wave reflection amplitude in the case of total reflection is more difficult (Budden (1961), Sections 16.12 to 16.17) because of the relative complexity of q_b^2 (given by (88)) compared to q^2.

6-8 Two turning points, and tunnelling

The penetration of wavemotion into a region where $q^2 < 0$ leads to the possibility of *tunnelling*, where the classically forbidden region is traversed by a small portion of the wave, which exits into a region where $q^2 > 0$ and propagates on. Such transmission via tunnelling involves an even number of turning points. We will restrict our consideration to two turning points z_1 and z_2, defined by $q^2(z_1) = 0 = q^2(z_2)$. Our aim is to derive short wave approximations for the reflection and transmission amplitudes in this case. But first we will show an analytically solvable example of tunnelling, provided by the sech^2 profile

$$\varepsilon(z) = \varepsilon_0 + \Delta\varepsilon \,\mathrm{sech}^2(z/a), \tag{130}$$

which was discussed in Section 4-3. The s wave equation $\mathrm{d}^2E/\mathrm{d}z^2 + q^2E = 0$ has

$$q^2(z) = q_0^2 + \Delta\varepsilon \frac{\omega^2}{c^2} \mathrm{sech}^2(z/a), \tag{131}$$

where

$$q_0^2 = \varepsilon_0 \frac{\omega^2}{c^2} - K^2 = \varepsilon_0 \frac{\omega^2}{c^2} \cos^2\theta \tag{132}$$

gives the common value at $\pm\infty$ of the wavevector component q_0 perpendicular to the interface, and θ is the common value of the angles of incidence and refraction. Thus q^2 is given by

$$q^2(z) = \varepsilon_0 \frac{\omega^2}{c^2} \left[\cos^2\theta + \frac{\Delta\varepsilon}{\varepsilon_0} \mathrm{sech}^2(z/a)\right]. \tag{133}$$

When $\Delta\varepsilon$ is positive there are no turning points (it is assumed $\varepsilon_0 > 0$), but when $\Delta\varepsilon$ is negative a pair of symmetrically placed turning points come into existence for

$$\theta > \theta_t = \arccos(-\Delta\varepsilon/\varepsilon_0)^{1/2}. \tag{134}$$

The zeros of $q^2(z)$ are at $\pm z_0$, with

$$z_0 = a \log\left\{\frac{\cos\theta_t + (\cos^2\theta_t - \cos^2\theta)^{1/2}}{\cos\theta}\right\}. \tag{135}$$

(When $\Delta\varepsilon < -\varepsilon_0$ the dielectric function becomes negative near the origin, and there are two turning points at any angle of incidence. Their location is still given by (134) and (135), with θ_t now a pure imaginary). The variation of q^2 with z and angle of incidence is shown in Figure 6-10.

The reflection properties of the sech^2 profile are characterized by two dimensionless parameters α and β, or s and β, where

$$\alpha = \Delta\varepsilon(\omega a/c)^2, \quad \beta = q_0 a, \quad s = \tfrac{1}{2}[-1 + (1 + 4\alpha)^{1/2}]. \tag{136}$$

For negative α there will be tunnelling at a large enough angle of incidence. The transition from no tunnelling to tunnelling (in quantum particle language, from *over* the potential barrier to *through* the barrier) takes place when $\beta = (-\alpha)^{1/2}$ if $-\varepsilon_0 < \Delta\varepsilon < 0$. We saw in Section 4-3 that the reflectivity takes different analytic

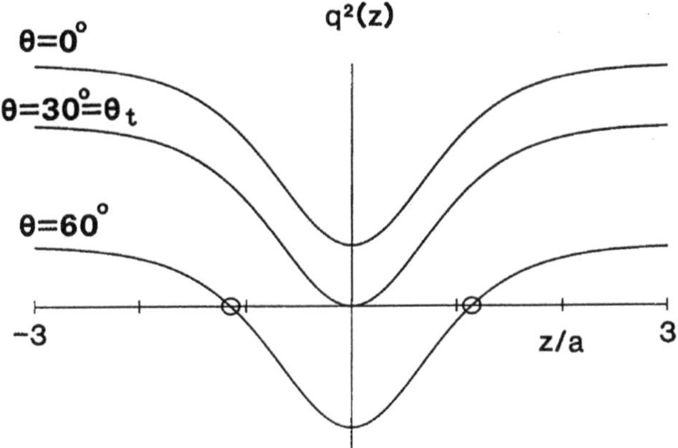

Figure 6-10. The dependence of $q^2(z)$, for the sech^2 profile, on angle of incidence. The curves are drawn for $\varepsilon_0 = 1$, $\Delta\varepsilon = -3/4$; then $\theta_t = 30°$. The two turning points for $\theta = 60°$ are circled.

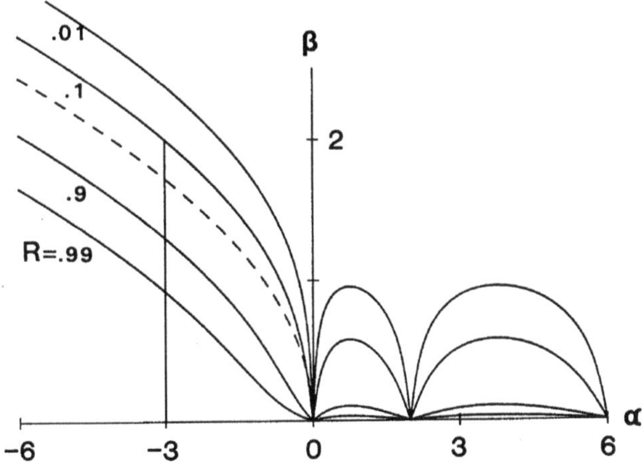

Figure 6-11. Contours of constant reflectivity for the sech^2 profile. Grazing incidence (corresponding fo $\beta \to 0$) generally has high reflectivity. For positive α (a positive $\Delta\varepsilon$ or a negative ΔV in the particle case) there are resonance zeros of R_s when $\alpha = n(n+1)$, $n = 0, 1, 2, \ldots$ (see (4.35)). The tunnelling region is below the dashed curve on the left, given by $\beta = (-\alpha)^{1/2}$, which is asymptotic to the $R_s = 1/2$ contour for large α, β. The three q^2 curves of Figure 6-10 at $\theta = 0$, 30° and 60° correspond to $\beta/(-\alpha)^{1/2} = 2/\surd 3$, 1 and $1/\surd 3$. The vertical line at $\alpha = -3$ relates to Figure 6-12.

forms according as $\alpha > -1/4$ (s is real), in which case $R_s = |r_s|^2$ is given by (4.35), or $\alpha < -1/4$, in which case s is complex:

$$s = -\tfrac{1}{2} + i\sigma, \qquad \sigma = \tfrac{1}{2}(4|\alpha| - 1)^{1/2}, \tag{137}$$

and

$$R_s = \frac{\cosh^2 \pi\sigma}{\cosh^2 \pi\sigma + \sinh^2 \pi\beta}. \tag{138}$$

Reflectivity contours for the sech^2 profile in the α, β plane are shown in Figure 6-11.

We note the rapid rise in the reflectivity on passage deeper into the tunnelling region (below the $\beta = (-\alpha)^{1/2}$ curve on the left), and the rapid fall on passage out of it. On the right-hand side of the figure, corresponding to positive $\Delta\varepsilon$ or a potential well in the particle case, the contour pattern is caused by the interplay of resonance reflectivity zeros with strong reflectivity at grazing incidence ($\beta \to 0$).

We shall now derive approximate expressions for r_s and t_s when there are two turning points, and then apply these to the sech^2 profile, comparing the reflectivity with (138). Away from the turning points, now denoted by z_1 and z_2, the waveforms will be approximated by $q^{-1/2}\,e^{\pm i\phi}$ for real q, and by $|q|^{-1/2}\,e^{\pm\Phi}$ for imaginary q (i.e. in between the turning points). The real and imaginary parts of the phase integral, ϕ and Φ, are defined by

$$\begin{aligned} \phi(z) &= \int_{z_1}^{z} d\zeta\, q(\zeta) + \phi_1 && (z \leqslant z_1) \\ \Phi(z) &= \int_{z_1}^{z} d\zeta |q(\zeta)| + \Phi_1 && (z_1 \leqslant z \leqslant z_2) \\ \phi(z) &= \int_{z_2}^{z} d\zeta\, q(\zeta) + \phi_1 && (z \geqslant z_2). \end{aligned} \tag{139}$$

Note that the real part of ϕ has the same value (ϕ_1) at z_1 and z_2, being continuous across the tunnelling interval (only the imaginary part changes). The limiting forms of the real part of ϕ are given by

$$q_0 z + \phi_- \leftarrow \phi(z) \to q_0 z + \phi_+ \tag{140}$$

as $z \to \pm\infty$ (for simplicity we consider the $\varepsilon_1 = \varepsilon_2$ case).

The method used to obtain the reflection and transmission amplitudes is similar to that used in Section 6-7, namely matching the approximate wave functions across the turning points. Since the approximate wave functions fail at z_1 and z_2, the matching is via the accurate Airy function solutions across z_1 and z_2. Near z_1 the dielectric function is decreasing with z; we approximate q^2 by

$$q^2(z) \simeq (z - z_1)\frac{\omega^2}{c^2}\left(\frac{d\varepsilon}{dz}\right)_1, \tag{141}$$

and set

$$\zeta_1 = (z - z_1)\left[\frac{\omega^2}{c^2}\left(-\frac{d\varepsilon}{dz}\right)_1\right]^{1/3}. \tag{142}$$

The accurate solutions in the neighbourhood of z_1 are then $Ai(\zeta_1)$ and $Bi(\zeta_1)$. Near z_2 the dielectric function is increasing with z; we set

$$q^2(z) \simeq (z - z_2)\frac{\omega^2}{c^2}\left(\frac{d\varepsilon}{dz}\right)_2, \tag{143}$$

$$\zeta_2 = (z - z_2)\left[\frac{\omega^2}{c^2}\left(\frac{d\varepsilon}{dz}\right)_2\right]^{1/3}, \tag{144}$$

and the accurate solutions in the neighbourhood of z_2 are $Ai(-\zeta_2)$ and $Bi(-\zeta_2)$. We again assume the existence of regions of overlap, where both the approximate solutions and the asymptotic forms of the Airy functions hold simultaneously. The condition for the existence of such regions was discussed in the last section. We use the asymptotic forms (125), the relations

$$\begin{aligned}
\tfrac{2}{3}(-\zeta_1)^{3/2} &\simeq \phi_1 - \phi && (z < z_1)\\
\tfrac{2}{3}(\zeta_1)^{3/2} &\simeq \Phi - \Phi_1 && (z > z_1)\\
\tfrac{2}{3}(-\zeta_2)^{3/2} &\simeq \Phi_2 - \Phi && (z < z_2)\\
\tfrac{2}{3}(\zeta_2)^{3/2} &\simeq \phi - \phi_1 && (z > z_2),
\end{aligned} \tag{145}$$

and match at four places (to the left and to the right of both z_1 and z_2). After removal of all common factors, the four matchings give

$$\begin{aligned}
e^{i(\phi-\phi_-)} + r_s\, e^{-i(\phi-\phi_-)} &= A_1 \sin(\phi_1 - \phi + \pi/4) + B_1 \cos(\phi_1 - \phi + \pi/4)\\
\tfrac{1}{2}A_1\, e^{\Phi_1-\Phi} + B_1\, e^{\Phi-\Phi_1} &= A\, e^{-\Phi} + B\, e^{\Phi}\\
A\, e^{-\Phi} + B\, e^{\Phi} &= \tfrac{1}{2}A_2\, e^{\Phi-\Phi_2} + B_2\, e^{\Phi_2-\Phi}\\
A_2 \sin(\phi - \phi_1 + \pi/4) + B_2 \cos(\phi - \phi_1 + \pi/4) &= t_s\, e^{i(\phi-\phi_+)}.
\end{aligned} \tag{146}$$

At each point we equate the coefficients of $e^{\pm i\phi}$ or $e^{\pm\Phi}$. Thus we have eight conditions to determine the eight coefficients r_s, A_1, B_1, A, B, A_2, B_2, t_s. The result of solving for r_s and t_s is

$$r_s = e^{2i(\phi_1-\phi_--\pi/4)} \tanh(\Delta\Phi + \log 2), \tag{147}$$

$$t_s = e^{i(\phi_+-\phi_-)} \operatorname{sech}(\Delta\Phi + \log 2), \tag{148}$$

where

$$\Delta\Phi = \Phi_2 - \Phi_1 = \int_{z_1}^{z_2} dz|q(z)| \tag{149}$$

(the log 2 comes from the factor $\frac{1}{2}$ multiplying A_1 and A_2 in the tunnelling region; this in turn comes from the first equation in (125)). Note that $1 - \tanh^2 x = \operatorname{sech}^2 x$, so that the conservation law $1 - |r_s|^2 = |t_s|^2$ of Section 2-1 is satisfied. When $\Delta\Phi$ is large,

$$|r_s|^2 \to 1 - e^{-2\Delta\Phi}, \qquad |t_s|^2 \to e^{-2\Delta\Phi}. \tag{150}$$

For the sech^2 profile, $\Delta\Phi$ may be evaluated analytically. We have, from (133) and (136),

$$\Delta\Phi = \int_{z_1}^{z_2} dz\, |q(z)| = 2\int_0^{x_0} dx\, [|\alpha| \operatorname{sech}^2 x - \beta^2]^{1/2}, \tag{151}$$

with x_0 being the value of $x = z/a$ for which the integrand is zero; this corresponds to z_0 as given by (135). The substitution $|\alpha| \operatorname{sech}^2 x = \beta^2 y^2$ reduces (151) to elementary form, leading to

$$\Delta\Phi = \pi(|\alpha|^{1/2} - \beta). \tag{152}$$

This expression holds for negative α, with $-\alpha \geqslant \beta^2$. When $-\alpha = \beta^2$ ($\theta = \theta_t$, see (134)) $\Delta\Phi$ is zero, the turning points having merged at the origin. The approximate treatment given above then fails, since it was based on the assumption of well separated points.

When $|\alpha|$ and β are large, the exact reflectivity formula (138) gives

$$R_s \to \frac{1}{1 - e^{-2\Delta\Phi}}, \tag{153}$$

which is in agreement with (105) provided $\Delta\Phi$ is large. When $|\alpha|$ is large but β small (grazing incidence), (138) leads to

$$R_s \to \frac{1}{1 + (2\pi\beta)^2\, e^{-2\pi|\alpha|^{1/2}}}. \tag{154}$$

This is not in agreement with (150), the latter having been based on short wave approximate waveforms which fail at grazing incidence. Thus a large $\Delta\Phi$ is not a guarantee of the accuracy of (150). The approximate reflectivity

$$R_s \simeq \tanh^2(\Delta\Phi + \log 2) \tag{155}$$

is compared with the exact reflectivity (138) in Figure 6-12.

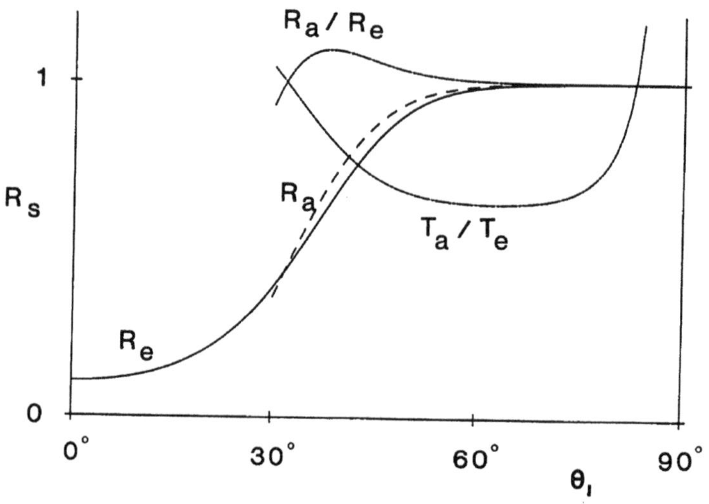

Figure 6-12. Reflectivity of the sech² profile as a function of the angle of incidence, shown for $\varepsilon_0 = 1$, $\Delta\varepsilon = -3/4$ and $\omega a/c = 2$. These dielectric function values are as for Figure 6-10, and together with $\omega a/c = 2$ correspond to the vertical line shown in Figure 6-11, with angle of incidence increasing downwards. The exact reflectivity R_e (equation (138), solid curve) is shown for all angles, while the approximate reflectivity R_a given by (152) and (155) (dashed curve) is shown in its range of applicability, $\theta \geqslant \theta_t = 30°$. The ratios R_a/R_e and $T_a/T_e = (1 - R_a)/(1 - R_e)$ are also shown. The latter demonstrates the poor accuracy obtained for the tunnelling probability, especially near grazing incidence.

References

References quoted in the text

F. W. J. Olver (1964) "Bessel functions of integer order" (Chapter 9 of 'Handbook of mathematical functions', edited by M. Abramowitz and I. A. Stegun, NBS Applied Math. Series **55**).

J. Liouville (1837) "Sur le développement des fonctions ou parties de fonctions en séries . . . ", J. Math. Pures Appl. **2**, 16–35.

G. Green (1837) "On the motion of waves in a variable canal of small depth and width", Trans. Camb. Phil. Soc. **6**, 457–462.
J. W. S. Rayleigh (1912) "On the propagation of waves through a stratified medium with special reference to the question of reflection", Proc. Roy. Soc., **A86**, 207–266.
R. Gans (1915) "Fortpflanzung des Lichts durch ein inhomogenes Medium", Ann. Physik **47**, 709–736.
J. Heading (1962) "An introduction to phase-integral methods", Methuen.
H. Jeffreys (1924) "On certain approximate solutions of linear differential equations of the second order", Proc. London Math. Soc. **23**, 428–436.
G. Wentzel (1926) "Eine Verallgemeinerung der Quantenbedingungen für die Zwecke der Wellenmechanik", Z. Physik **38**, 518–529.
H. A. Kramers (1926) "Wellenmechanik und halbzahlige Quantisierung", Z. Physik **39**, 828–840.
L. Brillouin (1926) "Remarques sur la méchanique ondulatoire", J. Phys. Radium **7**, 353–368.
F. W. J. Olver (1974) "Asymptotics and special functions", Academic Press, p. 228.
M. V. Berry and K. E. Mount (1972) "Semiclassical approximations in wave mechanics", Repts. Prog. in Phys. **35**, 315–397.
V. L. Pokrovskii, S. K. Savvinykh and F. R. Ulinich (1958), "Reflection from a barrier in the quasi-classical approximation", Soviet Physics JETP **34**, 879–882, and 1119–1120.
H. Bremmer (1951) "The W. K. B. approximation as the first term of a geometric-optical series", Commun. Pure and App. Math. **4**, 105–115.
R. Bellman and R. Kalaba (1959) "Functional equations, wave propagation and invariant imbedding", J. Math and Mech. **8**, 683–704.
F. V. Atkinson (1960) "Wave propagation and the Bremmer series", J. Math. Anal. and Applic. **1**, 255–276.
K. G. Budden (1961) "Radio waves in the ionosphere", Cambridge.
K. G. Budden (1985) "The propagation of radio waves", Cambridge.
C. Kittel (1966) "Introduction to solid state physics" (third edition), John Wiley and Sons.
D. R. Hartree (1931) "Optical and equivalent paths in a stratified medium treated from a wave standpoint", Proc. Roy. Soc. **A131**, 428–450.

Short wave perturbation theory (Section 6-5) is also developed in

H. Levine (1978) "Unidirectional wave motions", North-Holland.

Additional references for Section 6-7

V. L. Ginzburg (1964) "The propagation of electromagnetic waves in plasmas", Pergamon, Section 30;
L. D. Landau and E. M. Lifshitz (1965) "Quantum mechanics" (second edition), Pergamon, Section 52;
J. Heading (1975) "Ordinary differential equations, theory and practice", Elek Science, Section 7.12.

The results on tunnelling date from the early days of quantum mechanics. See

E. C. Kemble (1937) "The fundamental principles of quantum mechanics", McGraw-Hill, Section 21j.

Alternative derivations are given by Ginzburg (1964), Section 33, Landau and Lifshitz (1965), Section 50, and

N. Fröman and P. O. Fröman (1965) "JWKB approximation, contributions to the theory", North-Holland, Section 9.1;
E. Merzbacher (1970) "Quantum mechanics" (second edition), Wiley, Section 7.4.

7

Anisotropy

Up till now we have assumed the electrodynamics of a non-magnetic stratified system to be characterized by one dielectric function $\varepsilon(z)$. This is often a very good approximation: for example in the case of monatomic fluids, where a liquid–vapour interface needs two dielectric functions $\varepsilon_x(z)$ and $\varepsilon_z(z)$ for specification of the electrodynamics, the difference between these is small (Lekner 1983). On the other hand molecular systems can have strong anisotropy due to orientation of the molecules, extreme examples being liquid crystals. Other cases of interest are anisotropic crystals, and anisotropy in ionospheric propagation of radio waves due to the earth's magnetic field. Simple examples of reflection in the presence of anisotropy will be discussed here, with emphasis on the interplay of anisotropy and stratification in their effect on reflectivities and ellipsometric measurements.

7-1 Anisotropy with azimuthal symmetry

When the reflecting system is symmetric with respect to rotation about the normal to the interface (azimuthal symmetry), the electrodynamics is characterized by $\varepsilon_x(z, \omega)$ and $\varepsilon_z(z, \omega)$, corresponding to the electric field vector aligned respectively along and perpendicular to the interface. (The convention used throughout this book is that the interface lies in the xy plane, and propagation is in the zx plane; systems with azimuthal symmetry have $\varepsilon_x = \varepsilon_y$). Examples of such systems are: uniaxial crystals with the optic axis along the surface normal, and molecular fluids with molecular orientation not having a preferred direction along the surface.

Symmetry with respect to rotation about the surface normal conserves the s and p wave characterizations: these two polarizations, with $\mathbf{E} = (0, E_y, 0)$ and $\mathbf{B} = (0, B_y, 0)$ respectively, are together sufficient to represent any plane wave incident on such an anisotropic plane stratified medium. To derive equations for the s and p waves we repeat the analysis of Sections 1-1 and 1-2, with (1.1) unchanged, and the dielectric function ε in (1.2) now to be interpreted as the diagonal tensor

$$\varepsilon = \begin{pmatrix} \varepsilon_x & 0 & 0 \\ 0 & \varepsilon_x & 0 \\ 0 & 0 & \varepsilon_z \end{pmatrix}. \tag{1}$$

For the s wave (1.1) and (1.2) give

$$i\frac{\omega}{c}B_x = -\frac{\partial E_y}{\partial z}, \qquad B_y = 0, \qquad i\frac{\omega}{c}B_z = \frac{\partial E_y}{\partial x}, \tag{2}$$

$$-i\frac{\omega}{c}\varepsilon_x E_y = \frac{\partial B_x}{\partial z} - \frac{\partial B_z}{\partial x}. \tag{3}$$

On eliminating B_x and B_z from (2) and (3) we find

$$\frac{\partial^2 E_y}{\partial z^2} + \frac{\partial^2 E_y}{\partial x^2} + \varepsilon_x \frac{\omega^2}{c^2} E_y = 0. \tag{4}$$

Since the system retains invariance with respect to translation in the x or y directions, the x dependence of E_y is contained in the factor e^{iKx} as before,

$$E_y(z, x, t) = e^{i(Kx-\omega t)} E(z). \tag{5}$$

Substitution of (5) into (4) gives the usual form for the s wave equation,

$$\frac{d^2 E}{dz^2} + q_s^2 E = 0, \qquad q_s^2 = \varepsilon_x \frac{\omega^2}{c^2} - K^2. \tag{6}$$

Thus all the results we have derived in the last six chapters for the s wave apply to the s wave in the presence of azimuthally symmetric anisotropy, with the replacement of $\varepsilon(z)$ by $\varepsilon_x(z)$.

The p wave is more complicated, since it samples (at a general angle of incidence) both ε_x and ε_z. The Maxwell equation (1.1) gives

$$i\frac{\omega}{c}B_y = \frac{\partial E_x}{\partial z} - \frac{\partial E_z}{\partial x} \tag{7}$$

as before, but (1.2) now implies

$$i\frac{\omega}{c}\varepsilon_x E_x = \frac{\partial B_y}{\partial z}, \qquad E_y = 0, \qquad -i\frac{\omega}{c}\varepsilon_z E_z = \frac{\partial B_y}{\partial x}. \tag{8}$$

Elimination of E_x and E_z from (7) and (8) gives

$$\frac{\partial}{\partial z}\left(\frac{1}{\varepsilon_x}\frac{\partial B_y}{\partial z}\right) + \frac{\partial}{\partial x}\left(\frac{1}{\varepsilon_z}\frac{\partial B_y}{\partial x}\right) + \frac{\omega^2}{c^2} B_y = 0. \tag{9}$$

The substitution

$$B_y(z, x, t) = e^{i(Kx-\omega t)} B(z) \tag{10}$$

gives us a modified p wave equation,

$$\frac{\mathrm{d}}{\mathrm{d}z}\left(\frac{1}{\varepsilon_x}\frac{\mathrm{d}B}{\mathrm{d}z}\right) + \left(\frac{\omega^2}{c^2} - \frac{K^2}{\varepsilon_z}\right) B = 0. \tag{11}$$

This differs fundamentally from the isotropic case: (11) contains both dielectric functions, and thus results previously obtained for the p wave cannot be used directly to obtain results even for this restricted anisotropy.

Anisotropy implies birefringence (double refraction): consider the example of an electromagnetic wave incident onto a uniform azimuthally symmetric anisotropic material. The incident and reflected waves have the z dependence $e^{\pm iq_1 z}$ as usual, but the s and p transmitted waves have the z dependence $e^{iq_s z}$ and $e^{iq_p z}$, where q_s is given in (6) and (from 11))

$$q_p^2 = \varepsilon_x \frac{\omega^2}{c^2} - \frac{\varepsilon_x}{\varepsilon_z} K^2. \tag{12}$$

Since q_p differs from q_s, the angles of refraction θ_p and θ_s are different: from $K = q_1 \tan\theta_1 = q_s \tan\theta_s = q_p \tan\theta_p$, and (6) and (12),

$$\varepsilon_1 \sin^2\theta_1 = \varepsilon_x \sin^2\theta_s = \frac{\varepsilon_x \varepsilon_z \sin^2\theta_p}{\varepsilon_x \sin^2\theta_p + \varepsilon_z \cos^2\theta_p} \tag{13}$$

These angles give the wavevector directions in the anisotropic medium. For the s (or ordinary) wave the wavevector direction is the same as the ray direction, given by the energy flux or Poynting vector $\mathbf{E} \times \mathbf{B}$. For the p (or extraordinary) wave, the wavevector and $\mathbf{E} \times \mathbf{B}$ directions are different. The latter is given by

$$\tan\theta_p' = \frac{\varepsilon_x K}{\varepsilon_z q_p} = \frac{\varepsilon_x}{\varepsilon_z}\tan\theta_p \tag{14}$$

The s and p reflection amplitudes for a sharp interface between an isotropic medium and an anisotropic medium, characterized by ε_x and ε_z as above, are (for

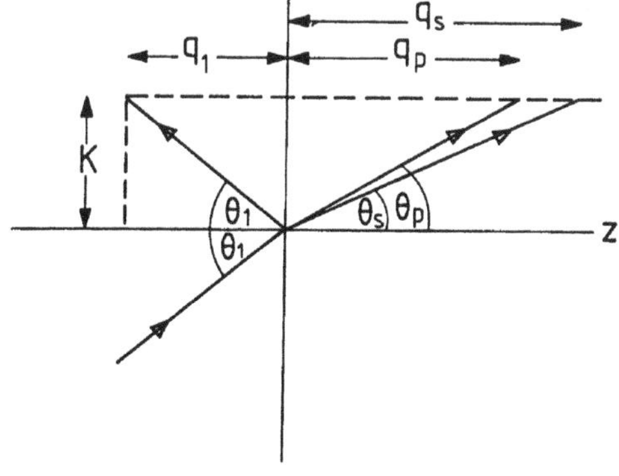

Figure 7-1. Double refraction in an azimuthally symmetric anisotropic system, such as a uniaxial crystal with its optic axis coincident with the z-axis

an interface at $z = 0$)

$$r_s = \frac{q_1 - q_s}{q_1 + q_s}, \qquad -r_p = \frac{Q_1 - Q}{Q_1 + Q}, \tag{15}$$

where q_s is given by (6), $Q_1 = q_1/\varepsilon_1$, and $Q = q_p/\varepsilon_x$. The p wave reflection amplitude follows from the continuity of B and $(1/\varepsilon_x)(\mathrm{d}B/\mathrm{d}z)$ at the boundary (compare (11)). The Brewster angle θ_B, at which $r_p = 0$, is given by $Q_1 = Q$, which leads to

$$\theta_B = \arctan\left\{\frac{\varepsilon_z(\varepsilon_x - \varepsilon_1)}{\varepsilon_1(\varepsilon_z - \varepsilon_1)}\right\}^{1/2}. \tag{16}$$

7-2 Ellipsometry off a thin film on an isotropic substrate

Ellipsometry determines the complex number r_p/r_s. In section 3-5, equation (3.52), we saw that this ratio is given to lowest order in the interface thickness/wavelength expansion by

$$r_{s0}\left(\frac{r_p}{r_s}\right) = r_{p0} - \frac{2iQ_1K^2/\varepsilon_1\varepsilon_2}{(Q_1 + Q_2)^2}\mathscr{I}_1 + \ldots, \tag{17}$$

where the invariant $\mathscr{I}_1$ is

$$\mathscr{I}_1 = \int_{-\infty}^{\infty} \mathrm{d}z\, \frac{(\varepsilon_1 - \varepsilon)(\varepsilon - \varepsilon_2)}{\varepsilon} = \int_{-\infty}^{\infty} \mathrm{d}z\left\{\varepsilon_1 + \varepsilon_2 - \frac{\varepsilon_1\varepsilon_2}{\varepsilon} - \varepsilon\right\}. \tag{18}$$

For the case of an anisotropic film characterized by ε_x and ε_z (with cylindrical symmetry about the surface normal) and resting on an isotropic substrate with dielectric constant ε_2, we shall show that the formula (17) remains valid, with

$$\mathscr{I}_1 = \int_{-\infty}^{\infty} \mathrm{d}z\left\{\varepsilon_1 + \varepsilon_2 - \frac{\varepsilon_1\varepsilon_2}{\varepsilon_z} - \varepsilon_x\right\}. \tag{19}$$

This generalization was given (without proof) by Buff (1966) and independently by Beaglehole (1980), the latter crediting Abelès (1976). The proof given here follows Lekner (1983), Appendix A. We begin by deriving an anisotropic generalization of the comparison identity (3.42). Let $\varepsilon_0(z)$ be the step function $\varepsilon_0 = \varepsilon_1$ for $z < 0$, $\varepsilon_0 = \varepsilon_2$ for $z > 0$, and $B_0(z)$ the solution of

$$\frac{\mathrm{d}}{\mathrm{d}z}\left(\frac{1}{\varepsilon_0}\frac{\mathrm{d}B_0}{\mathrm{d}z}\right) + \left(\frac{\omega^2}{c^2} - \frac{K^2}{\varepsilon_0}\right)B_0 = 0, \tag{20}$$

with

$$B_0(z) = \begin{cases} e^{iq_1z} - r_{p0}\, e^{-iq_1z} & (z < 0) \\ \left(\dfrac{\varepsilon_2}{\varepsilon_1}\right)^{1/2} t_{p0}\, e^{iq_2z} & (z > 0). \end{cases} \tag{21}$$

$B(z)$ is the solution of (11), with limiting forms

$$e^{iq_1z} - r_p\, e^{-iq_1z} \leftarrow B(z) \rightarrow \left(\frac{\varepsilon_2}{\varepsilon_1}\right)^{1/2} t_p\, e^{iq_2z}. \tag{22}$$

We multiply (11) by $B_0(z)$, (20) by $B(z)$, and subtract. The result is

$$\frac{d}{dz}(B_0C - BC_0) = K^2\left(\frac{1}{\varepsilon_z} - \frac{1}{\varepsilon_0}\right)BB_0 - (\varepsilon_x - \varepsilon_0)CC_0, \tag{23}$$

where

$$C_0 = \frac{1}{\varepsilon_0}\frac{dB}{dz}, \qquad C = \frac{1}{\varepsilon_x}\frac{dB}{dz}. \tag{24}$$

Integration of (23) from z_1 (deep in medium 1) to z_2 (deep in medium 2) and use of (21) and (22) then gives the identity

$$r_p = r_{p0} + \frac{1}{2iQ_1}\int_{-\infty}^{\infty} dz\left\{\left(\frac{1}{\varepsilon_0} - \frac{1}{\varepsilon_z}\right)K^2BB_0 + (\varepsilon_x - \varepsilon_0)CC_0\right\}. \tag{25}$$

To lowest order in the interface thickness we may replace B by $B_0(0)$ and C by $C_0(0)$, as given by (3.43). The result is

$$r_p = r_{p0} - \frac{2iQ_1}{(Q_1 + Q_2)^2}\left\{K^2\int_{-\infty}^{\infty} dz\left(\frac{1}{\varepsilon_0} - \frac{1}{\varepsilon_z}\right) - Q_2^2\int_{-\infty}^{\infty} dz\,(\varepsilon_x - \varepsilon_0)\right\} + \dots. \tag{26}$$

The s wave reflection amplitude is given by (3.18) on replacing ε by ε_x:

$$r_s = r_{s0} + \frac{2iq_1\omega^2/c^2}{(q_1 + q_2)^2}\int_{-\infty}^{\infty} dz\,(\varepsilon_x - \varepsilon_0) + \dots. \tag{27}$$

The ellipsometric ratio r_p/r_s may now be found from (26), (27) and (3.45):

$$r_{s0}\left(\frac{r_p}{r_s}\right) = r_{p0} - \frac{2iQ_1K^2}{(Q_1 + Q_2)^2}\left\{\int_{-\infty}^{\infty} dz\left(\frac{1}{\varepsilon_0} - \frac{1}{\varepsilon_z}\right) - \frac{1}{\varepsilon_1\varepsilon_2}\int_{-\infty}^{\infty} dz\,(\varepsilon_x - \varepsilon_0)\right\} + \dots. \tag{28}$$

The result (17, 18) follows from generalization of identity (3.37):

$$\varepsilon_1\varepsilon_2\int_{-\infty}^{\infty} dz\left(\frac{1}{\varepsilon_0} - \frac{1}{\varepsilon_z}\right) - \int_{-\infty}^{\infty} dz\,(\varepsilon_x - \varepsilon_0) = \int_{-\infty}^{\infty} dz\left\{\varepsilon_1 + \varepsilon_2 - \frac{\varepsilon_1\varepsilon_2}{\varepsilon_z} - \varepsilon_x\right\}. \tag{29}$$

Another two equivalent ways of writing $\mathscr{I}_1$ (given by (29)) are

$$\begin{aligned}\mathscr{I}_1 &= \int_{-\infty}^{\infty} dz\,\frac{(\varepsilon_1 - \varepsilon_z)(\varepsilon_z - \varepsilon_2)}{\varepsilon_z} + \int_{-\infty}^{\infty} dz\,(\varepsilon_z - \varepsilon_x)\\ &= \int_{-\infty}^{\infty} dz\,\frac{(\varepsilon_1 - \varepsilon_x)(\varepsilon_x - \varepsilon_2)}{\varepsilon_x} + \varepsilon_1\varepsilon_2\int_{-\infty}^{\infty} dz\left(\frac{1}{\varepsilon_x} - \frac{1}{\varepsilon_z}\right).\end{aligned} \tag{30}$$

The first form of $\mathscr{I}_1$ in (18) shows that $\mathscr{I}_1$ is positive when ε lies between ε_1 and ε_2. This is not necessarily so in the presence of anisotropy, even if both ε_x and ε_z lie between ε_1 and ε_2. For example, if on average ε_z is smaller than ε_x, the first form of (30) shows that $\mathscr{I}_1$ may be negative for sufficiently large anisotropy. In general, if $\varepsilon_z < \varepsilon_x$ on average, the anisotropy will give the appearance of a thinner film (a smaller $\mathscr{I}_1$), or can even make $\mathscr{I}_1$ negative. Conversely, if $\varepsilon_z > \varepsilon_x$ on average, the anisotropy will increase $\mathscr{I}_1$, giving the same signal as a thicker isotropic film.

For a *uniform* anisotropic film we can be more definite: if Δz is the film thickness, $\mathscr{I}_1/\Delta z = \varepsilon_1 + \varepsilon_2 - \varepsilon_x - \varepsilon_1\varepsilon_2/\varepsilon_z$, and is positive provided the sum of ε_x and $\varepsilon_1\varepsilon_2/\varepsilon_z$ is less than the sum of ε_1 and ε_2. The contours of constant $\mathscr{I}_1/\Delta z$ are lines of slope -1 in the ε_x, $\varepsilon_1\varepsilon_2/\varepsilon_z$ plane. In the same plane, the contours of fixed anisotropy $\varepsilon_z - \varepsilon_x$ are also shown in Figure 7-2.

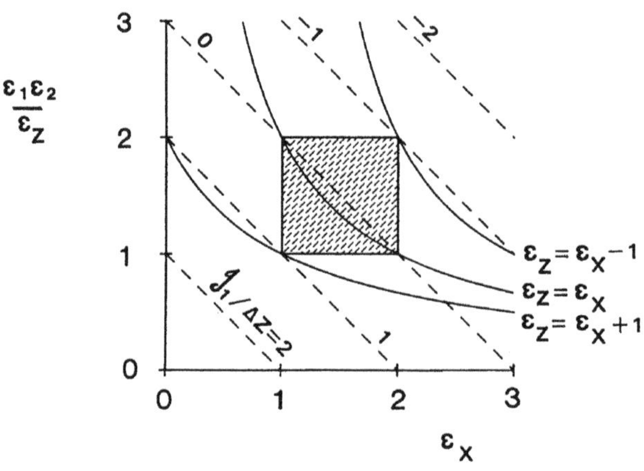

Figure 7-2. Lines of fixed $\mathscr{I}_1/\Delta z$ and contours of fixed anisotropy $\varepsilon_z - \varepsilon_x$ in the ε_x, $\varepsilon_1\varepsilon_2/\varepsilon_z$ plane, drawn for $\varepsilon_1 = 1$, $\varepsilon_2 = 2$. The ε_x and ε_z values lie between ε_1 and ε_2 in the shaded box.

We see that $\mathscr{I}_1/\Delta z$ increases with anisotropy. The $\varepsilon_z = \varepsilon_x$ contour has a maximum value of $\mathscr{I}_1/\Delta z$ equal to $(\sqrt{\varepsilon_2} - \sqrt{\varepsilon_1})^2$. (This is reached when the common value of ε_x and ε_z is $\sqrt{\varepsilon_1\varepsilon_2}$.) Thus if $\mathscr{I}_1$ is measured, Δz for the uniform film is known independently, and $\mathscr{I}_1/\Delta z$ is found to be bigger than $(\sqrt{\varepsilon_2} - \sqrt{\varepsilon_1})^2$, the anisotropy must be positive ($\varepsilon_z > \varepsilon_x$).

For a uniform anisotropic film the exact reflection amplitudes may be found by the methods of Section 2-4. The s wave takes the same form as in the isotropic film, with q replaced by q_s (defined in (6)). For a homogeneous layer located between z_1 and $z_1 + \Delta z$, the s wave reflection amplitude is

$$r_s = e^{2iq_1z_1}\frac{q_s(q_1 - q_2)c + i(q_s^2 - q_1q_2)s}{q_s(q_1 + q_2)c - i(q_s^2 + q_1q_2)s}, \tag{31}$$

where $c = \cos q_s\Delta z$ and $s = \sin q_s\Delta z$. For the p wave, the solutions within the film are $e^{\pm iq_pz}$, with q_p given by (12), and the boundary conditions are the continuity of B and $C = (1/\varepsilon_x)(dB/dz)$ at z_1 and z_2. We find that r_p has the same form as (2.68):

$$-r_p = e^{2iq_1z_1}\frac{Q(Q_1 - Q_2)c + i(Q^2 - Q_1Q_2)s}{Q(Q_1 + Q_2)c - i(Q^2 + Q_1Q_2)s}, \tag{32}$$

where now $c = \cos q_p\Delta z$, $s = \sin q_p\Delta z$, and $Q = q_p/\varepsilon_x$. From (31) and (32) we can verify that r_p/r_s takes the form (17) with $\mathscr{I}_1 = (\varepsilon_1 + \varepsilon_2 - \varepsilon_x - \varepsilon_1\varepsilon_2/\varepsilon_z)\Delta z$.

7-3 Thin film on an anisotropic substrate

We now include the possibility of substrate anisotropy, still keeping azimuthal symmetry in both stratified surface and in the uniform substrate (and thus retaining the s and p characterization of electromagnetic waves). The electromagnetic response of the system is determined by the three dielectric constants ε_1, ε_{2x}, ε_{2z} (the latter two for the substrate) and the two dielectric functions $\varepsilon_x(z)$, $\varepsilon_z(z)$, with

$$\varepsilon_1 \leftarrow \varepsilon_x(z) \rightarrow \varepsilon_{2x}, \qquad \varepsilon_1 \leftarrow \varepsilon_z(z) \rightarrow \varepsilon_{2z}. \tag{33}$$

We now need to define two step functions

$$\varepsilon_{0x}(z),\ \varepsilon_{0z}(z) = \begin{cases} \varepsilon_1 & (z < 0) \\ \varepsilon_{2x},\ \varepsilon_{2z} & (z > 0). \end{cases} \tag{34}$$

The results for the s wave need only the modification $\varepsilon_0 \rightarrow \varepsilon_{0x}$ (for example in (27)). For the p wave we use $B_0(z)$, the solution of

$$\frac{\mathrm{d}}{\mathrm{d}z}\left(\frac{1}{\varepsilon_{0x}}\frac{\mathrm{d}B_0}{\mathrm{d}z}\right) + \left(\frac{\omega^2}{c^2} - \frac{K^2}{\varepsilon_{0z}}\right) B_0 = 0. \tag{35}$$

The modified version of (23) is

$$\frac{\mathrm{d}}{\mathrm{d}z}(B_0 C - BC_0) = K^2\left(\frac{1}{\varepsilon_z} - \frac{1}{\varepsilon_{0z}}\right) - (\varepsilon_x - \varepsilon_{0x})CC_0, \tag{36}$$

with

$$C_0 = \frac{1}{\varepsilon_{0x}}\frac{\mathrm{d}B_0}{\mathrm{d}x}, \qquad C = \frac{1}{\varepsilon_x}\frac{\mathrm{d}B}{\mathrm{d}x}, \tag{37}$$

and leads to the comparison identity

$$r_p = r_{p0} + \frac{1}{2iQ_1}\int_{-\infty}^{\infty} \mathrm{d}z \left\{\left(\frac{1}{\varepsilon_{0z}} - \frac{1}{\varepsilon_z}\right) K^2 BB_0 + (\varepsilon_x - \varepsilon_{0x})CC_0\right\}. \tag{38}$$

The appropriate values of $B_0(0)$ and $C_0(0)$ are now

$$B_0(0) = \frac{2Q_1}{Q_1 + Q_2}, \qquad C_0(0) = \frac{2iQ_1Q_2}{Q_1 + Q_2}, \tag{39}$$

where $Q_1 = q_1/\varepsilon_1$, and $Q_2 = q_{2p}/\varepsilon_{2x}$, with

$$q_{2p}^2 = \varepsilon_{2x}\frac{\omega^2}{c^2} - \frac{\varepsilon_{2x}}{\varepsilon_{2z}}K^2. \tag{40}$$

To lowest order in the interface thickness,

$$r_p = \frac{Q_2 - Q_1}{Q_2 + Q_1} - \frac{2iQ_1}{(Q_1 + Q_2)^2}\left\{K^2 \int_{-\infty}^{\infty} dz\left(\frac{1}{\varepsilon_{0z}} - \frac{1}{\varepsilon_z}\right) - Q_2^2 \int_{-\infty}^{\infty} dz(\varepsilon_x - \varepsilon_{0x})\right\}. \tag{41}$$

On combining this with the (27) modified by replacing ε_0 by ε_{0x}, we find the generalization of (28):

$$r_{s0}\left(\frac{r_p}{r_s}\right) = r_{p0} - \frac{2iQ_1 K^2}{(Q_1 + Q_2)^2}\left\{\int_{-\infty}^{\infty} dz\left(\frac{1}{\varepsilon_{0z}} - \frac{1}{\varepsilon_z}\right)\right.$$
$$\left. - \left(\frac{1}{\varepsilon_1} - \frac{1}{\varepsilon_{2z}}\right)(\varepsilon_{2x} - \varepsilon_1)^{-1} \int_{-\infty}^{\infty} dz\,(\varepsilon_x - \varepsilon_{0x})\right\} + \ldots . \tag{42}$$

The expression in braces may be written as

$$\frac{1}{\varepsilon_1 \varepsilon_{2z}} \int_{-\infty}^{\infty} dz \left[\frac{\varepsilon_{2x}\varepsilon_{2z} - \varepsilon_1^2}{\varepsilon_{2x} - \varepsilon_1} - \frac{\varepsilon_{2z} - \varepsilon_1}{\varepsilon_{2x} - \varepsilon_1}\varepsilon_x - \frac{\varepsilon_{2z}\varepsilon_1}{\varepsilon_z}\right], \tag{43}$$

and thus the form of (17) is retained:

$$r_{s0}\left(\frac{r_p}{r_s}\right) = r_{p0} - \frac{2iQ_1 K^2/\varepsilon_1\varepsilon_{2z}}{(Q_1 + Q_2)^2}\mathscr{I}_1 + \ldots, \tag{44}$$

with $\mathscr{I}_1$ now being given by the integral in (43).

For a uniform anisotropic film on an anisotropic substrate, the formulae (31) and (32) remain valid, with q_2 being interpreted as q_{2s}, and Q_2 as q_{2p}/ε_{2x}. For a uniform film of thickness Δz, $\mathscr{I}_1/\Delta z$ is equal to the content of the square bracket in (43).

7-4 General results for anisotropic stratifications with azimuthal symmetry

In the previous two sections we have derived results for the ellipsometric properties of *thin* anisotropic films on isotropic and anisotropic substrates, respectively. Here we examine general properties of reflection by anisotropic stratified media, still keeping the constraint of cylindrical symmetry about the normal to the stratification (the z axis). As in Section 2-2 we take the inhomogeneous or interfacial region to be within the interval $[z_1, z_2]$; the results for such finite-ranged interfaces can be extended to continuously varying unbounded interfaces by a limiting process.

For the s wave the results of Section 2-2 follow directly, since anisotropy modifies the s wave equations only by the replacement of $\varepsilon(z)$ by $\varepsilon_x(z)$. Thus the general expressions for the reflection and transmission amplitudes, (2.25) and (2.26), remain valid, with q_1 and q_2 being the values of $q_s = (\varepsilon_x\omega^2/c^2 - K^2)^{1/2}$ for $z \leqslant z_1$ and $z \geqslant z_2$. In consequence, the result $|r_s| \leqslant 1$, the conservation law $q_1(1 - |r_s|^2) = q_2|t_s|^2$, and the reciprocity laws all follow. The result that $r_s \to -1$ at grazing incidence (Section 2-3) also holds, as does the inequality $|r_s| \leqslant |r_{s0}|$ for monotonic profiles (Section 5-4).

The p wave case involves both dielectric functions $\varepsilon_x(z)$ and $\varepsilon_z(z)$, which take the values ε_1 for $z \leqslant z_1$ and ε_{2x}, ε_{2z} for $z \geqslant z_2$. $B(z)$, the solution of (11), now has the limiting forms

$$B(z) = e^{iq_1 z} - r_p e^{-iq_1 z} \qquad (z \leqslant z_1)$$
$$B(z) = \left(\frac{\varepsilon_{2x}}{\varepsilon_1}\right)^{1/2} t_p e^{iq_{2p}z} \qquad (z \geqslant z_2). \tag{45}$$

The sign of r_p and the factor $(\varepsilon_{2x}/\varepsilon_1)^{1/2}$ multiplying t_p are chosen to make r_p and r_s, and t_p and t_s all apply to electric field components, and to agree at normal incidence. The electric field components for the p wave are found from (8); the effect of anisotropy of the substrate is the replacement of ε_2 by ε_{2x} in the square root of the ratio multiplying t_p (compare with Section 1-2).

For profiles which have ε_x continuous at z_1 and z_2, the equations (2.40) and (2.41) remain valid, with q_2 replaced by q_{2p} and $(\varepsilon_2/\varepsilon_1)^{1/2}$ by $(\varepsilon_{2x}/\varepsilon_1)^{1/2}$. The Wronskian of two solutions of (11) is now proportional to ε_x (compare with (2.47) and (2.48)). Again $|r_p|^2 \leqslant 1$ and the reciprocity relations remain valid, with q_2 replaced by q_{2p}. The conservation law for the p wave now reads

$$q_1(1 - |r_p|^2) = q_{2p}|t_p|^2. \tag{46}$$

The range of validity of the inequality $|r_p|^2 \leqslant |r_{p0}|^2$ will be examined in the next section.

7-5 Differential equations for the reflection amplitudes

We shall briefly examine some of the consequences of the non-linear first order differential equations of Chapter 5. The s wave need not be considered in detail, since all results remain valid on the replacement of $\varepsilon(z)$ by $\varepsilon_x(z)$. In the p wave case we set

$$\frac{1}{\varepsilon_x}\frac{dB}{dz} = C, \qquad \frac{dC}{dz} = -\frac{q_p^2}{\varepsilon_x} B, \tag{47}$$

in analogy with (5.32). This pair of coupled first order equations is equivalent to (11), with

$$q_p^2 = \varepsilon_x \frac{\omega^2}{c^2} - \frac{\varepsilon_x}{\varepsilon_z} K^2. \tag{48}$$

The anisotropic version of (5.33) is

$$B = F + G, \qquad C = i\frac{q_p}{\varepsilon_x}(F - G). \tag{49}$$

Elimination of B and C gives equations of the same form as (5.34) and (5.35),

$$F' = iq_p F - \frac{Q'}{2Q}(F - G), \tag{50}$$

$$G' = -iq_p G + \frac{Q'}{2Q}(F - G), \tag{51}$$

where now $Q = q_p/\varepsilon_x$. The reflection *coefficient* $\varrho = G/F$ (as distinct from the reflection *amplitude* to be discussed shortly) satisfies the equation

$$\varrho' + 2iq_p\varrho - \frac{Q'}{2Q}(1 - \varrho^2) = 0. \tag{52}$$

We write $\varrho = |\varrho|\, e^{i\theta}$; the absolute magnitude $|\varrho|$ satisfies

$$|\varrho|' = \frac{Q'}{2Q}(1 - |\varrho|^2)\cos\theta. \tag{53}$$

Integration of (53) gives the exact result

$$\log\frac{1 + |r_p|}{1 - |r_p|} = -\int_{-\infty}^{\infty} dz\, \frac{Q'}{Q}\cos\theta, \tag{54}$$

and for monotonic Q the inequality

$$R_p \leqslant \left(\frac{Q_1 - Q_2}{Q_1 + Q_2}\right)^2 \tag{55}$$

follows, with $Q_1 = q_1/\varepsilon_1$ and $Q_2 = q_{2p}/\varepsilon_{2x}$. We have

$$Q^2 = \frac{q_p^2}{\varepsilon_x^2} = \frac{1}{\varepsilon_x}\frac{\omega^2}{c^2} - \frac{K^2}{\varepsilon_x\varepsilon_z}. \tag{56}$$

and whether Q is monotonic or not depends on the variation of both ε_x and ε_z with z, as well as on the angle of incidence.

In Chapter 5 the useful Rayleigh approximations were obtained from differential equations for the reflection *amplitude*. We will give their anisotropic generalization here. There are now two phase integrals:

$$\phi_s = \int^z d\zeta\, q_s(\zeta), \qquad \phi_p = \int^z d\zeta\, q_p(\zeta). \tag{57}$$

The differential equation for the s wave reflection amplitude is (5.76) with ϕ replaced by ϕ_s and q by q_s. The equation for the p wave reflection amplitude is (5.81) with ϕ_p replacing ϕ and $Q = q_p/\varepsilon_x$. These equations lead to the Rayleigh or weak reflection approximations

$$r_s \simeq r_s^R = -\int_{-\infty}^{\infty} dz\, \frac{q_s'}{2q_s}\, e^{2i\phi_s}, \tag{58}$$

$$r_p \simeq r_p^R = \int_{-\infty}^{\infty} dz\, \frac{Q'}{2Q}\, e^{2i\phi_p}, \tag{59}$$

in parallel with (5.85) and (5.86).

This concludes our limited discussion of the optical aspects of reflection by stratified anisotropic media. Only the very simplest form of anisotropy has been treated: for more general cases (but restricted to systems with sharp boundaries) the reader is referred to Landau and Lifshitz (1960, Chapter 11), Born and Wolf (1970, Chapter 14), and Azzam and Bashara (1977). We next turn to anisotropy in ionospheric radio propagation.

7-6 Reflection from the ionosphere

In the days before satellite communication systems, radio propagation round the earth, using the ionosphere as a reflecting layer, was the only form of long distance "wireless" communication. The simplest model of the ionosphere, that of a plasma of free electrons in a neutralizing background of ions, leads to the dielectric function (Budden, 1985)

$$\varepsilon(z, \omega) = 1 - \frac{\omega_p^2(z)}{\omega^2}. \tag{60}$$

The height dependence of ω_p^2, the square of the angular plasma frequency, arises through its proportionality to the electron density. We noted in Section 6-7 that for waves radiated at θ_1 to the vertical, this model gives a turning point at height z_0 given by $\omega_p(z_0) = \omega \cos \theta_1$. For fixed θ_1 and ionospheric electron density profile, with maximum $\omega_p(z)$ equal to $\omega_p^{\max}$, frequencies below $\omega_p^{\max}/\cos \theta_1$ will be strongly reflected, while those above this value will be weakly reflected. When (for example) the electron density can be approximated by $\operatorname{sech}^2(z - h)/a$, the resulting reflectivity is that given in Section 6-7, with reflectivity contours shown in Figure 6-11.

The above assumes absence of electron collisions, and neglect of the earth's magnetic field. The effect of dissipation resulting from electron collisions will be discussed in the next chapter, while the anisotropy resulting from propagation in the earth's magnetic field will be briefly treated here. As in the case of anisotropic dielectrics, there is double refraction. The magneto-ionic case is more complex however, since for neither of the two polarizations which propagate unchanged do the wave normal and ray diffractions coincide (Ratcliffe (1959), Budden (1964)).

The simplest example of anisotropy arises for wave propagation along the direction of the earth's magnetic field, $\mathbf{B}_0$. This is referred to as the longitudinal case. Jackson (1962, Section 7.9) gives a simple argument which shows that for transverse electromagnetic waves propagating along $\mathbf{B}_0$ the two waves which propagate unchanged are left or right circularly polarized, with effective dielectric constants

$$\varepsilon_\pm = 1 - \frac{\omega_p^2}{\omega(\omega \pm \omega_B)}. \tag{61}$$

Here ω_B is the frequency of electron gyration round the magnetic field (the gyro-frequency), and is proportional to B_0. The gyro and plasma frequencies can have comparable magnitudes; Figure 7-3 shows ε_+ and ε_- for the case where $\omega_p = \omega_B$.

We see from the figure that there is a dramatic difference between the two circular polarizations. The + wave (with positive helicity) is strongly reflected for

$$\omega < \tfrac{1}{2}\{(\omega_B^2 + 4\omega_p^2 \sec^2\theta_1)^{1/2} - \omega_B\}, \tag{62}$$

while the − wave is strongly reflected in the interval

$$\omega_B < \omega < \tfrac{1}{2}\{(\omega_B^2 + 4\omega_p^2 \sec^2\theta_1)^{1/2} + \omega_B\}. \tag{63}$$

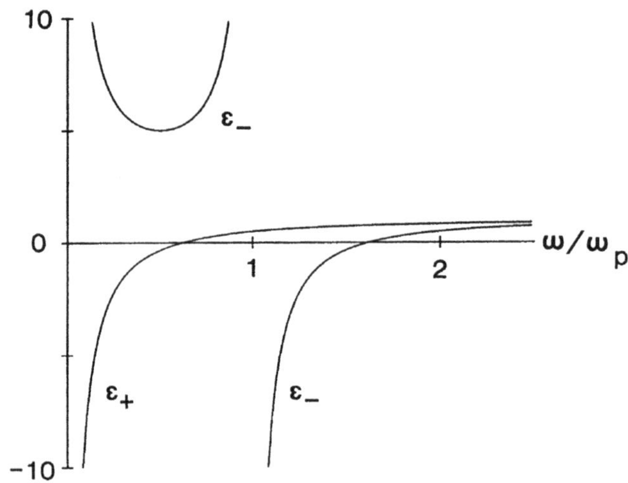

Figure 7-3 Dielectric functions $\varepsilon_{\pm}$ for the two circular polarizations, shown as a function of frequency when $\omega_B = \omega_p$.

(In these formulae ω_p stands for $\omega_p^{\max}$, and ω_B for the value attained within the region of maximum electron density.) The physical reason for the difference in the propagation of the two polarizations is that one reinforces and the other opposes the precessional motion of the electrons in the earth's magnetic field. Heading (1975, Sections 2.5 and 9.8) gives analytic results for reflection by an exponential ionosphere, in the longitudinal case.

There are also interesting effects in the transverse case, where propagation is perpendicular to the earth's magnetic field. Barber and Crombie (1959) have shown that the reflection from the ionosphere is greater for v.l.f. waves travelling west to east around the magnetic equator, than for those travelling from east to west. Exact solutions for this case have been given by Westcott (1970).

The general case, with arbitrary angle between the direction of propagation and the earth's magnetic field, is discussed by Heading and Whipple (1952), Heading (1955), Ratcliffe (1959), Heading (1963), Ginzburg (1964), Booker (1984), and Budden (1985).

References

References quoted in the text

J. Lekner (1983) "Anisotropy of the dielectric function within a liquid-vapour interface", Mol. Phys. **49**, 1385–1400.

F. P. Buff (1966) "Saline water conversion report", U.S. Govt. Printing Office, 26–27.

D. Beaglehole (1980) "Ellipsometric study of the surface of simple liquids", Physics **100B**, 163–174.

F. Abelès (1976) "Optical properties of very thin films", Thin Solid Films **34**, 291–302.

L. D. Landau and E. M. Lifshitz (1960) "Electrodynamics of continuous media", Pergamon.

M. Born and E. Wolf (1970) "Principles of optics", 4th edition, Pergamon.

R. M. A. Azzam and N. M. Bashara (1977) "Ellipsometry and polarized light", North Holland.

K. G. Budden (1985) "The propagation of radio waves", Cambridge.

J. A. Ratcliffe (1959) "The magneto-ionic theory and its applications to the ionosphere", Cambridge.

K. G. Budden (1964) "Lectures on the magnetoionic theory", Gordon and Breach.

J. D. Jackson (1962) "Classical electrodynamics", Wiley.

J. Heading (1975) "Ordinary differential equations, theory and practice", Elek Science.

N. F. Barber and D. D. Crombie (1959) "V.L.F. reflections from the ionosphere in the presence of a transverse magnetic field", J. Atm. Terr. Phys. **16**, 37–45.

B. S. Westcott (1970) "Exact solutions for vertically polarized electromagnetic waves in a horizontally stratified magneto-plasma", Proc. Camb. Phil. Soc. **67**, 491–501.

J. Heading and R. T. P. Whipple (1952) "THe oblique reflexion of long wireless waves from the ionosphere at places where the earth's magnetic field is regarded as vertical", Phil. Trans. Roy. Soc. **244A**, 469–503.

J. Heading (1955) "The reflexion of vertically-incident long radio waves from the ionosphere when the earth's magnetic field is oblique", Proc. Roy. Soc. **A231**, 414–435.

J. Heading (1963) "Composition of reflection and transmission formulae", J. Res. Nat. Bur. Stand. **67D**, 65–77.

V. L. Ginzburg (1964) "The propagation of electromagnetic waves in plasmas", Pergamon.

H. G. Booker (1984) "Cold plasma waves", Martinus Nijhoff.

Sections 7-1 and 7-2 are based on Lekner (1983) and

J. Lekner (1986) "Reflection of light by a non-uniform film between like media", J. Opt. Soc. Amer. **A3**, 9–15.

This paper also gives a variational theory for the anisotropic case.

8

Absorption

This chapter deals with the effect of absorption on reflection properties. The absorption, or dissipation of electromagnetic energy within the medium, can be due to conductivity (as in metals, and in the ionosphere). However, good insulators can also be absorbers at high frequencies, where the electromagnetic field energy is converted to heat via molecular or electronic excitations. The absorption is included in the Maxwell equation (1.2) by allowing the dielectric function ε to take complex values. In general, the curl of $\mathbf{B}$ is the sum of terms proportional to $\partial \mathbf{E}/\partial t$ and to the total current density. For non-magnetic media, and fields with the time variation $e^{-i\omega t}$, the form of (1.2) is retained, with the imaginary part of ε now proportional to the conductivity divided by the frequency (Born and Wolf, 1970, Section 13.1). The simplest model for conducting media is that of an electron gas, with mean free time between collisions τ. This leads to the dielectric function (see for example Kittel 1966, Booker 1984, Budden 1985)

$$\varepsilon(\omega, z) = 1 - \frac{\omega_p^2}{\omega^2 + i\omega/\tau}, \tag{1}$$

where ω_p is the plasma frequency. In the ionosphere, for example, ε is a function of height z through the proportionality of ω_p^2 to the electron density, as well as through the dependence of τ on the electron, ion, and neutral species densities.

We will represent the real and imaginary parts of physical variables such as ε by the subscripts r and i:

$$\varepsilon = \varepsilon_r + i\varepsilon_i. \tag{2}$$

The real and imaginary parts of ε are directly related to the electronic properties of the material under study. Either ε_r, ε_i, or the real and imaginary parts of the square root of ε (the complex refractive index) can be used in writing the reflectivity formulae. We shall use both, with the refractive index notation being particularly convenient at normal incidence. The relationship between the two is found from

$$\varepsilon_r + i\varepsilon_i = (n_r + in_i)^2, \tag{3}$$

giving

$$\varepsilon_r = n_r^2 - n_i^2, \qquad \varepsilon_i = 2n_r n_i. \tag{4}$$

The real and imaginary parts of ε are related in their frequency dependence by the Kramers–Krönig relations: $\varepsilon_r(\omega) - 1$ and $\varepsilon_i(\omega)$ are Hilbert transforms of each other, because the response of any system to an arbitrary signal must be causal (see, for example, Landau and Lifshitz, 1960, Section 62).

8-1 Fresnel reflection formulae for an absorbing medium

For the s wave, with $\mathbf{E} = (0, E_y, 0)$ for propagation in the zx plane,

$$E_y(z, x, t) = e^{i(Kx-\omega t)} E(z), \tag{5}$$

with

$$\frac{d^2E}{dz^2} + q^2E = 0, \qquad q^2(z) = \varepsilon(z)\frac{\omega^2}{c^2} - K^2. \tag{6}$$

The separation of variables constant K is the component of the wavevector along the interface, and its invariance leads to Snell's Law:

$$\sqrt{\varepsilon_1}\,\frac{\omega}{c}\sin\theta_1 = K = \sqrt{\varepsilon_2}\,\frac{\omega}{c}\sin\theta_2. \tag{7}$$

Here we consider radiation incident from a non-absorbing medium (real ε_1) onto an absorbing medium (complex ε_2). Thus the angle of refraction is complex, and has a formal meaning only. The behaviour of the refracted wave is found from its waveform

$$E_y(z, x, t) = e^{i(Kx+q_2z-\omega t)}. \tag{8}$$

We write $\varepsilon_2 = \varepsilon_r + i\varepsilon_i$; the real and imaginary parts of q_2 are found from

$$q_2^2 = \varepsilon_2\frac{\omega^2}{c^2} - K^2 = \frac{\omega^2}{c^2}(\varepsilon_r + i\varepsilon_i - \varepsilon_1\sin^2\theta_1). \tag{9}$$

Setting $q_2 = q_r + iq_i$, so that $q_2^2 = q_r^2 - q_i^2 + 2iq_rq_i$, we have

$$q_r^2 - q_i^2 = \frac{\omega^2}{c^2}(\varepsilon_r - \varepsilon_1\sin^2\theta_1), \qquad 2q_rq_i = \frac{\omega^2}{c^2}\varepsilon_i. \tag{10}$$

Thus, for $\varepsilon_i \neq 0$,

$$(cq_r/\omega)^2 = \tfrac{1}{2}\{\varepsilon_r - \varepsilon_1\sin^2\theta_1 + [(\varepsilon_r - \varepsilon_1\sin^2\theta_1)^2 + \varepsilon_i^2]^{1/2}, \tag{11}$$

$$cq_i/\omega = \frac{\varepsilon_i/2}{cq_r/\omega}. \tag{12}$$

(When $\varepsilon_i = 0$ we have either q_i or $q_r = 0$, depending on whether $\theta_1 < \theta_c$ or $\theta_1 > \theta_c$.) The waveform in the absorbing medium is

$$E_y(z, x, t) = e^{-q_iz}\, e^{i(Kx+q_rz-\omega t)}. \tag{13}$$

Thus q_i must be non-negative, which implies that ε_i and n_i must be non-negative. Surfaces of constant amplitude are planes parallel to the interface (z = constant), while surfaces of constant real phase are the planes $Kx + q_r z$ = constant. The normal to the surfaces of constant phase is at an angle θ_2' to the normal to the interface, where

$$\tan^2\theta_2' = \frac{K^2}{q_r^2} = \frac{2\varepsilon_1 \sin^2\theta_1}{\varepsilon_r - \varepsilon_1 \sin^2\theta_1 + [(\varepsilon_r - \varepsilon_1 \sin^2\theta_1)^2 + \varepsilon_i^2]^{1/2}}. \tag{14}$$

The real angle θ_2' and the angle θ_2 (in general complex) coincide only for real ε_2, or at normal incidence.

For a sharp boundary between media 1 and 2, represented by a step dielectric function at $z = 0$, the continuity of E and $\mathrm{d}E/\mathrm{d}z$ at the boundary (implied by (6)) give the reflection amplitude

$$r_s = \frac{q_1 - q_2}{q_1 + q_2} = \frac{q_1 - q_r - iq_i}{q_1 + q_r + iq_i}. \tag{15}$$

The s reflectivity is thus

$$R_s = \frac{(q_1 - q_r)^2 + q_i^2}{(q_1 + q_r)^2 + q_i^2}. \tag{16}$$

At normal incidence this reduces to

$$R_n = \frac{(n_1 - n_r)^2 + n_i^2}{(n_1 + n_r)^2 + n_i^2}, \tag{17}$$

since then $q_r = n_r\,\omega/c$ and $q_i = n_i\,\omega/c$.

The p wave, which has $\mathbf{B} = (0, B_y, 0)$, again has

$$B_y(z, x, t) = \mathrm{e}^{i(Kx - \omega t)}\, B(z), \tag{18}$$

with the same separation of variables constant K as the s wave. The equation satisfied by B is

$$\frac{\mathrm{d}}{\mathrm{d}z}\left(\frac{1}{\varepsilon}\frac{\mathrm{d}B}{\mathrm{d}z}\right) + \left(\frac{\omega^2}{c^2} - \frac{K^2}{\varepsilon}\right) B = 0. \tag{19}$$

At a sharp boundary between two media, B and $\mathrm{d}B/\varepsilon\mathrm{d}z$ are continuous; the reflection amplitude is thus

$$-r_p = \frac{Q_1 - Q_2}{Q_1 + Q_2} = \frac{Q_1 - Q_r - iQ_i}{Q_1 + Q_r + iQ_i}, \tag{20}$$

where $Q_1 = q_1/\varepsilon_1$ and $Q_2 = q_2/\varepsilon_2$. The latter is the ratio of the two complex quantities $q_r + iq_i$ and $\varepsilon_r + i\varepsilon_i$, and thus has the real and imaginary parts

$$Q_r = \frac{\varepsilon_r q_r + \varepsilon_i q_i}{\varepsilon_r^2 + \varepsilon_i^2}, \qquad Q_i = \frac{\varepsilon_r q_i - \varepsilon_i q_r}{\varepsilon_r^2 + \varepsilon_i^2}. \tag{21}$$

We note that $r_s \to -1$ and $r_p \to +1$ at grazing incidence, as in the case of nonabsorbing media. The p reflectivity is

$$R_p = \frac{(Q_1 - Q_r)^2 + Q_i^2}{(Q_1 + Q_r)^2 + Q_i^2}, \tag{22}$$

and reduces to (17) at normal incidence, where the real and imaginary parts of Q takes the values

$$Q_r = \frac{n_r\omega/c}{n_r^2 + n_i^2}, \qquad Q_i = \frac{-n_i\omega/c}{n_r^2 + n_i^2}. \tag{23}$$

We shall see later in this Section that R_p is never greater than R_s for a step profile.

In the absense of absorption, the p wave reflectivity is zero when $Q_1 = Q_2$ (at the Brewster angle arctan $(\varepsilon_2/\varepsilon_1)^{1/2}$). The condition $Q_1 = Q_2$ cannot be satisfied for absorbing reflectors, since this would imply $Q_1 = Q_r$ and $Q_i = 0$. The latter condition is satisfied at angle of incidence θ_1 such that

$$\sin^2\theta_1 = \frac{\varepsilon_r^2 + \varepsilon_i^2}{2\varepsilon_1\varepsilon_r}, \tag{24}$$

and is thus possible if $\varepsilon_r > 0$ and $\varepsilon_r^2 + \varepsilon_i^2 < 2\varepsilon_1\varepsilon_r$. But when (24) holds, the first condition $Q_1 = Q_r$ could be satisfied only if $(\varepsilon_1 - \varepsilon_r)^2 + \varepsilon_i^2 = 0$. Thus perfect absorption at a single sharp boundary is not possible. When a dielectric layer is placed over an absorbing medium, zero reflectivity *is* possible, for both polarizations (at different angles), as we shall see in Section 8-3.

The s and p reflectivities are shown in Figure 8-1, for a metal (Al) and a semiconductor (Si) at the visible He–Ne laser wavelength, $\lambda_0 = 0.633\,\mu$m.

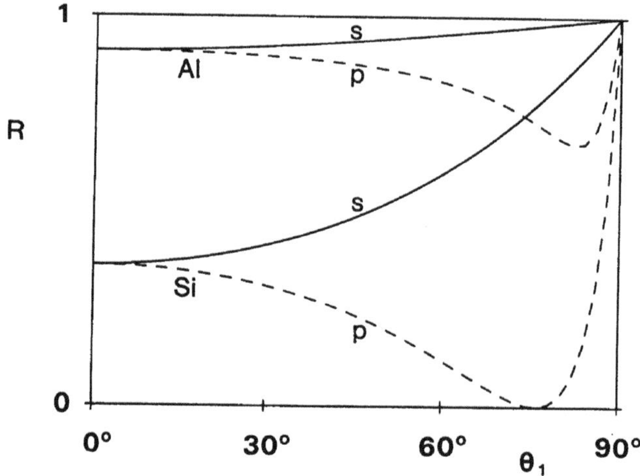

Figure 8-1. Reflectivity as a function of the angle of incidence, for the s and p polarizations at 0.633 μm. The refractive indices are Al: 1.566 + 7.938i, Si: 4.0 + 0.12i (the Al values are for bulk metal; vapour-deposited values are different: see Allen, 1976). The corresponding dielectric functions are Al: −60.56 + 24.86i, Si: 16.0 + 0.96i.

We note the high metallic reflectivities, which are due to wavenumber mismatch: q_1 is real and $q_2 = q_r + iq_i$ has a large imaginary part. An example of this real/imaginary type of mismatch was seen in total internal reflection in dielectric materials, where for $\theta_1 >$ arcsin $(\varepsilon_1/\varepsilon_2)^{1/2}$ the wavenumber component q_2 is pure imaginary, giving total reflection for both polarizations. We see from (11) that when $\varepsilon_i \neq 0$ total reflection is not possible for the s wave (except at grazing incidence, when $q_1 \to 0$). The same result follows for the p wave from (21), (11) and (12).

The ellipsometric quantity r_p/r_s is obtained from (15) and (20). It has the real and imaginary parts

$$\mathrm{Re}\left(\frac{r_p}{r_s}\right) = -\frac{\{(Q_1^2 - Q_r^2 - Q_i^2)(q_1^2 - q_r^2 - q_i^2) + 4Q_1Q_iq_1q_i\}}{[(Q_1 + Q_r)^2 + Q_i^2][(q_1 - q_r)^2 + q_i^2]}, \quad (25)$$

$$\mathrm{Im}\left(\frac{r_p}{r_s}\right) = 2\frac{Q_1Q_i(q_1^2 - q_r^2 - q_i^2) - q_1q_i(Q_1^2 - Q_r^2 - Q_i^2)}{[(Q_1 + Q_r)^2 + Q_i^2][(q_1 - q_r)^2 + q_i^2]}. \quad (26)$$

The computation of these quantities is simplified by the identity

$$Q_r^2 + Q_i^2 = \frac{q_r^2 + q_i^2}{\varepsilon_r^2 + \varepsilon_i^2}. \quad (27)$$

Equivalent and somewhat simpler formulae (in terms of q_1, q_r, q_i and K) are given in Section 9-1. The trajectories of r_p/r_s in the complex plane for variable angle of incidence are shown in Figure 8-2 for Al and Si at 0.633 μm.

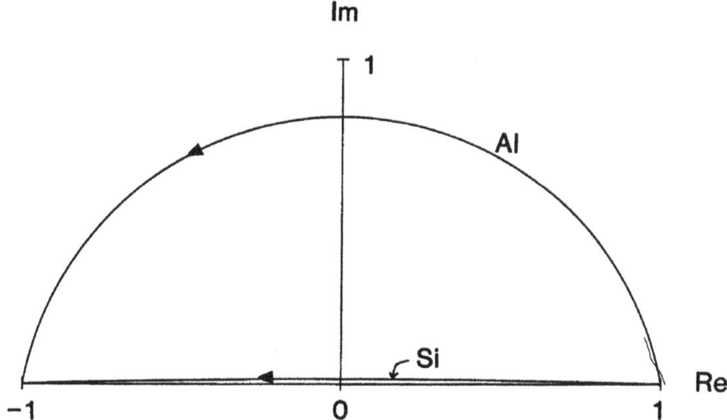

Figure 8-2. The ellipsometric ratio r_p/r_s in the complex plane; trajectories for Al and Si are shown. The refractive indices are for 0.633 μm, as in Figure 8-1. For a perfect dielectric (no absorption) the trajectory is the real axis from 1 (at 0°) to −1 (at 90°).

The trajectory of r_p/r_s always lies within the upper half of the unit circle for an arbitrary absorbing medium with a sharply defined surface. To see this, it is convenient to define a complex angle of refraction, $\theta_2 = \theta_r + i\theta_i$, via

$$q_2 = \varepsilon_2^{1/2}(\omega/c)\cos\theta_2 \quad (28)$$

(this definition is consistent with (6) and (7)). Then (15) and (20) may be written in the Fresnel forms (1.14) and (1.32),

$$r_s = \frac{\sin(\theta_2 - \theta_1)}{\sin(\theta_2 + \theta_1)}, \qquad r_p = \frac{\tan(\theta_2 - \theta_1)}{\tan(\theta_2 + \theta_1)}, \quad (29)$$

and the ellipsometric ratio as $\cos(\theta_2 + \theta_1)/\cos(\theta_2 - \theta_1)$, or

$$\frac{r_p}{r_s} = \frac{\cos(\theta_1 + \theta_r)\cosh\theta_i - i\sin(\theta_1 + \theta_r)\sinh\theta_i}{\cos(\theta_1 - \theta_r)\cosh\theta_i + i\sin(\theta_1 - \theta_r)\sinh\theta_i}. \quad (30)$$

The fact that $R_p \leqslant R_s$ follows from $\sinh^2\theta_i < \cosh^2\theta_i$. The sign of Im (r_p/r_s) is opposite to that of θ_i. From (28) we have

$$\left(\frac{c}{\omega}\right)(q_r + iq_i) = (n_r + in_i)(\cos\theta_r \cosh\theta_i - i\sin\theta_r \sinh\theta_i). \tag{31}$$

From the real and imaginary parts of (31) we obtain

$$\cos\theta_r \cosh\theta_i = \frac{c}{\omega}(n_r q_r + n_i q_i)/(n_r^2 + n_i^2), \tag{32}$$

$$\sin\theta_r \sinh\theta_i = \frac{c}{\omega}(n_i q_r - n_r q_i)/(n_r^2 + n_i^2). \tag{33}$$

These relations may in turn be used to find θ_r and θ_i as a function of the angle of incidence, θ_1. Here we are interested mainly in the sign of θ_i, which is that of $n_i q_r - n_r q_i$. We note (see (13)) that q_r and q_i are non-negative. Thus ε_i is non-negative also (from (10)), and so is n_i (both n_r and n_i are $\geqslant 0$, since $q_r = n_r\omega/c$, $q_i = n_i\omega/c$ at normal incidence). It thus follows from (11), (12) and (33) that θ_i is never positive, so that r_p/r_s always stays in the upper half of the unit circle.

For non-absorbing dielectrics R_p is zero at the Brewster angle. In the presence of absorption the reflectivity ratio R_p/R_s has a minimum at what is known as the pseudo-Brewster angle. The extraction of the optical constants n_r and n_i (or ε_r and ε_i) from measurements of this angle and of the minimum reflectivity ratio is discussed by Potter (1969).

8-2 General results for reflection by absorbing media

In Section 2-1 we derived the conservation law

$$q_1(1 - |r|^2) = q_2|t|^2 \tag{34}$$

(valid for s, p and quantum particle waves in the absence of absorption). This relation represents conservation of energy in the electromagnetic case, and conservation of the probability density current in the particle case. In the presence of absorption the conservation law is no longer valid, since energy or particles are removed by the absorbing medium. This was noted in Section 2-1; it is mathematically more explicit in the approach of Section 6-2, where it is clear that the derivation of (34) depends on the reality of $q^2(z)$.

The quantity $T_{12} = (q_2/q_1)|t_{12}|^2$ is called the transmittance (see the discussion following (2.8); equivalently one may take the ratio of Im $(\psi^* d\psi/dz)$ for $\psi_2 = t_{12}\,e^{iq_2 z}$ and $\psi_1 = e^{iq_1 z}$, as in (2.9)). For an arbitrary inhomogeneous and absorbing layer between the nonabsorbing media 1 and 2, we showed in Section 2-1 that the reciprocity relation $q_2 t_{12} = q_1 t_{21}$ holds. Thus the transmittances for propagation in either direction through an absorbing layer are equal:

$$T_{12} = \frac{q_2}{q_1}|t_{12}|^2 = \frac{q_1}{q_2}|t_{21}|^2 = T_{21}. \tag{35}$$

The corresponding result for reflectivities, $R_{12} = R_{21}$, holds only in the absence of absorption (Section 2-1, equation 18)).

The result (34) may be written as $1 - R = T$. For an absorbing layer between nonabsorbing media, the ratio $(1 - R)/T$ is greater than unity, since the conservation law $1 = R + T$ is replaced by $1 = R + T + A$, where A represents absorption. Abelès (1950) has shown that if an arbitrary nonabsorbing layer is inserted in the front of the absorbing layer, causing the reflectance to change to R' and the transmittance to T', the ratio of $1 - R$ to T is unaltered:

$$\frac{1 - R}{T} = \frac{1 - R'}{T'}. \tag{36}$$

This result is proved by matrix methods in Section 12-6.

The general formulae for r_s, t_s, r_p and t_p given in Section 2-2 remain true in the presence of absorption, with the solutions F, G and C, D now complex. Thus it remains true that $r_s \rightarrow -1$ and $r_p \rightarrow 1$ at grazing incidence. Since $r_p = r_s$ at normal incidence, it follows that the trajectory of r_p/r_s still starts at $+1$ and ends at -1 in the complex plane, and consequently there always exists an ellipsometric Brewster angle where $\text{Re}\,(r_p/r_s) = 0$ (in general there can be an odd number of such angles, as noted in Section 2-3).

8-3 Dielectric layer on an absorbing substrate

The reflection from a uniform dielectric layer on a transparent substrate was discussed in Section 2-4. When the substrate is absorbing (typically it is metallic), the formulae for the reflection amplitudes derived in Section 2-4 remain valid, with $q_2 = q_r + iq_i$ and $Q_2 = Q_r + iQ_i$. The fact that q_2 and Q_2 are complex changes the reflectivity properties markedly. Of particular interest is the design of *reflection polarizers*, in which the reflectance of one of the components of polarization is extinguished by interface effects, while that of the other is not. See, for example, Ruiz-Urbieta and Sparrow (1972), Bennett and Bennett (1978), and Azzam (1985).

Consider a transparent uniform film, of thickness Δz, located between z_1 and $z_1 + \Delta z$. The light is incident at angle θ_1 from medium 1, of dielectric constant ε_1. The film has dielectric constant ε, and the refracted ray within it makes an angle θ to the normal. Snell's Law (the invariance of K^2) gives $\varepsilon_1 \sin^2\theta_1 = \varepsilon \sin^2\theta$. The substrate has dielectric constant $\varepsilon_2 = \varepsilon_r + i\varepsilon_i$. The normal components of the wavevector in the three media are q_1, q, and $q_2 = q_r + iq_i$, where the real and imaginary parts of q_2 are given by (11) and (12). The s wave reflection amplitude for this system is given by (2.58):

$$r_s = e^{2iq_1z_1} \frac{r + r'\, e^{2iq\Delta z}}{1 + rr'\, e^{2iq\Delta z}}. \tag{37}$$

where

$$r = \frac{q_1 - q}{q_1 + q}, \qquad r' = \frac{q - q_2}{q + q_2}, \tag{38}$$

are the reflection amplitudes (without phase factors associated with location) for the ambient-film and film-substrate interfaces. From (37), r_s will be zero when $r' = -r\,e^{-2iq\Delta z}$; on equating the real and imaginary part we find

$$\begin{aligned} \frac{q_r^2 + q_i^2 - q^2}{(q + q_r)^2 + q_i^2} &= r \cos 2q\Delta z, \\ \frac{-2qq_i}{(q + q_r)^2 + q_i^2} &= r \sin 2q\Delta z. \end{aligned} \tag{39}$$

The angle of incidence at which zero reflection occurs (and the corresponding wavevector components to be inserted into (39) to determine the appropriate values of Δz) is found from $|r'|^2 = r^2$, which leads to

$$(q^2 - q_1 q_r)(q_r - q_1) = q_1 q_i^2. \tag{40}$$

This equation is solved numerically. For metallic substrates the solution lies near grazing incidence; for example, for a layer of Al_2O_3 of refractive index 1.6 on aluminium (with the optical parameters used in Figures 8-1 and 8-2), zero reflection for the s wave occurs at 87.93°. Since the reflectivity is always unity at grazing incidence, its variation with angle of incidence is necessarily rapid between the polarizing angle and 90°.

Kitajima, Fujita and Cizmic (1984) give numerical and experimental examples of the reflectivity near the extinction point as a function of film thickness. The dependence is strong, so observation of the oblique incidence reflectance during film deposition is a sensitive thickness monitor.

For the p wave the reflection amplitude is given by (2.70):

$$-r_p = e^{2iq_1 z_1} \frac{r + r' e^{2iq\Delta z}}{1 + rr' e^{2iq\Delta z}}, \tag{41}$$

where now

$$r = \frac{Q_1 - Q}{Q_1 + Q}, \qquad r' = \frac{Q - Q_2}{Q + Q_2}, \tag{42}$$

with $Q_1 = q_1/\varepsilon_1$, $Q = q/\varepsilon$ and $Q_2 = q_2/\varepsilon_2$. The real and imaginary parts of Q_2 are given by (21). The condition for zero p reflectivity is $r' = -r\,e^{-2iq\Delta z}$, which is equivalent to the equations derived for the s wave with Q's replacing q's except in the oscillatory functions of $2q\Delta z$. The equation analogous to (40),

$$(Q^2 - Q_1 Q_r)(Q_r - Q_1) = Q_1 Q_i^2, \tag{43}$$

again has a solution close to grazing incidence: for the Al_2O_3 on Al case, extinction of the p wave occurs at $\theta_1 \simeq 88.67^0$. Azzam (1985) has used the fact that zero reflection occurs near 90° to obtain approximate but explicit solutions of (40) and (43).

8-4 Absorbing film on a transparent substrate

The derivation of the reflection and transmission amplitudes for a uniform layer between two uniform media given in Section 2-4 remains valid when the layer, the substrate, or both, are absorbing. Here we examine the case when the layer is absorbing, and the substrate is not. An example is a metallic film on glass. The s wave results may be obtained from (2.52) and (2.53) or from (2.58) and (2.59). The latter are more convenient when ε (the dielectric constant of the layer) is complex. We set $\varepsilon = \varepsilon_r + i\varepsilon_i$, $q^2 = \varepsilon\omega^2/c^2 - K^2 = (q_r + iq_i)^2$, to obtain

$$(cq_r/\omega)^2 = \tfrac{1}{2}\{\varepsilon_r - \varepsilon_1 \sin^2\theta_1 + [(\varepsilon_r - \varepsilon_1 \sin^2\theta_1)^2 + \varepsilon_i^2]^{1/2}\}, \tag{44}$$

$$cq_i/\omega = \frac{\varepsilon_i/2}{cq_r/\omega}. \tag{45}$$

(These formulae are the same as (11) and (12), but here ε_r, ε_i and q_r, q_i refer to a film rather than to a bulk medium.) The reflection amplitude is given formally by (37) and (38), with $q = q_r + iq_i$ now complex. To simplify the analysis we write

$$r = \varrho\, e^{i\delta}, \qquad r'\, e^{2iq\Delta z} = \varrho'\, e^{i\delta'}, \tag{46}$$

with ϱ' including the exponential decay factor $e^{-2q_i\Delta z}$, and δ' the $2q_r\Delta z$ phase increment. Then r_s takes the simple form

$$r_s = e^{2iq_1z_1} \frac{\varrho\, e^{i\delta} + \varrho'\, e^{i\delta'}}{1 + \varrho\varrho'\, e^{i(\delta+\delta')}}, \tag{47}$$

and gives the reflectivity

$$|r_s|^2 = \frac{\varrho^2 + 2\varrho\varrho' \cos(\delta - \delta') + (\varrho')^2}{1 + 2\varrho\varrho' \cos(\delta + \delta') + (\varrho\varrho')^2}. \tag{48}$$

The transmission amplitude is found from (2.59):

$$t_s = e^{i(q_1z_1 - q_2z_2)} \frac{(1 + \varrho\, e^{i\delta})(e^{iq\Delta z} + e^{-iq\Delta z}\,\varrho'\, e^{i\delta'})}{1 + \varrho\varrho'\, e^{i(\delta+\delta')}}. \tag{49}$$

We set

$$e^{iq\Delta z} = e^{-q_i\Delta z}\, e^{iq_r\Delta z} = f\, e^{i\phi}, \tag{50}$$

and obtain

$$|t_s|^2 = \frac{(1 + 2\varrho \cos\delta + \varrho^2)(f^2 + 2\varrho' \cos(2\phi - \delta) + (\varrho'/f)^2)}{1 + 2\varrho\varrho' \cos(\delta + \delta') + (\varrho\varrho')^2}. \tag{51}$$

When the film is "thick", by which we mean here that the absorption within it is large, with $f = e^{-q_i\Delta z} \ll 1$, the reflection properties reduce to those of Section 8-1 for waves incident on a semi-infinite absorbing medium. The transmitted flux is then proportional to $e^{-2q_i\Delta z}$. In such a film the effect of interference of multiply reflected light (see Figure 2-5) is negligible because of the decrease in the amplitude due to absorption.

In the absence of absorption the above formulae reduce to those of Section 2-4, with $|r_s|^2$ and $|t_s|^2$ periodic functions of Δz of period $\pi/q = \lambda/2$, λ being the wavelength within the film associated with motion in the z direction (perpendicular to the surface). An example was shown in Figure 2-6. Absorption destroys the

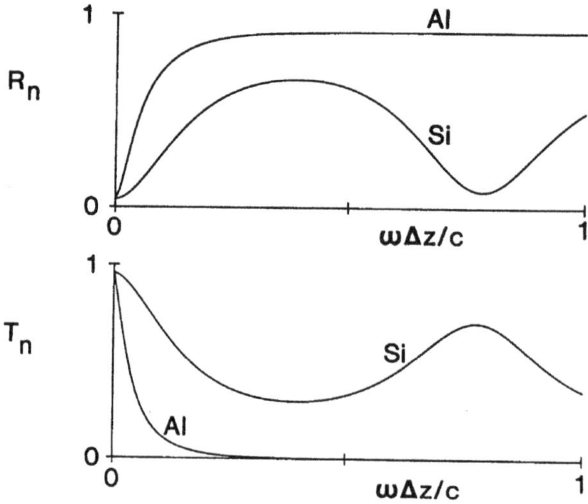

Figure 8-3. Reflectivity $R = |r|^2$ and transmittance $T = (q_2/q_1)|t|^2$ for light of wavelength 0.633 μm incident normally on films of Si ($n = 4.0 + 0.12i$) and vapour-deposited Al ($n = 1.212 + 6.924i$) on glass ($n = 1.5$). At normal incidence $q_i = n_i\omega/c$; for large $q_i\Delta z$ the transmitted intensity varies as $e^{-2q_i\Delta z}$. When $\Delta z = c/\omega = \lambda_0/2\pi$, this factor is approximately 10^{-6} for Al; the thickness is then about 0.1 μm.

periodicity, and strong absorption removes the oscillations altogether. Figure 8-3 shows the reflectivity and transmittance for Al and Si films on glass.

The p wave formulae may be obtained from (2.70) and (2.71), with q and Q now complex and given by (44, 45) and (21). The results are completely analogous to the s wave case, with the exception of the $(\varepsilon_2/\varepsilon_1)^{1/2}$ factor multiplying t_p in (2.71).

8-5 Thin non-uniform absorbing films

So far we have considered uniform media, and uniform layers of arbitrary thickness. We now specialize to thin films or interfaces ("thin" meaning that the film thickness times ω/c is small), which however may have arbitrary depth dependence in the real and imaginary parts of the dielectric function, $\varepsilon_r(z)$ and $\varepsilon_i(z)$.

The presence of absorption within the interface and/or substrate has major effects on the reflection and ellipsometric properties. We saw in Chapter 3 that for nonabsorbing media the reflectivities are unchanged to first order in the interface thickness/wavelength expansion: the interfacial profile characteristics appear only in the second order. For absorbing media, the wavenumber components q and $Q = q/\varepsilon$ are complex. The s and p reflection amplitudes to first order in the interface thickness are given by (3.23) and (3.44):

$$r_s = r_{s0} + \frac{2iq_1\omega^2/c^2}{(q_1 + q_2)^2}\lambda_1 + \ldots, \tag{52}$$

$$r_p = r_{p0} - \frac{2iQ_1}{(Q_1 + Q_2)^2}\left\{\frac{K^2\Lambda_1}{\varepsilon_1\varepsilon_2} - Q_2^2\lambda_1\right\} + \ldots. \tag{53}$$

The integrals λ_1 and Λ_1 are given by

$$\lambda_1 = \int_{-\infty}^{\infty} dz\,(\varepsilon - \varepsilon_0), \qquad \Lambda_1 = \varepsilon_1\varepsilon_2 \int_{-\infty}^{\infty} dz \left(\frac{1}{\varepsilon_0} - \frac{1}{\varepsilon}\right), \tag{54}$$

where the step function $\varepsilon_0(z)$, which takes the values ε_1 for $z < 0$ and ε_2 for $z > 0$, may now be complex, since ε_2 may be complex.

We will consider in detail the transparent substrate case (ε_2 real). The integrals λ_1 and Λ_1 are still complex, since the interface is absorbing, with complex $\varepsilon(z)$. The s and p reflectivities thus contain first order correction terms proportional to the imaginary parts of λ_1 and Λ_1:

$$R_s = R_{s0} - \frac{4q_1(q_1 - q_2)\omega^2/c^2}{(q_1 + q_2)^3}\,\text{Im}\,\lambda_1 + \ldots, \tag{55}$$

$$R_p = R_{p0} - \frac{4Q_1(Q_1 - Q_2)}{(Q_1 + Q_2)^3}\left\{\frac{K^2}{\varepsilon_1\varepsilon_2}\,\text{Im}\,\Lambda_1 - Q_2^2\,\text{Im}\,\lambda_1\right\} + \ldots \tag{56}$$

(these formulae and the following discussion apply to the transparent substrate case only). The step function ε_0 is real when ε_2 is real, so

$$\text{Im}\,\lambda_1 = \int_{-\infty}^{\infty} dz\,\varepsilon_i(z), \qquad \frac{1}{\varepsilon_1\varepsilon_2}\,\text{Im}\,\Lambda_1 = \int_{-\infty}^{\infty} dz\,\frac{\varepsilon_i}{\varepsilon_r^2 + \varepsilon_i^2}. \tag{57}$$

The reflectivity corrections are proportional to integrals over the imaginary part of ε through the absorbing region, as may be expected. At normal incidence both (55) and (56) reduce to

$$R_n = R_{n0} - \frac{4n_1(n_1 - n_2)}{(n_1 + n_2)^3}\,\frac{\omega}{c}\,\text{Im}\,\lambda_1 + \ldots, \tag{58}$$

where $n_1 = \varepsilon_1^{1/2}$ and $n_2 = \varepsilon_2^{1/2}$ are the real refractive indices of the media bounding the inhomogeneous region.

For passive media the absorption term $\text{Im}\,\lambda_1$ is non-negative, and so (to first order in the film thickness) absorption in the film *increases* the system reflectivity at normal incidence if $n_1 < n_2$, and *decreases* it if $n_1 > n_2$. This statement remains true at all angles of incidence for the s wave, but not for the p wave, for which the correction term in (56) changes sign at the Brewster angle arctan (n_2/n_1) (at which $Q_1 = Q_2$), and also when $\sin^2\theta_1 = (\varepsilon_1/\varepsilon_2 + \text{Im}\,\Lambda_1/\text{Im}\,\lambda_1)^{-1}$. At these two angles for the p wave there is no contribution to the reflectivity in first order in the film thickness. The first order term is also absent for absorbing films between like media ($\varepsilon_1 = \varepsilon_2$), for both polarizations and at all angles of incidence.

In all cases there is however a first order effect in the transmission: for the s wave we have from (2.15) that

$$t_s = t_{s0} + \frac{2iq_1\omega^2/c^2\lambda_1}{(q_1 + q_2)^2} + \ldots, \tag{59}$$

which gives the transmittance

$$T_s = \frac{q_2}{q_1}|t_s|^2 = \frac{4q_1q_2}{(q_1 + q_2)^2}\left\{1 - \frac{2\omega^2/c^2\,\text{Im}\,\lambda_1}{q_1 + q_2} + \ldots\right\}. \tag{60}$$

The p wave result is a little more complicated. To obtain an identity similar to (2.15) we start with two p wave equations, with dielectric functions ε and $\tilde{\varepsilon}$, and incident from media 1 and 2, respectively:

$$\frac{dC_{12}}{dz} + \left(\frac{\omega^2}{c^2} - \frac{K^2}{\varepsilon}\right) B_{12} = 0, \qquad e^{iq_1 z} - r_{12}\, e^{iq_1 z} \leftarrow B_{12} \rightarrow \tau_{12}\, e^{iq_2 z}, \tag{61}$$

$$\frac{d\tilde{C}_{21}}{dz} + \left(\frac{\omega^2}{c^2} - \frac{K^2}{\tilde{\varepsilon}}\right) \tilde{B}_{21} = 0, \qquad \tilde{\tau}_{21}\, e^{-iq_1 z} \leftarrow \tilde{B}_{21} \rightarrow e^{-iq_2 z} - \tilde{r}_{21}\, e^{iq_2 z}. \tag{62}$$

Here C stands for $dB/\varepsilon dz$, τ_{12} for $(\varepsilon_2/\varepsilon_1)^{1/2} t_{12}$. On multiplying (61) by $\tilde{B}_{21}$ and (62) by B_{12}, subtracting, and integrating from $-\infty$ to $+\infty$, we get the identity

$$2i(Q_1\tilde{\tau}_{21} - Q_2\tau_{12}) = \int_{-\infty}^{\infty} dz \left\{K^2\left(\frac{1}{\tilde{\varepsilon}} - \frac{1}{\varepsilon}\right) B_{12}\tilde{B}_{21} + (\varepsilon - \tilde{\varepsilon})C_{12}\tilde{C}_{21}\right\}. \tag{63}$$

This holds for any pair of profiles ε and $\tilde{\varepsilon}$ (with the same limiting values ε_1 and ε_2), and thus also for $\tilde{\varepsilon} = \varepsilon$, when the right-hand side is zero. Thus $Q_1\tau_{21} = Q_2\tau_{12}$ (equivalent to $q_1 t_{21} = q_2 t_{12}$, equation (2.14)), and (63) may be rewritten as

$$\tau_{12} = \tilde{\tau}_{12} - \frac{1}{2iQ_2}\int_{-\infty}^{\infty} dz \left\{K^2\left(\frac{1}{\tilde{\varepsilon}} - \frac{1}{\varepsilon}\right) B_{12}\tilde{B}_{21} + (\varepsilon - \tilde{\varepsilon})C_{12}\tilde{C}_{21}\right\}. \tag{64}$$

We now set $\tilde{\varepsilon} = \varepsilon_0$, the step function profile. To lowest order in the film thickness, it suffices to replace B and C by the values taken by B_0 and C_0 at the origin:

$$B_{12} \rightarrow \frac{2Q_1}{Q_1 + Q_2}, \quad \tilde{B}_{21} \rightarrow \frac{2Q_2}{Q_1 + Q_2}, \quad C_{12} \rightarrow \frac{2iQ_1Q_2}{Q_1 + Q_2}, \quad \tilde{C}_{21} \rightarrow -\frac{2iQ_1Q_2}{Q_1 + Q_2}. \tag{65}$$

Thus

$$\tau = \tau_0 + \frac{2iQ_1}{(Q_1 + Q_2)^2}\left[\frac{K^2}{\varepsilon_1\varepsilon_2}\Lambda_1 + Q_1Q_2\lambda_1\right] + \ldots. \tag{66}$$

The corresponding p wave transmittance is, on using $\tau_0 = 2Q_1/(Q_1 + Q_2)$,

$$\begin{aligned} T_p &= \frac{q_2}{q_1}|t_p|^2 = \frac{Q_2}{Q_1}|\tau|^2 \\ &= \frac{4Q_1Q_2}{(Q_1 + Q_2)^2}\left\{1 - \frac{2}{Q_1 + Q_2}\left[\frac{K^2}{\varepsilon_1\varepsilon_2}\operatorname{Im}\Lambda_1 + Q_1Q_2\operatorname{Im}\lambda_1\right] + \ldots\right\}. \end{aligned} \tag{67}$$

The imaginary parts of λ_1 and Λ_1 are both positive, so (60) and (67) show that, to first order in the film thickness, the transmission through a film is always decreased by absorption within the film (in contrast to the reflection, which we saw could be either decreased or increased by absorption). At normal incidence both (60) and (67) reduce to

$$T_n = \frac{4n_1n_2}{(n_1 + n_2)^2}\left\{1 - \frac{2}{n_1 + n_2}\frac{\omega}{c}\operatorname{Im}\lambda_1 + \ldots\right\}. \tag{68}$$

The conservation law $R + T = 1$ for non-dissipative media can be generalized to $R + T + A = 1$, where A represents absorption within the system, and is non-negative for passive media. From (55), (56), (60) and (67) we find the absorptance for the two polarizations:

$$A_s = \frac{4q_1}{(q_1 + q_2)^2} \frac{\omega^2}{c^2} \operatorname{Im} \lambda_1 + \ldots, \tag{69}$$

$$A_p = \frac{4Q_1}{(Q_1 + Q_2)^2} \left\{ \frac{K^2}{\varepsilon_1 \varepsilon_2} \operatorname{Im} \Lambda_1 + Q_2^2 \operatorname{Im} \lambda_1 \right\} + \ldots. \tag{70}$$

We now turn to the ellipsometric characterization of thin absorbing films. The derivation given in Section 3-4 remains valid for complex ε. To first order in the film thickness we have

$$r_{s0} \left(\frac{r_p}{r_s} \right) = r_{p0} - \frac{2iQ_1 K^2/\varepsilon_1 \varepsilon_2}{(Q_1 + Q_2)^2} \mathscr{I}_1 + \ldots, \tag{71}$$

where the integral invariant $\mathscr{I}_1$ is given by

$$\mathscr{I}_1 = \int_{-\infty}^{\infty} \mathrm{d}z \frac{(\varepsilon - \varepsilon_1)(\varepsilon_2 - \varepsilon)}{\varepsilon} = \int_{-\infty}^{\infty} \mathrm{d}z \left(\varepsilon_1 + \varepsilon_2 - \frac{\varepsilon_1 \varepsilon_2}{\varepsilon} - \varepsilon \right). \tag{72}$$

We again consider the simplest case where only the film is absorbing, with ε_1 and ε_2 real. Then, with $\varepsilon(z) = \varepsilon_r(z) + i\varepsilon_i(z)$,

$$\mathscr{I}_1 = \int_{-\infty}^{\infty} \mathrm{d}z \left(\varepsilon_1 + \varepsilon_2 - \frac{\varepsilon_1 \varepsilon_2 \varepsilon_r}{\varepsilon_r^2 + \varepsilon_i^2} - \varepsilon_r \right) + i \int_{-\infty}^{\infty} \mathrm{d}z \left(\frac{\varepsilon_1 \varepsilon_2}{\varepsilon_r^2 + \varepsilon_i^2} - 1 \right) \varepsilon_i. \tag{73}$$

Since ε_r has the limiting values ε_1 and ε_2, and ε_i is zero outside the absorbing region, both integrands go to zero at the end-points. For non-absorbing films the ellipsometric Brewster angle θ'_B, at which Re $(r_p/r_s) = 0$, differs in second order in the film thickness from $\theta_B = \arctan (\varepsilon_2/\varepsilon_1)^{1/2}$ (determined by $Q_1 = Q_2$), as we saw in Section 3-4. When the film is absorbing there is a first order correction: from (71) we find

$$\Delta\theta_B = \theta'_B - \theta_B = \frac{(\omega/c) \operatorname{Im} \mathscr{I}_1}{(\varepsilon_1 + \varepsilon_2)^{1/2} \left(\frac{\varepsilon_1}{\varepsilon_2} \right)^{1/2} \left(\frac{\varepsilon_1}{\varepsilon_2} - \frac{\varepsilon_2}{\varepsilon_1} \right)} + \ldots. \tag{74}$$

This difference between the Brewster angles is proportional to Im Λ_1 − Im λ_1, and may be large even for thin films if $\varepsilon_1 \simeq \varepsilon_2$. The case $\varepsilon_1 = \varepsilon_2$ (absorbing film between identical media) requires special consideration, since then both r_{s0} and r_{p0} are zero. The leading term in the ellipsometric ratio depends on the ratio of Λ_1 to λ_1: from (52) and (53),

$$\frac{r_p}{r_s} = \frac{r_{p1}}{r_{s1}} + \cdots = \cos^2\theta_0 - \frac{\Lambda_1}{\lambda_1} \sin^2\theta_0 + \ldots, \tag{75}$$

where θ_0 is the common angle of incidence and refraction. The zero-thickness Brewster angle is $\theta_B = \pi/4$, while the ellipsometric Brewster angle at which $\mathrm{Re}\,(r_p/r_s) = 0$ is given by

$$\cot^2\theta_B' = \mathrm{Re}\left(\frac{\Lambda_1}{\lambda_1}\right) + \cdots = \frac{\Lambda_r\lambda_r + \Lambda_i\lambda_i}{\lambda_r^2 + \lambda_i^2} + \ldots . \tag{76}$$

Here $\Lambda_1 = \Lambda_r + i\Lambda_i$ and $\lambda_1 = \lambda_r + i\lambda_i$, and the real and imaginary parts may be extracted from (54), with $\varepsilon_1 = \varepsilon_2 = \varepsilon_0$. This differs in second order in ε_i from the angle at which a non-absorbing film has zero reflection of the p wave, given by (3.59).

8-6 Attenuated total reflection; surface waves

When light is incident from a dielectric of refractive index n_1 onto another dielectric of refractive index $n_2 < n_1$ (for example from glass to air) there will be total reflection when $\theta_1 > \theta_c = \arcsin(n_2/n_1)$. This holds for both polarizations, and irrespective of whether the transition between the dielectrics is sharp or gradual, *provided there is no absorption within the interface.* When an absorbing layer (typically a metal film) is deposited between the two dielectrics, the transmission is still zero (since $q_2 = (\varepsilon_2\omega^2/c^2 - K^2)^{1/2}$ is imaginary for $\theta_1 > \theta_c$) but the p wave reflectance can be very much less than total. A sharp resonance in the absorption can appear, and this is the basis of an experimental technique for the determination of the optical constants of metal and semiconductor films. The technique is due to Otto (1968) and Kretschmann and Raether (1968). Otto originally referred to "frustrated total reflection"; it is now called *attenuated total reflection.*

Other experimental configurations are also used, for example with a low refractive index material between the high refractive index dielectric and the metal, and a symmetric high/low/complex/low/high system. In the latter case the transmittance can be high. The basic configurations are illustrated in Figure 8-4.

We will consider the left and centre configurations in this chapter. The right-hand configuration will be treated in the chapter on matrix methods. For the dielectric/absorbing layer/dielectric case on the left, we can use the results of

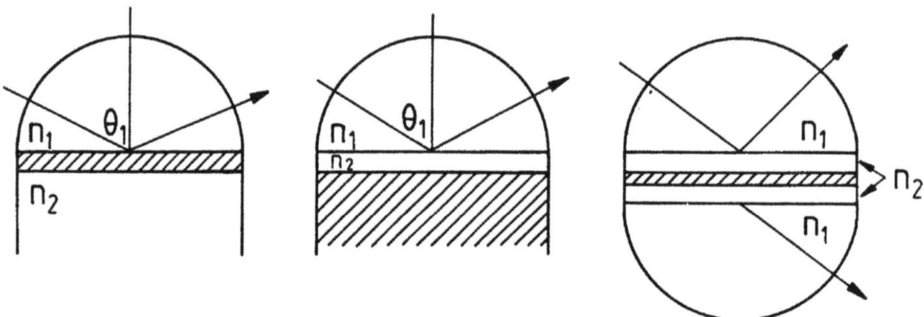

Figure 8-4. Three configurations for attenuated total reflection. The one on the right can show high transmittance as well. Prisms are often used instead of the half-cylinders illustrated. The shaded material is a conductor, with $n = n_r + in_i$.

Section 8-4, taking account of the fact that q_2 and Q_2 become imaginary for $\theta_1 > \theta_c = \arcsin(n_2/n_1)$. The s and p reflectivities have the form (48),

$$R_s, R_p = \frac{\varrho^2 + 2\varrho\varrho' \cos(\delta - \delta') + (\varrho')^2}{1 + 2\varrho\varrho' \cos(\delta + \delta') + (\varrho\varrho')^2}, \tag{77}$$

where ϱ, δ and ϱ', δ' are defined for the two polarizations by

$$\varrho_s\, \mathrm{e}^{i\delta_s} = \frac{q_1 - q}{q_1 + q}, \qquad \varrho_s'\, \mathrm{e}^{i\delta_s'} = \frac{q - q_2}{q + q_2}\, \mathrm{e}^{2iq\Delta z}, \tag{78}$$

$$\varrho_p\, \mathrm{e}^{i\delta_p} = \frac{Q_1 - Q}{Q_1 + Q}, \qquad \varrho_p'\, \mathrm{e}^{i\delta_p'} = \frac{Q - Q_2}{Q + Q_2}\, \mathrm{e}^{2iq\Delta z}. \tag{79}$$

The wavenumber normal components q and Q are complex, with real and imaginary parts given as before by (44, 45) and (21). We have

$$\varrho_s^2 = \frac{(q_1 - q_r)^2 + q_i^2}{(q_1 + q_r)^2 + q_i^2}, \qquad \varrho_p^2 = \frac{(Q_1 - Q_r)^2 + Q_i^2}{(Q_1 + Q_r)^2 + Q_i^2}. \tag{80}$$

We interpret ϱ_s and ϱ_p as the positive square roots of these expressions, and define atn (y, x) as the arctangent of y/x placed in the correct quadrant according to the signs of x and y. The corresponding phases are then

$$\begin{aligned} \delta_s &= \operatorname{atn}(-2q_1q_i,\ q_1^2 - q_r^2 - q_i^2), \\ \delta_p &= \operatorname{atn}(-2Q_1Q_i,\ Q_1^2 - Q_r^2 - Q_i^2). \end{aligned} \tag{81}$$

The primed variables take different forms depending on whether θ_1 is less or greater than θ_c. For $\theta_1 < \theta_c$, with the notation $f = \mathrm{e}^{-q_i\Delta z}$, $\phi = q_r\Delta z$,

$$\varrho_s' = \left\{\frac{(q_r - q_2)^2 + q_i^2}{(q_r + q_2)^2 + q_i^2}\right\}^{1/2} f^2, \qquad \varrho_p' = \left\{\frac{(Q_r - Q_2)^2 + Q_i^2}{(Q_r + Q_2)^2 + Q_i^2}\right\}^{1/2} f^2. \tag{82}$$

The positive square roots are again understood. The corresponding phases are

$$\begin{aligned} \delta_s' &= 2\phi + \operatorname{atn}(2q_iq_2,\ q_r^2 + q_i^2 - q_2^2), \\ \delta_p' &= 2\phi + \operatorname{atn}(2Q_iQ_2,\ Q_r^2 + Q_i^2 - Q_2^2). \end{aligned} \tag{83}$$

For $\theta_1 > \theta_c$ we set $q_2 = i|q_2|$ and $Q_2 = i|Q_2|$. Then

$$\varrho_s' = \left\{\frac{q_r^2 + (q_i - |q_2|)^2}{q_r^2 + (q_i + |q_2|)^2}\right\}^{1/2} f^2, \qquad \varrho_p' = \left\{\frac{Q_r^2 + (Q_i - |Q_2|)^2}{Q_r^2 + (Q_i + |Q_2|)^2}\right\}^{1/2} f^2, \tag{84}$$

$$\begin{aligned} \delta_s' &= 2\phi + \operatorname{atn}(-2q_r|q_2|,\ q_r^2 + q_i^2 - |q_2|^2), \\ \delta_p' &= 2\phi + \operatorname{atn}(-2Q_r|Q_2|,\ Q_r^2 + Q_i^2 - |Q_2|^2). \end{aligned} \tag{85}$$

Figure 8-5 shows the s and p reflectivities for a high refractive index glass/silver film/lithium fluoride system at $\lambda_0 = 546\,\text{nm}$. The refractive indices are $n_1 = 1.9018$, $n = 0.055 + 3.28i$, $n_2 = 1.392$ (these correspond to those used in Figure 13b of Otto, 1976).

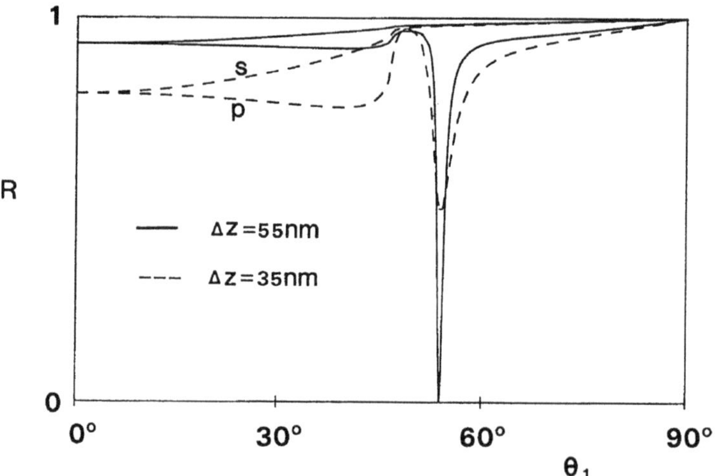

Figure 8-5. Attenuated total reflection for the left-hand configuration in Figure 8-4. The p reflectivity shows a marked minimum in the region where total reflection would occur in the absence of the metal film of thickness Δz (the critical angle is $\theta_c \simeq 47°$).

An estimate of the location of the minimum in the p reflectance can be obtained from the thin film formula (53). For $\theta_1 > \theta_c$ we have $Q_2 = i|Q_2|$, and (53) gives

$$R_p = 1 - \frac{4Q_1}{Q_1^2 + |Q_2|^2}\left[\frac{K^2}{\varepsilon_1 \varepsilon_2} \operatorname{Im} \Lambda_1 + |Q_2|^2 \operatorname{Im} \lambda_1\right] + \ldots. \tag{85}$$

For a uniform metallic film of thickness Δz this becomes (on using (57))

$$R_p = 1 - \frac{4Q_1}{Q_1^2 + |Q_2|^2}\left(\frac{K^2}{\varepsilon_r^2 + \varepsilon_i^2} + |Q_2|^2\right)\varepsilon_i \Delta z + \ldots. \tag{86}$$

This has a minimum when

$$\left(\frac{cK}{\omega}\right)^2 = \varepsilon_1 \sin^2\theta_1 = \tfrac{1}{2}\{3v - u + [(3v - u)^2 - 4uv + 8\varepsilon_1(u - v)]^{1/2}\}, \tag{87}$$

where

$$u = \frac{(\varepsilon_r^2 + \varepsilon_i^2)\varepsilon_2}{\varepsilon_r^2 + \varepsilon_i^2 + \varepsilon_2^2}, \qquad v = \frac{\varepsilon_1 \varepsilon_2}{\varepsilon_1 + \varepsilon_2}. \tag{88}$$

For the parameters of Figure 8-5 these expressions locate the minimum at about 59°. The actual minima are at smaller angles: the 55 nm metal film (for which $(\omega/c)\Delta z \simeq 0.633$) has a reflectance minimum at about 54°. One reason for the lack of precision in this estimate is that the thin film formulae contain no direct information about the complex wavenumber component $q = q_r + iq_i$ within the absorber. For metals, ε_r can be large and negative, and q_i is then much larger than q_r, making the effect of the exp $(-2q_i\Delta z)$ factor in the reflection formulae very strong. The same comments apply to (74), which is accurate for metallic films only when these are unrealistically thin.

The ellipsometric quantity r_p/r_s shows remarkable behaviour in the vicinity of strong attenuated total reflection. It is given by

$$\frac{r_p}{r_s} = -\frac{\varrho_p\,e^{i\delta_p} + \varrho_p'\,e^{i\delta_p'}}{1 + \varrho_p\varrho_p'\,e^{i(\delta_p+\delta_p')}} \cdot \frac{1 + \varrho_s\varrho_s'\,e^{i(\delta_s+\delta_s')}}{\varrho_s\,e^{i\delta_s} + \varrho_s'\,e^{i\delta_s'}}, \tag{89}$$

where the magnitudes and phases of the component amplitudes are given by (78) to (85). For thick metal layers the trajectories r_p/r_s tend to those of Figure 8-2; for very thin metal films the trajectories approach those of Figure 2-9. The layers of intermediate thickness which show strongly attenuated total reflection have a variety of trajectories between these two limiting cases. Two examples are shown in Figure 8-6: note the very rapid variation with angle in the vicinity of the reflection minimum.

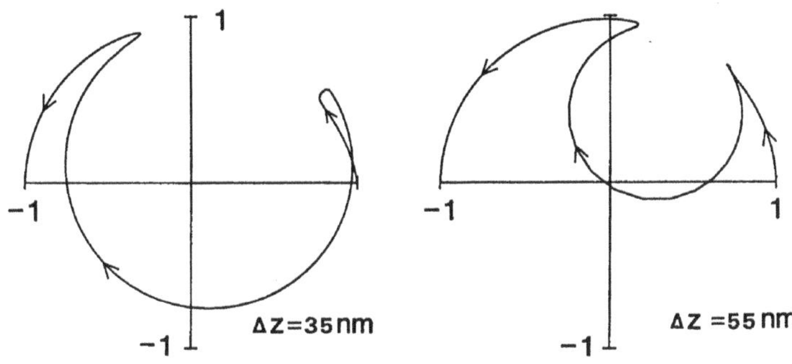

Figure 8-6. The ellipsometric ratio r_p/r_s in the complex plane, for the prism |metal layer| dielectric configuration illustrated in the left-hand diagram of Figure 8-4. The parameters used are those of Figure 8-5. The values of $(\omega/c)\Delta z$ for the two cases are 0.403 and 0.633. The 35 nm case has a single ellipsometric Brewster angle $\theta_B' \simeq 53.4°$, while the 55 nm thick film shows triple Brewster angles ($\theta_B' \simeq 53.9°, 55.1°, 63.4°$). (The possibility of multiple Brewster angles was discussed in Section 2-3.) As the thickness of silver is increased the indentation near 54° diminishes, and eventually the trajectory becomes a simple arc in the upper half plane, with $\theta_B' \simeq 65°$.

We mentioned in the discussion following (88) why the thin film formula (74) for the location of the Brewster angle might give accurate results only for unrealistically thin metal films. In fact (74) predicts a small *negative* shift $\Delta\theta_B = \theta_B' - \theta_B$ for the silver film case illustrated in Figures 8-5 and 8-6. For vanishing thickness of silver $\theta_B' = \theta_B \simeq 36°$, while for thick silver layers $\theta_B' \simeq 65°$, an *increase* of nearly 30°. This large increase is almost complete when $(\omega/c)\Delta z \simeq 1$, and swamps the small predicted decrease even for monolayers of silver.

The phenomenon of attenuated total reflection is due to the generation of electromagnetic surface waves in situations where total reflection would occur in the absence of the metal layer. Consider the simplest possible case of an idealized conductor with negative dielectric function ε (e.g., $\varepsilon = 1 - \omega_p^2/\omega^2$ with $\omega < \omega_p$), bounded by a dielectric with $\varepsilon_2 > 0$. A surface wave solution for the p polarization is possible: for an interface in the $z = 0$ xy plane, let

$$B_y(z, x, t) = \begin{cases} e^{i(Kx-\omega t)}\,e^{|q|z} & (z < 0) \\ e^{i(Kx-\omega t)}\,e^{-|q_2|z} & (z > 0). \end{cases} \tag{90}$$

This function satisfies (1.18) and its consequent boundary conditions, namely the continuity of B_y and $\partial B_y/\varepsilon\partial z$ at $z = 0$, provided

$$K^2 - |q|^2 = \varepsilon\frac{\omega^2}{c^2}, \quad K^2 - |q_2|^2 = \varepsilon_2\frac{\omega^2}{c^2}, \quad \frac{|q|}{\varepsilon} = -\frac{|q_2|}{\varepsilon_2} \tag{91}$$

(note that a surface wave solution for s polarization is not possible, since this would require $|q| = -|q_2|$). The first two equations may be regarded as the usual relation between the tangential and normal components of the wave-number, $K^2 + q^2 = \varepsilon\omega^2/c^2$, except that q and q_2 are imaginary and give exponential decay away from the surface rather than propagation in the z direction. The last relation in (91) can be satisfied if $|\varepsilon| > \varepsilon_2$, since $|q| = (K^2 + |\varepsilon|\omega^2/c^2)^{1/2} > (K^2 - \varepsilon_2\omega^2/c^2)^{1/2} = |q_2|$. On eliminating $|q|$ and $|q_2|$ from (91) we find for the wavevector component K (for propagation along the interface) the dispersion relation

$$K^2 = \frac{|\varepsilon|\varepsilon_2}{|\varepsilon| - \varepsilon_2}\frac{\omega^2}{c^2}. \tag{92}$$

The electromagnetic surface wave described here has no real normal component of its wavevector and thus cannot be coupled into by an incident plane wave. When incidence is from an optically denser medium, so as to produce total reflection in the absence of the metal film, strong coupling is possible for a special combination of angle of incidence and thickness of metal. The modulus of the magnetic field is shown within the silver layer and the lithium fluoride substrate in Figure 8-7.

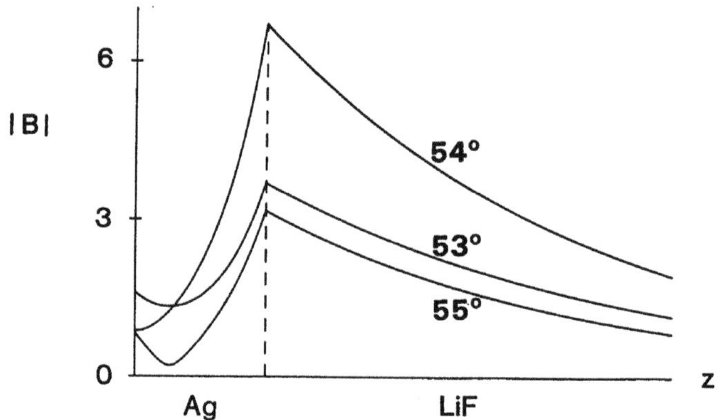

Figure 8-7. The modulus of B as a function of z, for the prism|silver|lithium fluoride configuration, with the Ag layer being 55 nm thick. Note the sensitivity to angle of incidence: in this case even a 1° shift is sufficient to decrease the peak amplitude by a factor of 2.

We see that near the angle for minimum reflectivity (here about 53.9°) the fields peak at the interface between the metal and the second dielectric, as in the idealized metal|dielectric case discussed above. The fields decay exponentially into the second dielectric, and at resonance *increase* (approximately exponentially) into the metal, as opposed to the usual decay away from the illuminated surface.

The conditions for minimum reflection, and thus maximum absorption within the metal film, can be seen from (77). For R_p to be minimum we need ϱ_p and ϱ_p' approximately equal, and $\cos(\delta_p - \delta_p')$ near -1. Since ϱ_p' contains the factor

$\exp(-2q_i\Delta z)$ it would normally be much smaller than ϱ_p, especially as q_i is large for metals. But by varying the angle it is possible to make $Q_i + |Q_2|$ zero or very small, Q_i being usually negative for metals (note that $\theta_1 > \theta_c$ is needed here). Since Q_r is small, this makes the factor multiplying f^2 in ϱ_p' large, and $\varrho_p' \simeq \varrho_p$. Both angle and metal thickness adjustment are involved in attaining approximate equality of ϱ_p and ϱ_p'. The other condition, $\cos(\delta_p - \delta_p') \simeq -1$, depends mainly on angle, since here the metal thickness enters as $q_r\Delta z$, and q_r is small.

These rough arguments can be made more precise. For example, we can ask whether a certain combination of prism|metal|dielectric can give zero reflection (total absorption) at some angle and thickness combination. This amounts to satisfying $\varrho_p = \varrho_p'$ and $\cos(\delta_p - \delta_p') = -1$ simultaneously. On eliminating Δz between these two relations, we find the condition

$$\operatorname{atn}(-2Q_1Q_i,\, Q_1^2 - Q_r^2 - Q_i^2) - \operatorname{atn}(-2Q_r|Q_2|,\, Q_r^2 + Q_i^2 - |Q_2|^2) - \frac{1}{2}\frac{q_r}{q_i}\log\left\{\frac{Q_r^2 + (Q_i - |Q_2|)^2}{Q_r^2 + (Q_i + |Q_2|)^2}\cdot\frac{(Q_1 + Q_r)^2 + Q_i^2}{(Q_1 - Q_r)^2 + Q_i^2}\right\} = (2m + 1)\pi, \tag{93}$$

where m is a positive or negative integer, or zero. For the case illustrated in Figures 8-5 to 8-7 we find that (93) is satisfied (with $m = 0$) at $\theta_0 \simeq 53.9°$, and gives the optimum thickness $\Delta z = (4q_i)^{-1}\log\{\ \}_0 \simeq 55.6\,\text{nm}$, where $\{\ \}_0$ denotes the contents of the braces in (93) evaluated at θ_0. For vapour-deposites Al at 633 nm (refractive index $1.212 + 6.924i$, as in Figure 8-3) between the same two dielectrics, (93) gives $\theta_0 \simeq 49.5°$ (again with $m = 0$), and an optimum thickness of 13 nm.

For perfectly attenuated total reflection the trajectory of r_p/r_s passes through the origin at θ_0. The trajectory shown on the right in Figure 8-6 was for a 55 nm thickness of silver, and thus not quite a perfect absorber at θ_0.

We now turn to the middle configuration in Figure 8-4, in which the prism is followed by a low-index material (or an air gap) and then by a metallic substrate. In the absence of the metal we would have exponential decay of the fields into the second dielectric for $\theta_1 > \arcsin(n_2/n_1)$, since $q_2 = i|q_2|$ is then imaginary. When the metal is present, both exponential increase and decrease are possible, these going as $\exp(\pm|q_2|z)$. Attenuated total reflection occurs when the increase dominates, producing large fields at the second dielectric|metal boundary, and thus large absorption.

The reflection amplitudes may be obtained as before. We have, for the $n_1|n_2|n_r + in_i$ configuration with boundaries at $z = z_1$ and $z + \Delta z$,

$$r_s = e^{2iq_1z_1}\frac{r + r'\,e^{2iq_2\Delta z}}{1 + rr'\,e^{2iq_2\Delta z}}, \tag{94}$$

where

$$r = \frac{q_1 - q_2}{q_1 + q_2}, \qquad r' = \frac{q_2 - q}{q_2 + q}, \tag{95}$$

and $q_1^2 = \varepsilon_1\omega^2/c^2 - K^2 = \varepsilon_1(\omega^2/c^2)\cos^2\theta_1$, $q_2^2 = \varepsilon_2\omega^2/c^2 - K^2$, $q^2 = \varepsilon\omega^2/c^2 - K^2$. The real and imaginary parts of q are given by (11) and (12) or (44) and (45). We again set $r = \varrho\, e^{i\delta}$ and $r'\, e^{2iq_2\Delta z} = \varrho'\, e^{i\delta'}$; when q_2 is imaginary ϱ' includes the exponential factor $e^{-2|q_2|\Delta z}$. The p wave reflection amplitude is

$$-r_p = e^{2iq_1z_1} \frac{r + r'\, e^{2iq_2\Delta z}}{1 + rr'\, e^{2iq_2\Delta z}}, \tag{96}$$

where now

$$r = \frac{Q_1 - Q_2}{Q_1 + Q_2}, \qquad r' = \frac{Q_2 - Q}{Q_2 + Q}, \tag{97}$$

and $Q_1 = q_1/\varepsilon_1$, $Q_2 = q_2/\varepsilon_2$, and $Q = q/\varepsilon$ with real and imaginary parts given by (21). The amplitudes ϱ, ϱ' and the phases δ and δ' are defined as for the s wave. The reflectivities then take the form (77). For $\theta_1 < \theta_c = \arctan(n_2/n_1)$ we have

$$\varrho_s = \frac{q_1 - q_2}{q_1 + q_2}, \qquad \varrho_p = \frac{Q_1 - Q_2}{Q_1 + Q_2},$$

$$\varrho_s' = \left\{\frac{(q_2 - q_r)^2 + q_i^2}{(q_2 + q_r)^2 + q_i^2}\right\}^{1/2}, \quad \delta_s' = \operatorname{atn}(-2q_2q_i, q_2^2 - q_r^2 - q_i^2) + 2q_2\Delta z,$$

$$\varrho_p' = \left\{\frac{(Q_2 - Q_r)^2 + Q_i^2}{(Q_2 + Q_r)^2 + Q_i^2}\right\}^{1/2}, \quad \delta_p' = \operatorname{atn}(-2Q_2Q_i, Q_2^2 - Q_r^2 - Q_i^2) + 2q\Delta z. \tag{97}$$

(In the $\theta_1 < \theta_c$ case it is convenient to set $\delta_p = 0$ and allow ϱ_p to carry the change of sign at the Brewster angle $\theta_B = \arctan(n_2/n_1)$, where $Q_1 = Q_2$.) For angle of incidence greater than the critical angle, $q_2 = i|q_2|$, $Q_2 = i|Q_2|$, and

$$\varrho_s = 1, \qquad \delta_s = \operatorname{atn}(-2q_1|q_2|, q_1^2 - |q_2|^2),$$

$$\varrho_p = 1, \qquad \delta_p = \operatorname{atn}(-2Q_1|Q_2|, Q_1^2 - |Q_2|^2), \tag{98}$$

$$\varrho_s' = \left\{\frac{q_r^2 + (|q_2| - q_i)^2}{q_r^2 + (|q_2| + q_i)^2}\right\}^{1/2} e^{-2|q_2|\Delta z}, \quad \delta_s' = \operatorname{atn}(2|q_2|q_r, |q_2|^2 - q_r^2 - q_i^2),$$

$$\varrho_p' = \left\{\frac{Q_r^2 + (|Q_2| - Q_i)^2}{Q_r^2 + (|Q_2| + Q_i)^2}\right\}^{1/2} e^{-2|q_2|\Delta z}, \quad \delta_p' = \operatorname{atn}(2|Q_2|Q_r, |Q_2|^2 - Q_i^2 - Q_r^2).$$

The p wave reflectivity can be zero if $\varrho_p = \varrho_p'$ and $\cos(\delta_p - \delta_p') = -1$ are satisfied simultaneously. The angle θ_0 at which this can happen is (for $\theta_1 > \theta_c$) found from

$$\operatorname{atn}(-2Q_1|Q_2|, Q_1^2 - |Q_2|^2) - \operatorname{atn}(2|Q_2|Q_r, |Q_2|^2 - Q_i^2 - Q_r^2) = (2m + 1)\pi. \tag{99}$$

The thickness of the second dielectric which gives perfectly attenuated total reflection (that is, total absorption) is given by

$$\Delta z = \frac{1}{4|q_2|} \log \left\{\frac{Q_r^2 + (|Q_2| - Q_i)^2}{Q_r^2 + (|Q_2| + Q_i)^2}\right\}, \tag{100}$$

evaluated at θ_0.

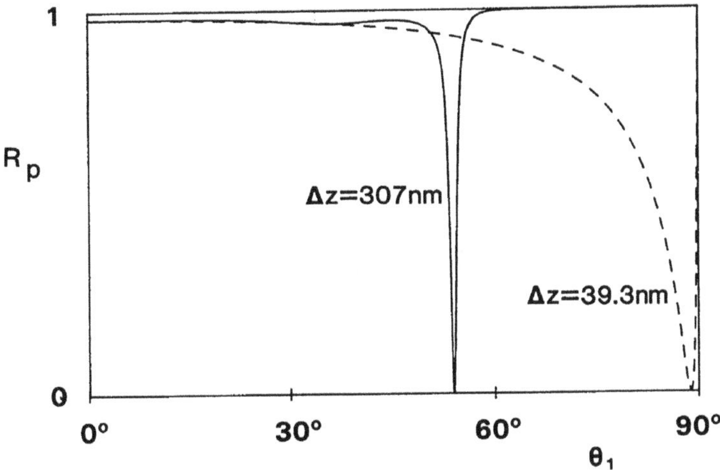

Figure 8-8. The p reflectivity as a function of angle of incidence, for the center configuration in Figure 8-4, for the two thicknesses of the second dielectric (LiF) which give total absorption of the p wave at one angle. The thicknesses and angles are 307 nm, 54° and 39.3 nm, 89° (at $\lambda_0 = 546$ nm).

For the high refractive index prism/lithium fluoride/silver system at a wavelength of 546 nm, with the refractive indices 1.9018, 1.392, $0.055 + 3.28i$, (99) gives $\theta_0 \simeq 54°$ (with $m = -1$) and (100) gives $\Delta z \simeq 307$ nm. There is also zero reflection of the p wave near grazing incidence, with $\theta_0 \simeq 89°$ (again $m = -1$) and $\Delta z \simeq 39.3$ nm: compare the discussion of reflection polarizers consisting of a dielectric layer on a metal substrate in Section 8-3. The reflectivities are given by (77) and r_p/r_s by (89). Figure 8-8 shows R_p for the above combination, at the two optimum thicknesses of the second dielectric, which correspond to $(\omega/c)\Delta z = 3.53$ and 0.452. Figure 8-9 shows the corresponding r_p/r_s curves.

There is wide variety in both the reflectance and the ellipsometric curves as the thickness of the second dielectric varies. For thin layers of the dielectric the curves tend to those of Section 8-1.

The phenomenon of attenuated total reflection has been treated here purely by classical electrodynamics. It was seen to be an interference-attenuation effect,

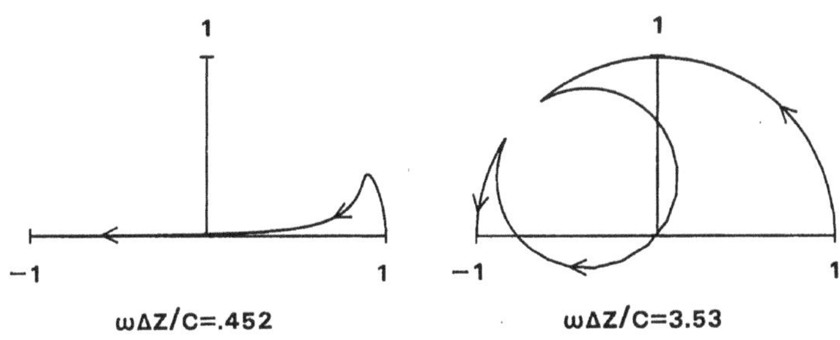

Figure 8-9. The ratio r_p/r_s is the complex plane, for the two dielectric thicknesses which give perfectly attenuated reflection at one angle. The thicknesses are 39.3 nm and 307 nm for the left and right diagrams; the corresponding curves pass through the origin at about 89° and 54° respectively. The diagram on the right shows the triple Brewster angle phenomenon, as in Figure 8-6.

linked to the excitation of electromagnetic surface waves. These are coupled into by means of the exponential decay or growth in the second dielectric which is possible for $\theta_1 > \theta_c$. In radio physics the electromagnetic surface waves sometimes go by the names of Zenneck or Sommerfeld–Zenneck (Barlow and Brown 1961), or ground waves (Budden 1985). In solid state physics the terms surface polariton, surface plasmon, or sometimes surface polariton-plasmon or phonon-polariton are used. There the phenomenon of attenuated total reflection has many applications: for example the determination of the optical constants of metals and semiconductors (Otto 1976), and the study of adsorbates (McIntyre 1976). The literature on the solid state aspects of surface wave phenomena is very large; see for example the collections of papers edited by Burstein and DeMartini (1974), Seraphin (1976), Boardman (1982), and Agranovich and Mills (1982).

8-7 Reflection by a diffuse absorbing interface: the tanh profile

In Section 2-5 we considered the reflection properties of the hyperbolic tangent profile

$$\varepsilon(z) = \tfrac{1}{2}(\varepsilon_1 + \varepsilon_2) - \tfrac{1}{2}(\varepsilon_1 - \varepsilon_2)\tanh z/2a = \frac{\varepsilon_1 + \varepsilon_2\, e^{z/a}}{1 + e^{z/a}}. \tag{101}$$

The s wave reflectivity was shown to be

$$R_s = \left\{\frac{\sinh \pi a(q_1 - q_2)}{\sinh \pi a(q_1 + q_2)}\right\}^2 \tag{102}$$

for real q_2, and unity for imaginary q_2 ($q_2 = i|q_2|$ for $\theta_1 > \theta_c = \arcsin(\varepsilon_2/\varepsilon_1)^{1/2}$). Here we shall discuss reflection by the profile (101) when ε_2 is complex, $\varepsilon_2 = \varepsilon_r + i\varepsilon_i$. The particular example we have in mind is reflection by an ionospheric layer in which the electron density approximately takes the functional form (101). The electron gas dielectric function (1) has the real and imaginary parts

$$\varepsilon_r = 1 - \frac{\omega_p^2}{\omega^2 + 1/\tau^2}, \qquad \varepsilon_i = \frac{\omega_p^2/\omega\tau}{\omega^2 + 1/\tau^2}. \tag{103}$$

Since ω_p^2 is proportional to the electron density, both ε_r and ε_i take the form (101), and so does $\varepsilon(z)$ with $\varepsilon_2 = \varepsilon_r + i\varepsilon_i$, if the variation of τ through the inhomogeneity can be neglected.

The theory leading to r_s as given by (2.84) remains valid when ε_2 is complex, with q_2 being replaced by $q_r + iq_i$. The reflection amplitude in the absorbing case is thus given by (with $y_1 = q_1 a$ as before and $q_2 a = y_r + iy_i$)

$$r_s = -\frac{\Gamma(2iy_1)\Gamma(y_i - i(y_1 + y_r))\Gamma(-y_i - i(y_1 - y_r))\sinh \pi(y_1 - y_r - iy_i)}{\Gamma(-2iy_1)\Gamma(-y_i + i(y_1 + y_r))\Gamma(y_i + i(y_1 - y_r))\sinh \pi(y_1 + y_r + iy_i)}. \tag{104}$$

From (2.86) the ratio $\Gamma(2iy_1)/\Gamma(-2iy_1)$ has modulus unity, and so

$$R_s = |r_s|^2 = G\frac{\sinh^2\pi(y_1 - y_r) + \sin^2\pi y_i}{\sinh^2\pi(y_1 + y_r) + \sin^2\pi y_i}, \tag{105}$$

where G is the modulus squared of the gamma function ratios in (104):

$$G = \left|\frac{\Gamma(y_i - i(y_1 + y_r))\Gamma(-y_i - i(y_1 - y_r))}{\Gamma(-y_i + i(y_1 + y_r))\Gamma(y_i + i(y_1 - y_r))}\right|^2. \tag{106}$$

To evaluate G we consider the ratio $\Gamma(-z)/\Gamma(z)$. From the infinite product representation (2.85) we have

$$\frac{\Gamma(-z)}{\Gamma(z)} = -e^{2\gamma z}\prod_{n=1}^{\infty}\left(\frac{n+z}{n-z}\right)e^{-2z/n}, \tag{107}$$

and thus, with $z = x + iy$,

$$\left|\frac{\Gamma(-z)}{\Gamma(z)}\right|^2 = e^{4\gamma x}\prod_{n=1}^{\infty}\frac{(n+x)^2 + y^2}{(n-x)^2 + y^2}e^{-4x/n}. \tag{108}$$

Since both $\Gamma(-z)/\Gamma(z)$ ratios in G have the same real part of z, the exponential factors cancel and

$$G(y_1, y_r, y_i) = \prod_{n=1}^{\infty}\frac{(n - y_i)^2 + (y_1 + y_r)^2}{(n + y_i)^2 + (y_1 + y_r)^2}\cdot\frac{(n + y_i)^2 + (y_1 - y_r)^2}{(n - y_i)^2 + (y_1 - y_r)^2}. \tag{109}$$

Since q_1, q_r and q_i are all non negative, G is always greater than unity in the presence of absorption. G tends to 1 as $\varepsilon_i \to 0$, and also at grazing incidence where q_1 and y_1 tend to zero.

Figure 8-10 shows the s wave reflectivity for a tanh profile with $\varepsilon_r = 0.25$ and $\varepsilon_i = 0.001$, corresponding roughly to a frequency a bit above ($2/\sqrt{3}$ larger than) the maximum plasma frequency, with absorption typical of the ionospheric E layer. In the absence of absorption there would be total reflection for angle of incidence

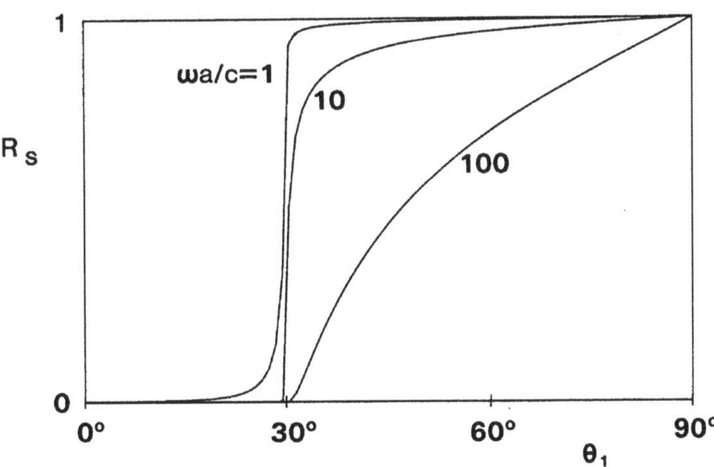

Figure 8-10. The s wave reflectivity as a function of the angle of incidence for the tanh profile with fixed absorption and varying thickness a ($\varepsilon_1 = 1$, $\varepsilon_r = 0.25$, $\varepsilon_i = 0.001$, $(\omega/c)a = 1$, 10 and 100).

greater than arcsin (1/2) = 30°, for any layer thickness. Electron collisions decrease the reflectivity, the decrease being greater for greater thickness of the transition, there being more penetration into the absorbing region.

An interesting phenomenon appears in the reflection at a gradual transition to negative ε_r. In the absence of absorption both s and p waves would be totally reflected. The turning point (where $q^2(z) = 0$) is given by $\varepsilon(z) = \varepsilon_1 \sin^2\theta_1$, and the s wave shows exponential decay beyond this point. The p wave however has a singularity at $\varepsilon(z) = 0$ (arising from the $\varepsilon^{-1}\,dB/dz$ term in the equation satisfied by B), and this leads to a logarithmic singularity in B, and infinities in E_x and E_z. Absorption removes the infinities, but the fields at the point where $\varepsilon_r(z) = 0$ can still be large. This problem is discussed by Landau and Lifshitz (1960, Section 68) for real ε; the effect of absorption is considered by Ginzburg (1964, Section 20) and Budden (1985, Section 15.6) who also give references to earlier work.

References

References quoted in the text

M. Born and E. Wolf (1970) "Principles of optics", 4th ed, Pergamon.

C. Kittel (1966) "Introduction to solid state physics", Wiley.

H. G. Booker (1984) "Cold plasma waves", Martinus Nijhoff.

K. G. Budden (1985) "The propagation of radio waves", Cambridge.

L. D. Landau and E. M. Lifshitz (1960) "Electrodynamics of continuous media", Pergamon.

T. H. Allen (1976) "Study of Al with a combined Auger electron spectrometer-ellipsometric system", J. Vac. Sci. Technol. **13**, 112–115.

F. Abelès (1950) "Recherches sur la propagation des ondes électromagnétiques sinusoidales dans les milieux stratifiés. Application aux couches minces", Annales de Physique **5**, 596–640.

M. Ruiz-Urbieta and E. M. Sparrow (1972) "Reflection polarization by a transparent-film-absorbing substrate system", J. Opt. Soc. Amer. **62**, 1188–1194.

J. M. Bennett and H. E. Bennett (1978) "Polarization"; Chapter 10 in the Handbook of Optics (W. G. Driscoll and W. Vaughan, eds.), McGraw Hill.

R. M. A. Azzam (1985) "Explicit equations for the polarizing angles of a high-reflectance substrate coated by a transparent thin film", J. Opt. Soc. Amer. **A2**, 480–482.

H. Kitajima, K. Fujita and H. Cizmic (1984) "Zero reflection from a dielectric film on a metal substrate at oblique angles of incidence", Applied Optics **23**, 1937–1939.

A. Otto (1968) "Excitation of nonradiative surface plasma waves in silver by the method of frustrated total reflection", Zeit. fur Physik **216**, 398–410.

E. Kretschmann and H. Raether (1968) "Radiative decay of non radiative surface plasmons by light", Zeit. für Naturforschung, **23a**, 2135–2136.

A. Otto (1976), "Spectroscopy of surface polaritons by attenuated total reflection" (Chapter 13 in the volume edited by Seraphin, listed below).

H. M. Barlow and J. Brown (1962) "Radio surface waves", Oxford.

J. D. E. McIntyre (1976) "Optical reflection spectroscopy of chemisorbed monolayers" (Chapter 11 in the volume edited by Seraphin, listed below.)

E. Burnstein and F. DeMartini (eds.) (1974) "Polaritons", Pergamon.

B. O. Seraphin (ed.) (1976) "Optical properties of solids: new developments", North Holland.

A. D. Boardman (ed.) (1982) "Electromagnetic surface modes", Wiley.

V. M. Agranovich and D. L. Mills (eds) (1982) "Surface polaritons: electromagnetic waves at surfaces and interfaces", North-Holland; see in particular F. Abelès and T. Lopez-Rios on "Surface polaritons at metal surfaces and interfaces".

V. L. Ginzburg (1964) "The propagation of electromagnetic waves in plasmas", Pergamon.

9

Inverse problems

The *direct* problem in reflection is the calculation of the reflection amplitudes (and thence the reflectivities and the ellipsometric ratio), given the characteristics of the reflecting profile. *Inverse* or *inversion* problems consist in the estimation of the profile characteristics, given some experimental reflection or ellipsometric data. In general, the wider the range of the experimental data (in polarization, angle, and frequency), the more can be said about the reflecting profile. But the information is never complete, and if sparse, can be ambiguous. For example: suppose we measure the s or p reflectivity from a uniform layer between two other uniform media, all three refractive indices being known. What can be said about the thickness of this layer? Only that it has one of an infinity of possible values, since the reflectivity is periodic in the thickness (see (2.66) and Figure 2.6). One measurement does not guarantee the evaluation of one parameter, even if it is the only unknown in the model. There can also be measurements which give no profile information whatever, serving only to verify experimental accuracy or calibrate the apparatus. An example is the reflectivity at grazing incidence, this being unity for either polarization, for arbitrary profiles with or without absorption (Section 2-3).

More surprising is the fact that null reflectivity at any given angle of incidence, from an interface between media of given dielectric constants ε_1 and ε_2, can be produced in a non-denumerable infinity of ways. The prescription is to pick any function $\varepsilon(z)$ which takes the values ε_1 at $z = -\infty$ and ε_2 at $z = +\infty$, and form $q^2(z) = \varepsilon(z)\omega^2/c^2 - K^2$ (K has the usual meaning, being given by $\varepsilon_1^{1/2}(\omega/c)\sin\theta_1$). Then the profile

$$\varepsilon_s(z) = \varepsilon(z) + \left(\frac{c}{\omega}\right)^2 \left\{\frac{q''}{2q} - \frac{3}{4}\left(\frac{q'}{q}\right)^2\right\} \tag{1}$$

will give zero reflection for the s wave. This was noted by Kofink (1947); the result follows from the fact that the Liouville–Green approximations to the s wave-functions both satisfy (6.26), and that ψ_1^+ tends to $e^{iq_1 z}$ as $z \to -\infty$, thus having zero component of the reflected wave (see Section 6-2). The analogous result for the p polarization is obtained from the equation satisfied by $q_b^{-1/2}\, e^{i\phi_b}$, $\phi_b' = q_b$, where q_b is given by (6.86).

After these cautionary notes we will examine some inverse problems relating to reflection, beginning with restricted and thus simple examples, and progressing to more general results. References to related inverse problems in other fields are given at the end of the chapter.

9-1 Reflection at a sharp boundary

The case of a sharp boundary between two uniform nonabsorbing media is straightforward: we have

$$R_s = \left(\frac{q_1 - q_2}{q_1 + q_2}\right)^2, \qquad R_p = \left(\frac{Q_1 - Q_2}{Q_1 + Q_2}\right)^2, \tag{2}$$

and thus

$$\frac{q_2}{q_1} = \frac{1 \pm R_s^{1/2}}{1 \mp R_s^{1/2}}, \qquad \frac{Q_2}{Q_1} = \frac{\varepsilon_1}{\varepsilon_2}\frac{q_2}{q_1} = \frac{1 \pm R_p^{1/2}}{1 \mp R_p^{1/2}}, \tag{3}$$

with the upper signs to be taken for $q_2 > q_1$ and $Q_2 > Q_1$, respectively. From the wavenumber ratios one can extract the dielectric constant ratio $\varepsilon_2/\varepsilon_1$ via $q_1^2 = \varepsilon_1(\omega^2/c^2) - K^2 = (\omega^2/c^2)\varepsilon_1 \cos^2\theta_1$, $q_2^2 = \varepsilon_2(\omega^2/c^2) - K^2 = (\omega^2/c^2)(\varepsilon_2 - \varepsilon_1 \sin^2\theta_1)$. This gives, for example,

$$\frac{\varepsilon_2}{\varepsilon_1} = \sin^2\theta_1 + \cos^2\theta_1 \left(\frac{1 \pm R_s^{1/2}}{1 \mp R_s^{1/2}}\right)^2. \tag{4}$$

The ellipsometric ratio r_p/r_s moves on the real axis from $+1$ at normal incidence to -1 at grazing incidence, passing through the origin at $\theta_B = \arctan\,(\varepsilon_2/\varepsilon_1)^{1/2}$.

When the second medium is absorbing, $\varepsilon_2 = \varepsilon_r + i\varepsilon_i$, the wavevector normal components q_2 and Q_2 are also complex, and

$$R_s = \frac{(q_1 - q_r)^2 + q_i^2}{(q_1 + q_r)^2 + q_i^2}, \qquad R_p = \frac{(Q_1 - Q_r)^2 + Q_i^2}{(Q_1 + Q_r)^2 + Q_i^2}, \tag{5}$$

with q_r, q_i, Q_r, Q_i, being given by (8.11), (8.12) and (8.21). There is no known analytic inversion of (5) to obtain ε_r and ε_i at an arbitrary angle of incidence, but Potter (1969) has developed an inversion procedure based on the values of R_p/R_s and θ_1 at the minimum of R_p/R_s, this angle of incidence being known as the pseudo-Brewster angle. A simple explicit result for the real and imaginary parts of ε_2 in terms of the ellipsometric ratio, variously written as

$$\frac{r_p}{r_s} = \varrho = \tan\psi\, e^{i\Delta}, \tag{6}$$

is possible (see for example Vašíček 1960 or Aspnes 1976). We have

$$\varrho = -\frac{Q_1 - Q_2}{Q_1 + Q_2}\cdot\frac{q_1 + q_2}{q_1 - q_2} = \frac{q_1 q_2 - K^2}{q_1 q_2 + K^2}, \tag{7}$$

having used $Q_1 = q_1/\varepsilon_1$, $Q_2 = q_2/\varepsilon_2$ and

$$\varepsilon_2 q_1^2 - \varepsilon_1 q_2^2 = (\varepsilon_1 - \varepsilon_2)K^2. \tag{8}$$

Thus

$$\frac{1 + \varrho}{1 - \varrho} = \frac{q_1 q_2}{K^2}, \qquad \left(\frac{1 + \varrho}{1 - \varrho}\right)^2 = \frac{\varepsilon_2/\varepsilon_1 - \sin^2\theta_1}{\sin^2\theta_1 \tan^2\theta_1}, \tag{9}$$

and therefore

$$\frac{\varepsilon_2}{\varepsilon_1} = \sin^2\theta_1 + \sin^2\theta_1 \tan^2\theta_1 \left(\frac{1 + \varrho}{1 - \varrho}\right)^2. \tag{10}$$

This equation gives the real and imaginary parts of ε_2 in terms of the real and imaginary parts of $(1 + \varrho)^2/(1 - \varrho)^2$. If we write $\varrho = \varrho_r + i\varrho_i$, then

$$\frac{\varepsilon_r}{\varepsilon_1} = \sin^2\theta_1 + \sin^2\theta_1 \tan^2\theta_1 \frac{(1 - \varrho_r^2)^2 - 4\varrho_i^2}{[(1 - \varrho_r)^2 + \varrho_i^2]^2}, \tag{11}$$

$$\frac{\varepsilon_i}{\varepsilon_1} = \sin^2\theta_1 \tan^2\theta_1 \frac{4(1 - \varrho_r^2)\varrho_i}{[(1 - \varrho_r)^2 + \varrho_i^2]^2}. \tag{12}$$

In terms of the ellipsometric angles ψ and Δ, these formulae read

$$\frac{\varepsilon_r}{\varepsilon_1} = \sin^2\theta_1 + \sin^2\theta_1 \tan^2\theta_1 \frac{\cos^2 2\psi - \sin^2 2\psi \sin^2\Delta}{(1 - \sin 2\psi \cos\Delta)^2}, \tag{13}$$

$$\frac{\varepsilon_i}{\varepsilon_1} = \sin^2\theta_1 \tan^2\theta_1 \frac{\sin 4\psi \sin\Delta}{(1 - \sin 2\psi \cos\Delta)^2}. \tag{14}$$

The angle representation is ambiguous without the specification of the range of one of them; we take $0 \leqslant \psi \leqslant \pi/2$, in which case $\tan\psi = |\varrho|$. From (6) and (7) we have

$$\varrho_r = \tan\psi \cos\Delta = \frac{q_1^2(q_r^2 + q_i^2) - K^4}{(q_1 q_r + K^2)^2 + q_1^2 q_i^2}, \tag{15}$$

$$\varrho_i = \tan\psi \sin\Delta = \frac{2 q_1 q_i K^2}{(q_1 q_r + K^2)^2 + q_1^2 q_i^2}. \tag{16}$$

We showed in Section 8-1 that r_p/r_s always lies within the upper semicircle of unit radius (for a sharp boundary between a dielectric and an absorbing medium). Thus $\psi \leqslant \pi/4$; the value $\pi/4$ is attained at normal incidence and at grazing incidence. The angle Δ increases from 0 at normal incidence to π at grazing incidence. According to Aspnes (1976), the currently attainable precision in ψ and Δ is about $\delta\psi \simeq \delta\Delta/2 \simeq 1$ millidegree. Figure 9-1 shows an example of the uncertainty in ε_r and ε_i as a function of the angle at which measurement is carried out, assuming the much larger random scatter of up to 0.1° in ψ and 0.2° in Δ, at all angles of incidence. We see that the accuracy in ε_r and ε_i is best near the ellipsometric Brewster angle θ_B' (also known as the principal angle), at which $\varrho_r = 0$ and $\Delta = \pi/2$.

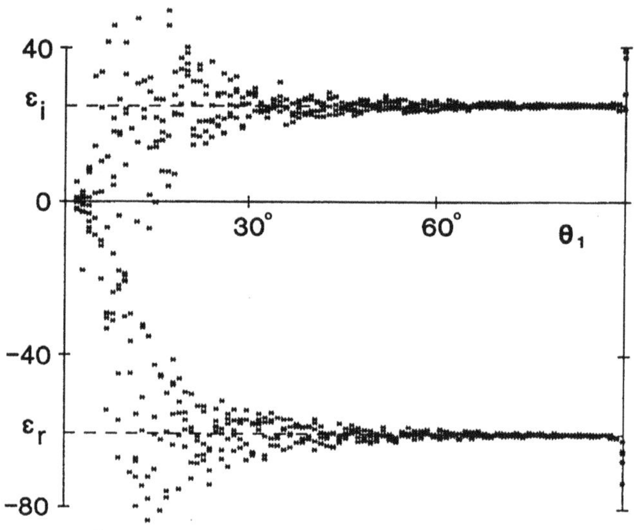

Figure 9-1. Scatter in ε_r and ε_i values deduced from (13) and (14), on the assumption of uniformly distributed random errors of up to 0.1° in ψ and 0.2° in Δ. The "true" values of ψ and Δ are calculated from (15) and (16), using the bulk Al parameters at 633 nm, $\varepsilon_r = -60.56$, $\varepsilon_i = 24.86$ (as in Figures 8-1 and 8-2), for which $\theta'_B \simeq 83°$.

The above ellipsometric extraction of ε_r and ε_i, carried out over a range of frequencies, gives $\varepsilon_r(\omega)$ and $\varepsilon_i(\omega)$, or $n_r(\omega)$ and $n_i(\omega)$. Another method is to measure the reflectivity at normal incidence (given in terms of n_r and n_i by (8.17)). This determines the modulus $|r(\omega)|$ of the reflection amplitude $r = |r|\, e^{i\delta}$. The phase $\delta(\omega)$ is found from a Kramers–Krönig relation between log $|r|$ and δ (extrapolation of measured reflectivity data is required), and finally $n_r(\omega)$ and $n_i(\omega)$ are found from $r = (n_1 - n_r - in_i)/(n_1 + n_r + in_i)$. Details of this procedure are given by Wooten (1972; Chapter 6 and Appendix G).

9-2 Uniform film between like media

An explicit inversion of ellipsometric data for a uniform absorbing film has not been found, except in the special case where the media bounding the film have the same dielectric constant. The solution for this case is due to Azzam (1983); both reflection and transmission ellipsometric coefficients are required. Consider a uniform film of thickness Δz and dielectric constant $\varepsilon_r + i\varepsilon_i$, in a medium of dielectric constant ε_0. From (2.58), (2.59) and (2.70), (2.71) we have

$$\varrho = \frac{r_p}{r_s} = \frac{p}{s} \cdot \frac{1 - s^2\, e^{2iq\Delta z}}{1 - p^2\, e^{2iq\Delta z}}. \tag{17}$$

$$\tau = \frac{t_p}{t_s} = \frac{1 - p^2}{1 - s^2} \cdot \frac{1 - s^2\, e^{2iq\Delta z}}{1 - p^2\, e^{2iq\Delta z}}, \tag{18}$$

where

$$s = \frac{q_0 - q}{q_0 + q}, \qquad -p = \frac{Q_0 - Q}{Q_0 + Q} \tag{19}$$

are the s and p reflection amplitudes at a step between media with dielectric constants ε_0 and ε, q_0 and q being the corresponding real and complex normal components of the wavevector. From (17) and (18) we see that the ratio of the transmission and reflection ellipsometric ratios is independent of the thickness of the layer:

$$\frac{\tau}{\varrho} = \frac{s}{p}\frac{1 - p^2}{1 - s^2}. \tag{20}$$

This equation is to be solved for $\varepsilon = \varepsilon_r + i\varepsilon_i$. The identity (Azzam 1979)

$$p = \frac{s(\cos 2\theta - s)}{1 - s\cos 2\theta}, \tag{21}$$

(which may be verified by solving (21) for $\cos 2\theta$ and using (19) and $q^2 = (\omega/c)^2(\varepsilon - \varepsilon_0 \sin^2\theta)$) serves to eliminate p from (20), which reduces to a quadratic for s in terms of the measured θ and τ/ϱ:

$$s^2 - 2\sigma s + 1 = 0, \qquad \sigma = \frac{\cos 2\theta + (\tau/2\varrho)(1 + \cos^2 2\theta)}{1 + (\tau/\varrho)\cos 2\theta}. \tag{22}$$

This has the solutions

$$s_\pm = \sigma \pm (\sigma^2 - 1)^{1/2}, \tag{23}$$

of which one needs to select the root with $|s| < 1$. That one and only one such root exists can be seen by writing $\sigma = \cosh \zeta$: then $s_\pm = e^{\pm\zeta}$. In general ζ is complex, and one takes s_+ or s_- according as $\mathrm{Re}\,\zeta < 0$ or $\mathrm{Re}\,\zeta > 0$. (In the special case where $\mathrm{Re}\,\zeta = 0$ and σ is equal to $\cos(\mathrm{Im}\,\zeta)$, both roots would have unit modulus. But for the absorbing media $|s|^2$ cannot be unity except at grazing incidence, where $q_0 \to 0$.) Having obtained the complex value of s, the dielectric function may be found from

$$\frac{\varepsilon}{\varepsilon_0} = \sin^2\theta + \cos^2\theta\left(\frac{1 - s}{1 + s}\right)^2 \tag{24}$$

(this relation is obtained by squaring $(1 - s)/(1 + s) = q/q_0$; compare (4)). It then remains to evaluate the thickness Δz. Since s and p (from the Azzam identity) are now known, $e^{2iq\Delta z}$ may be found from ϱ or τ as given by (17) and (18):

$$e^{2iq\Delta z} = \frac{1}{sp}\cdot\frac{p - \varrho s}{s - \varrho p}, \tag{25}$$

$$e^{2iq\Delta z} = \frac{1 - p^2 - \tau(1 - s^2)}{s^2(1 - p^2) - \tau p^2(1 - s^2)}. \tag{26}$$

If the right side of either of these is written as $e^{2i(\alpha + i\beta)}$, then

$$q_r\Delta z = \alpha + m\pi, \qquad q_i\Delta z = \beta, \tag{27}$$

where m is some positive or negative integer, and q_r and q_i have been found from (8.11) and (8.12). Only one of (25) and (26) and only the second relation in (27) need be used to obtain Δz; the others provide a check on the accuracy. Azzam

(1983) gives an example of the application of this technique to the determination of the thickness and optical parameters of a thin gold foil.

9-3 Synthesis of a profile from r as a function of wavenumber

A general solution to the inverse reflection problem (or inverse scattering problem in quantum mechanics) has been found by Gelfand and Levitan (1951/1955) and others. More references are given at the end of this chapter; here we shall give only a brief description of the theory, discuss some results which follow from it, and give an approximate but explicit solution which is simple enough to have practical application.

The inversion procedure assumes the knowledge of the reflection amplitude as a function of wavenumber (the latter ranging from zero to infinity) and in addition, coefficients relating to any bound states that may exist. Since experiments generally give $|r|^2$, not r, and that over a finite range of wavenumbers, we prefer the term *synthesis* to *inversion* in this case: a model reflection amplitude, complete with phase, can be constructed to have some desired properties, such as high reflectance in one wavenumber region and low reflectance in another region. The theory then gives a procedure for synthesising the refractive index profile which will give the desired reflectance. In the general theory, one constructs an integral equation from the Fourier transform of the reflection amplitude (analytically continued to negative wavenumbers). The solution of the integral equation then gives the refractive index profile. In special cases an explicit solution can be found, for example when the reflection amplitude is a rational function of the wavenumber (Kay 1960, Jordan 1980). Another special case is the construction of an infinite set of profiles which are transparent to the s wave, at fixed frequency but *for any angle of incidence* (Kay and Moses, 1956). The simplest of these is the sech^2 profile discussed in Section 4-3, for certain special values of its parameters. We saw there that for profile

$$\varepsilon(z) = \varepsilon_0 + \Delta\varepsilon \operatorname{sech}^2(z/a), \tag{28}$$

the s reflection amplitude was a function of two dimensionless parameters $\alpha = \Delta\varepsilon(\omega a/c)^2$ and $\beta = \varepsilon_0^{1/2}(\omega a/c)\cos\theta$, θ being the angle of incidence. When $\alpha \geqslant -1/4$,

$$R_s = \frac{\cos^2\left[\frac{\pi}{2}(1+4\alpha)^{1/2}\right]}{\cos^2\left[\frac{\pi}{2}(1+4\alpha)^{1/2}\right] + \sinh^2\pi\beta}. \tag{29}$$

This is zero when $\alpha = m(m+1)$, m an integer, for any θ, the angle of incidence appearing only through β.

We now give an approximate solution of a synthesis problem, due to Hirsch (1979). Only the essence of the method will be given, since we then show how a more general result can be obtained in a simpler way. The problem is that of constructing a refractive index profile $n(z) = \sqrt{\varepsilon(z)}$ so as to give a desired reflection amplitude at normal incidence, $r(k_1)$. Here the wave is incident from a medium of unit refractive index, and transmitted into a medium of refractive index n_2. The respective wavenumbers are $k_1 = \omega/c$ and $k_2 = n_2(\omega/c)$. The E field satisfies

$$\frac{d^2 E}{dz^2} + n^2 k_1^2 E = 0, \qquad e^{ik_1 z} + r\, e^{-ik_1 z} \leftarrow E \rightarrow t\, e^{ik_2 z}. \tag{30}$$

The geometric path increment dz is replaced by the optical path increment $dx = n dz$. Also E is replaced by the function $w = n^{1/2}E$. The resulting equation is

$$w'' + [k_1^2 - U(x)]w = 0, \qquad U(x) = \frac{1}{2}\frac{n''}{n} - \frac{1}{4}\left(\frac{n'}{n}\right)^2. \tag{31}$$

(Throughout this section primes will denote differentiation with respect to x.) Since E and w are proportional as z and x tend to $-\infty$, and if further the refractive index is taken to be equal to 1 for $z \leqslant 0$ and x is defined by

$$x = \int_0^z d\zeta\, n(\zeta), \tag{32}$$

then x and z are equal for $z \leqslant 0$ and $r(k_1)$ is the reflection amplitude for $w(x)$ as well as for $E(z)$. (What we have just done is ensure that the phase of r is the same for both.) From $r(k_1)$ and its analytic continuation to negative k_1 via $r(-k_1) = r^*(k_1)$, we form the Fourier transform

$$F(x) = \frac{1}{2\pi}\int_{-\infty}^{\infty} dk_1\, r(k_1)\, e^{-ik_1 x}. \tag{33}$$

The first two terms of $U(x)$ in a series expansion formally equivalent to the Gelfand–Levitan equation are (Moses 1956)

$$U_a(x) = -2\frac{d}{dx}F(2x) + 4F^2(2x). \tag{34}$$

The second relation in (31) may be written as

$$(n^{1/2})'' = n^{1/2}U. \tag{35}$$

When U is approximated by U_a, this differential equation for $n^{1/2}(x)$ may be integrated; the solution incorporating the boundary condition $n \rightarrow 1$ at $-\infty$ and $n \rightarrow$ finite constant at $+\infty$ is then

$$n_a(x) = \exp\left\{-2\int_{-\infty}^{2x} dy\, F(y)\right\}. \tag{36}$$

We note that $F(x)$ is real, and also that since r is the inverse Fourier transform of F,

$$r(k_1) = \int_{-\infty}^{\infty} dx\, F(x)\, e^{ik_1 x}, \tag{37}$$

the final value of the refractive index is approximately

$$n_a(\infty) = \exp\{-2r(0)\}. \tag{38}$$

As an example of these relations, consider the application of the inverse of (36),

$$F(2x) \simeq -\frac{1}{4}\frac{\mathrm{d}}{\mathrm{d}x}\log n(x). \tag{39}$$

We will use (39) to obtain the approximate reflection amplitude for the Rayleigh profile studied in Section 2-5, for which n^{-1} is linear in z:

$$n^{-1}(z) \equiv \eta(z) = \eta_1 + (\Delta\eta/\Delta z)z, \quad 0 \leqslant z \leqslant \Delta z. \tag{40}$$

For this profile we have, in the interval $0 \leqslant z \leqslant \Delta z$,

$$x = \frac{\Delta z}{\Delta\eta}\log\eta, \quad \frac{\mathrm{d}}{\mathrm{d}x}\log n(x) = -\frac{\Delta\eta}{\Delta z}, \quad F(2x) = \frac{1}{4}\frac{\Delta\eta}{\Delta z}, \tag{41}$$

and so from (37) the approximate reflection amplitude is

$$r_a(k_1) = \tfrac{1}{2}\log\frac{\eta_2}{\eta_1}\,\mathrm{e}^{ik_1\Delta x}\frac{\sin k_1\Delta x}{k_1\Delta x}, \quad \Delta x = \frac{\Delta z}{\Delta\eta}\log\left(\frac{\eta_2}{\eta_1}\right). \tag{42}$$

This is precisely the Rayleigh or weak reflection approximation result obtained in Section 5-8 for this profile (see (5.98) and Figure 5-4).

We will now show that this suprising accord is not accidental: the approximate solution of the synthesis problem given above is identical to the result obtained by inverting the Rayleigh approximation for the reflection amplitude. The latter was given in Section 5-7; at normal incidence (5.85) and (5.86) reduce to

$$r \simeq -\int_{-\infty}^{\infty}\mathrm{d}\phi\,\frac{\mathrm{d}n/\mathrm{d}\phi}{2n}\,\mathrm{e}^{2i\phi}, \qquad \phi = \int^{z}\mathrm{d}\zeta\,k(\zeta). \tag{43}$$

Thus the Rayleigh approximation gives the reflection amplitude at normal incidence as the Fourier transform in the ϕ variable of the logarithmic derivative of $n^{1/2}$. To keep common notation with the Hirsch inversion we set $\mathrm{d}\phi = k\mathrm{d}z = k_1\mathrm{d}x$. Then (43) reads

$$r(k_1) \simeq -\int_{-\infty}^{\infty}\mathrm{d}x\,\frac{n'}{2n}\,\mathrm{e}^{2ik_1x}, \tag{44}$$

and has the inverse (compare (33))

$$-\frac{n'}{2n} \simeq \frac{1}{2\pi}\int_{-\infty}^{\infty}\mathrm{d}k_1\,r(k_1)\,\mathrm{e}^{-2ik_1x} \equiv F(2x), \tag{45}$$

$$n(x) \simeq n_1\exp\left\{-2\int_{-\infty}^{2x}\mathrm{d}y\,F(y)\right\}. \tag{46}$$

(This equation is slightly more general than (36), in that $n_1 = 1$ has not been assumed). Discussion of the validity of the Rayleigh approximation, and error bounds on the resulting reflection amplitudes, may be found in Section 5-7. In general it is expected to break down when the reflection is strong.

Inversion of the Rayleigh approximation reflection amplitudes is possible for all angles of incidence for both polarizations. In the s wave case we have, from (5.85)

$d\phi = q_1\, dx$ or $x = q_1^{-1} \int^z d\zeta\, q(\zeta)$,

$$r_s(q_1) \simeq -\int_{-\infty}^{\infty} dx \frac{q'}{2q} e^{2iq_1 x}, \tag{47}$$

which has the Fourier inverse

$$-\frac{q'}{2q} \simeq \frac{1}{2\pi} \int_{-\infty}^{\infty} dq_1\, e^{-2iq_1 x}\, r_s(q_1) \equiv F_s(2x). \tag{48}$$

Thus

$$q(x) \simeq q_1 \exp\left(-2 \int_{-\infty}^{2x} dy\, F_s(y)\right), \tag{49}$$

or

$$\frac{\varepsilon(x)}{\varepsilon_1} \simeq \sin^2\theta_1 + \cos^2\theta_1 \exp\left(-4 \int_{-\infty}^{2x} dy\, F_s(y)\right) \tag{50}$$

(note the formal similarity with (4) and (24)). The p wave equation (5.86) is inverted using the same variable x:

$$r_p(q_1) \simeq \int_{-\infty}^{\infty} dx \frac{Q'}{2Q} e^{2iq_1 x}, \tag{51}$$

$$\frac{Q'}{2Q} \simeq \frac{1}{2\pi} \int_{-\infty}^{\infty} dq_1\, e^{2iq_1 x}\, r_p(q_1) \equiv F_p(2x), \tag{52}$$

$$Q(x) \simeq Q_1 \exp\left(2 \int_{-\infty}^{2x} dy\, F_p(y)\right). \tag{53}$$

Since $Q = q/\varepsilon$ and $(cq/\omega)^2 = \varepsilon - \varepsilon_1 \sin^2\theta_1$, (53) gives a quadratic for $\varepsilon(x)$. We take the root which agrees with (50) at normal incidence;

$$\frac{\varepsilon(x)}{\varepsilon_1} \simeq \frac{\exp\left(-4 \int_{-\infty}^{2x} dy\, F_p(y)\right)}{2\cos^2\theta_1} \times \left\{1 + \left[1 - \sin^2 2\theta_1 \exp\left(4 \int_{-\infty}^{2x} dy\, F_p(y)\right)\right]^{1/2}\right\}. \tag{54}$$

A measure of the accuracy of this inversion may be obtained by calculating $\varepsilon_2/\varepsilon_1$ from (50) and (54), letting x tend to $+\infty$, and using

$$\int_{-\infty}^{\infty} dx\, F_s(x) = r_s(0), \qquad \int_{-\infty}^{\infty} dx\, F_p(x) = r_p(0). \tag{55}$$

The long wave limits of the Rayleigh approximations for r_s and r_p are given by (5.87); when these are substituted in (50) and (54) the right-hand sides are equal to $\varepsilon_2/\varepsilon_1$, at all angles. The true long wave limits, given by (5.88), do not give agreement between the left and right hand sides. For example, at normal incidence where

$r_s(0) = r_p(0) = (n_1 - n_2)/(n_1 + n_2)$, (46), (50) and (54) all read

$$\frac{n_2}{n_1} \simeq \exp 2\left(\frac{n_2 - n_1}{n_2 + n_1}\right), \tag{56}$$

which has an error of the third order in $(n_2 - n_1)/(n_2 + n_1)$.

The inversion formulae (50) and (54) give $\varepsilon(x)$, not the required $\varepsilon(z)$, and thus need to be complemented by a functional relation between the physical coordinate z and the "optical" coordinate x. This is obtained from the given reflection amplitude r_s via its Fourier transform F_s by integrating $q\mathrm{d}z = q_1\mathrm{d}x$ using (49):

$$z(x) = \int_0^x \mathrm{d}x_1 \exp\left[2\int_{-\infty}^{2x_1} \mathrm{d}x_2\, F_s(x_2)\right]. \tag{57}$$

References

References quoted in the text

W. Kofink (1947) "Reflexion elektromagnetischer Wellen an einer inhomogenen Schicht", Annalen der Physik **1**, 119–132.

R. F. Potter (1969) "Pseudo-Brewster angle technique for determining optical constants", Chapter 16 of "Optical properties of solids", edited by S. Nudelman and S. S. Mitra, Plenum.

A. Vašíček (1960) "Optics of thin films", North-Holland, Section 5.2.

D. E. Aspnes (1976) "Spectroscopic ellipsometry of solids", Chapter 15 in "Optical properties of solids: new developments" edited by B. O. Seraphin, North Holland.

F. Wooten (1972) "Optical properties of solids", Academic Press.

R. M. A. Azzam (1983) "Ellipsometry of unsupported and embedded thin films", J. de Physique **C10**, 67–70.

R. M. A. Azzam (1979) "Direct relation between Fresnel's interface reflection coefficients for the parallel and perpendicular polarisations", J. Opt. Soc. Amer. **69**, 1007–1016.

I. M. Gelfand and B. M. Levitan (1951/1955) "On the determination of a differential equation by its spectral function", Amer. Math. Soc. Transl. (Ser. 2), **1**, 253–304.

I. Kay (1960) "The inverse scattering problem when the reflection coefficient is a rational function", Commun. Pure App. Math. **13**, 371–393.

A. K. Jordan (1980) "Inverse scattering theory: exact and approximate solutions", in "Mathematical methods and applications of scattering theory", edited by J. A. DeSanto, A. W. Sáenz and W. W. Zachary, Springer Lecture Notes in Physics No. 130.

I. Kay and H. E. Moses (1956) "Reflectionless transmission through dielectrics and scattering potentials". J. Appl. Phys. **27**, 1503–1508.

J. Hirsch (1979) "An analytic solution to the synthesis problem for dielectric thin-film layers", Optica Acta **26**, 1273–1279.

H. E. Moses (1956) "Calculation of the scattering potential from reflection coefficients", Phys. Rev. **102**, 559–567.

Introductory and review articles on inverse problems

M. Kac (1966) "Can one hear the shape of a drum?", Amer. Math. Monthly **73**, (Pt.II) 1–23.

R. G. Newton (1970) "Inverse problems in physics", SIAM Review **12**, 346–355.

J. B. Keller (1976) "Inverse problems", Amer. Math. Monthly **83**, 107–118.

F. J. Dyson (1976) "Old and new approaches to the inverse scattering problem", in "Studies in Mathematical Physics, Essays in Honor of Valentine Bargmann", edited by E. H. Lieb, B. Simon and A. S. Wightman, Princeton, pp. 151–167.

Papers and books on the inverse problem of scattering theory

V. Bargmann (1949) "Remarks on the determination of a central field of force from the elastic scattering phase shifts", Phys. Rev. **75**, 301–303.

V. Bargmann (1949) "On the connection between phase shifts and scattering potential", Rev. Mod. Phys. **21**, 488–493.

R. Jost and W. Kohn (1952) "Equivalent potentials", Phys. Rev. **88**, 382–385.

V. A. Marchenko (1955) "The construction of the potential energy from the phases of the scattered waves", Dokl. Akad. Nauk SSSR **104**, 695–698 [Math. Rev. **17**, 740 (1956)].

L. D. Faddeev (1963) "The inverse problem in the quantum theory of scattering", J. Math. Phys. **4**, 72–104.

Z. S. Agranovich and V. A. Marchenko (1963) "The inverse problem of scattering theory", Gordon and Breach.

K. Chadan and P. C. Sabatier (1977) "Inverse problems in quantum scattering theory", Springer.

Collections of papers on electromagnetic and optical inverse problems

H. P. Baltes, ed. (1978) "Inverse source problems in optics", Springer.

H. P. Baltes, ed. (1980) "Inverse scattering problems in optics", Springer.

W. M. Boemer, A. K. Jordan and I. W. Kay, eds. (1981), I.E.E.E. Transactions on Antennas and Propagation, special issue on "Inverse methods in electromagnetics", **AP-29**, No. 2, 185–417.

A. J. Devaney, ed. (1985) "Inverse problems in propagation and scattering", J. Opt. Soc. Amer. **A2**, 1901–2061. (In relation to Section 9-3, see especially the papers by H. D. Landouceur and A. K. Jordan, and by D. L. Jaggard and Y. Kim.)

Another important inversion problem arises in the extraction of the electron density as a function of height from the measured times of travel of nearly monochromatic radio pulses which are reflected from the ionosphere. The problem reduces to that of solving Abel's integral equation, and is related to several inverse problems in mechanics (Keller, 1976). The ionospheric case is considered in detail by

K. G. Budden (1961) "Radio waves in the ionosphere", Cambridge, Chapter 10.

K. G. Budden (1985) "The propagation of radio waves", Cambridge, Chapter 12.

10

Pulses, finite beams

The preceding chapters have dealt with the reflection of monochromatic plane waves from planar interfaces. Here we consider the reflection of pulses, and of finite beams. The theory of pulse reflection is simplest for those bounded in time but still having a plane wave spatial character. These are built up by a superposition of plane waves of differing frequencies. The simplest beams to consider are those bounded in space but still monochromatic. These are built up by a superposition of plane waves of differing propagation directions. We shall find, accordingly, that the reflection of pulses is determined by the frequency dependence of the reflection amplitude, while the reflection of beams depends on the angular dependence of the reflection amplitude. Particularly important is the case of total reflection, where all the frequency and angle dependence is contained in the phase of the reflection amplitude, since its modulus is then unity.

10-1 Reflection of pulses: the time delay

An incident pulse $E_i(t)$ may be written as a superposition of monochromatic waves by means of the Fourier integral:

$$E_i(t) \;=\; \int_{-\infty}^{\infty} \mathrm{d}\omega\, f(\omega)\, \mathrm{e}^{-i\omega t}. \tag{1}$$

The Fourier inverse of (1) is

$$f(\omega) \;=\; \frac{1}{2\pi}\int_{-\infty}^{\infty} \mathrm{d}t\, \mathrm{e}^{i\omega t}\, E_i(t). \tag{2}$$

For example, if the incident pulse is sinusoidal with amplitude $A(t)$

$$E_i(t) \;=\; A(t)\,\mathrm{e}^{-i\omega_0 t}, \qquad f(\omega) \;=\; \frac{1}{2\pi}\int_{-\infty}^{\infty} \mathrm{d}t\, A(t)\, \mathrm{e}^{i(\omega-\omega_0)t}. \tag{3}$$

If $A(t)$ varies slowly over most of its range, this represents a wavepacket which is nearly monochromatic. The simplest case is a truncated sine wave, for which

$A(t) = A_0$ when $-T/2 < t < T/2$, and $A(t) = 0$ otherwise; then

$$f(\omega) = A_0 \frac{\sin(\omega - \omega_0)T/2}{\pi(\omega - \omega_0)}. \tag{4}$$

Each Fourier component reflects with its own reflection amplitude $r(\omega)$. Thus if (1) represents the incident wave at (say) $z = 0$, the reflected wave at $z = 0$ will be given by

$$E_r(t) = \int_{-\infty}^{\infty} \mathrm{d}\omega\, r(\omega) f(\omega)\, \mathrm{e}^{-i\omega t}. \tag{5}$$

For a sinusoidal pulse this may be written as

$$E_r(t) = \frac{1}{2\pi} \int_{-\infty}^{\infty} \mathrm{d}\omega\, r(\omega)\, \mathrm{e}^{-i\omega t} \int_{-\infty}^{\infty} \mathrm{d}\tau\, A(\tau)\, \mathrm{e}^{i(\omega-\omega_0)\tau}. \tag{6}$$

We now specialize further to the case where the modulus of $r(\omega)$ is slowly varying compared to the phase, and the frequency variation of the phase is adequately approximated by the first term in its Taylor expansion about ω_0: $r(\omega) = |r(\omega)|\, \mathrm{e}^{i\delta(\omega)}$, with

$$|r(\omega)| \simeq |r(\omega_0)|, \qquad \delta(\omega) \simeq \delta_0 + (\omega - \omega_0)\delta_0'. \tag{7}$$

Here $\delta_0 = \delta(\omega_0)$ and δ_0' is the derivative $\mathrm{d}\delta/\mathrm{d}\omega$ evaluated at ω_0. (This approximation is particularly suited to the treatment of total reflection, where $|r(\omega)| = 1$.) On substituting (7) in (6), $(2\pi)^{-1}$ times the integral over ω becomes a delta function, which selects the time $\tau = t - \delta_0'$ in the τ integral. Thus

$$E_r(\omega) \simeq |r(\omega_0)|\, \mathrm{e}^{i\delta_0} A(t - \delta_0')\, \mathrm{e}^{-i\omega_0 t}. \tag{8}$$

The reflected pulse in this approximation is thus decreased in amplitude by $|r(\omega_0)|$ and phase shifted by δ_0. The pulse is unchanged in shape (it has the same time envelope A), but is delayed by the *group delay time*

$$\Delta t = \delta_0' = \left(\frac{\mathrm{d}\delta}{\mathrm{d}\omega}\right)_{\omega=\omega_0}. \tag{9}$$

We will consider some examples of the application of (9). The simplest case is that of reflection from a discontinuity in the refractive index. If the discontinuity occurs at z_1, the reflection amplitude at normal incidence is, from (1.15),

$$r_n = \mathrm{e}^{2in_1(\omega/c)z_1} \frac{n_1 - n_2}{n_1 + n_2}. \tag{10}$$

The phase is thus a constant (0 or $\pm\pi$, depending on the sign of $n_1 - n_2$) plus ω times $2n_1 z_1/c$, and the delay time is $2n_1 z_1/c$, this being the distance travelled from $z = 0$ to z_1 and back divided by the speed c/n_1.

The above example is special because the medium is uniform everywhere except at the discontinuity. In the general case of reflection by an inhomogeneous medium we showed in Section 6-7 that in the short wave limit, the phase shift on total reflection is given by

$$\delta \simeq 2(\phi_0 - \phi_- - \pi/4), \tag{11}$$

where

$$\phi_0 = \int^{z_0} d\zeta\, q(\zeta), \qquad \phi_- = \lim_{z\to-\infty} \left\{ \int^{z} d\zeta\, q(\zeta) - q_1 z \right\}. \tag{12}$$

The lower limit in the phase integral is arbitrary. It is convenient to set it equal to z_0 (the turning point, at which $q^2 = 0$). Thus $\phi_0 = 0$, and ϕ_- may be written as

$$\phi_- = \lim_{z\to-\infty} \int_{z_0}^{z} d\zeta\, (q - q_1) - q_1 z_0. \tag{13}$$

If $q(z) = q_1$ at and below the observation point $z = 0$, the phase of the reflection amplitude becomes

$$\delta \simeq 2 \int_0^{z_0} dz\, q(z,\omega) - \pi/2. \tag{14}$$

At normal incidence (vertical propagation in the case of pulses reflected from the ionosphere),

$$\delta \simeq 2 \frac{\omega}{c} \int_0^{z_0} dz\, n(z, \omega) - \pi/2. \tag{15}$$

The time delay is thus

$$\Delta t = \frac{d\delta}{d\omega} \simeq \frac{2}{c} \int_0^{z_0} dz \left[n(z, \omega) + \omega \frac{\partial n}{\partial \omega} \right] \tag{16}$$

(the turning point z_0 is also a function of ω, but its derivative is multiplied by $n(z_0, \omega)$, which is zero). Thus the pulse travels to the turning point and back at the *group velocity*

$$u(z, \omega) = \frac{c}{n(z, \omega) + \omega(\partial n/\partial \omega)}; \tag{17}$$

hence the name "group delay time" given to Δt. Equation (17) is equivalent to the usual definition of group velocity, $u = d\omega/dk$, since here $k = n\omega/c$. In the simplest model of the ionosphere,

$$\varepsilon(z, \omega) = n^2(z, \omega) = 1 - \frac{\omega_p^2(z)}{\omega^2}; \tag{18}$$

then the group velocity u and the phase velocity $v = c/n$ are related by

$$uv = c^2. \tag{19}$$

The above derivation of the group delay time is based, in part, on Ginzburg (1964, Section 21). An alternative treatment may be found in Budden (1961, Chapter 10, and 1985, Chapter 5); a general discussion of phase, group, signal and energy transport velocities is given by Brillouin (1960).

We have considered only the linear term in the Taylor expansion

$$\delta(\omega) = \delta_0 + (\omega - \omega_0)\delta_0' + \tfrac{1}{2}(\omega - \omega_0)^2 \delta_0'' + \dots. \tag{20}$$

The first order term leads to the time delay discussed above; the second and higher order terms cause pulse spreading and distortion. These effects are discussed by Budden and Ginzburg in the limit (common in optics and radio) where the pulse is nearly monochromatic. The opposite extreme is common in underwater acoustics and in seismology, where explosive sources or sudden crust movements give pulses which are strongly localized, and not at all harmonic. There is then no dominant frequency ω_0, and use of expansions such as (20) is not appropriate. A discussion of this case and further references may be found in Brekhovskikh (1980, Section 15).

10-2 Phase change on total internal reflection

In the next section we treat the reflection of bounded beams, with emphasis on the problem of beam shift. The latter depends on the variation of the phase of the reflection amplitude with the angle of incidence, and is greatest near the critical angle where this variation becomes infinite. In this section we give examples of the angular dependence of the phases of r_s and r_p, and then show that a square root singularity at the critical angle is universal for nonabsorbing profiles.

Reflection at a sharp boundary. The s and p reflection amplitudes for a step profile located at $z = 0$ are given by (1.13) and (1.31):

$$r_s = \frac{q_1 - q_2}{q_1 + q_2}, \qquad r_p = \frac{Q_2 - Q_1}{Q_1 + Q_2}. \tag{21}$$

When medium 1 is optically denser ($\varepsilon_1 > \varepsilon_2$), $q_1 > q_2$ and r_s has zero phase (all phases are modulo 2π) up to the critical angle θ_c, where $q_2^2 = (\omega^2/c^2)(\varepsilon_2 - \varepsilon_1 \sin^2\theta_1)$ passes through zero and q_2 changes from real to imaginary:

$$q_2 = i|q_2| = i\frac{\omega}{c}(\varepsilon_1 \sin^2\theta_1 - \varepsilon_2)^{1/2} \qquad (\theta_1 \geqslant \arcsin(\varepsilon_2/\varepsilon_1)^{1/2} = \theta_c) \tag{22}$$

Beyond this point $|r_s| = 1$ and

$$\delta_s = -2\arctan\frac{|q_2|}{q_1}, \tag{23}$$

where

$$\frac{|q_2|}{q_1} = \{\cos^2\theta_c \tan^2\theta_1 - \sin^2\theta_c\}^{1/2} = \left\{\left(1 - \frac{\varepsilon_2}{\varepsilon_1}\right)\tan^2\theta_1 - \frac{\varepsilon_2}{\varepsilon_1}\right\}^{1/2}. \tag{24}$$

We note the square root singularity at θ_c, which leads to an infinite value of $d\delta_s/d\theta_1$ at θ_c^+: in terms of $\alpha = \theta_1 - \theta_c$ this is

$$\delta_s = -2\left(\frac{4\varepsilon_2}{\varepsilon_1 - \varepsilon_2}\right)^{1/4}\alpha^{1/2} + 0(\alpha). \tag{25}$$

The s wave phase decreases monotonically from 0 at θ_c to $-\pi$ at grazing incidence, approaching $-\pi$ linearly in the grazing angle γ:

$$\delta_s = -\pi + 2\left(\frac{\varepsilon_1}{\varepsilon_1 - \varepsilon_2}\right)^{1/2}\gamma + 0(\gamma^2), \qquad \gamma = \pi/2 - \theta_1. \tag{26}$$

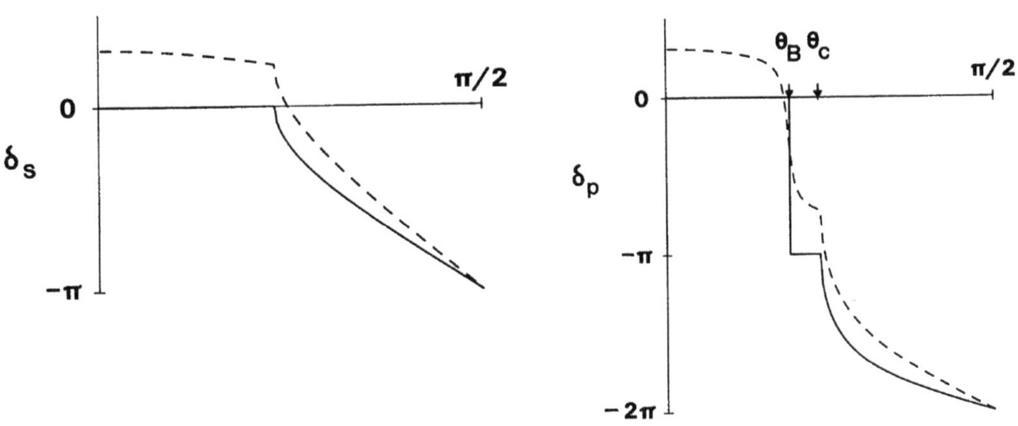

Figure 10-1. Phases of the s and p wave reflection amplitudes, for a sharp boundary between glass and air, refractive indices $\frac{3}{2}$ and 1 (solid curves). The dashed curves are for a uniform layer of water (refractive index $\frac{4}{3}$) between the glass and air, with $(\omega/c)\Delta z = \frac{1}{2}$.

The p wave phase is zero from normal incidence to the Brewster angle $\theta_B = \arctan(\varepsilon_2/\varepsilon_1)^{1/2}$, when $Q_1 = Q_2$ and r_p changes sign. In the interval $\theta_B < \theta_1 < \theta_c$ we can set δ_p equal to $+\pi$ or $-\pi$. We take $\delta_p = -\pi$, this choice being dictated by continuity of the phase as a function of interfacial thickness, as the next example will make clear. Beyond θ_c the p wave phase is (from (21) with $Q_2 = i|Q_2|$)

$$\delta_p = -\pi - 2\arctan\frac{|Q_2|}{Q_1} = -\pi - 2\arctan\frac{\varepsilon_1}{\varepsilon_2}\frac{|q_2|}{q_1}. \tag{27}$$

The strength of the square root singularity is thus larger for the p phase shift by the factor $\varepsilon_1/\varepsilon_2$:

$$\delta_p = -\pi - 2\frac{\varepsilon_1}{\varepsilon_2}\left(\frac{4\varepsilon_2}{\varepsilon_1 - \varepsilon_2}\right)^{1/4}\alpha^{1/2} + 0(\alpha). \tag{28}$$

The inverse factor applies as δ_p tends to -2π at grazing incidence:

$$\delta_p = -2\pi + 2\frac{\varepsilon_2}{\varepsilon_1}\left(\frac{\varepsilon_1}{\varepsilon_1 - \varepsilon_2}\right)^{1/2}\gamma + 0(\gamma^2). \tag{29}$$

Figure 10-1 shows δ_s and δ_p for the sharp boundary between two media, and also for a uniform layer between the same two media (the latter to be discussed shortly).

The ellipsometric ratio r_p/r_s is equal to $\exp i(\delta_p - \delta_s)$ for $\theta_1 > \theta_c$. The phase difference $\Delta = \delta_p - \delta_s$ is given by

$$\Delta = -\pi + 2\left(\arctan\frac{|q_2|}{q_1} - \arctan\frac{\varepsilon_1}{\varepsilon_2}\frac{|q_2|}{q_1}\right). \tag{30}$$

The phase difference has an extremum at the angle of incidence

$$\theta_m = \arctan\left(\frac{2\varepsilon_2}{\varepsilon_1 - \varepsilon_2}\right)^{1/2};$$

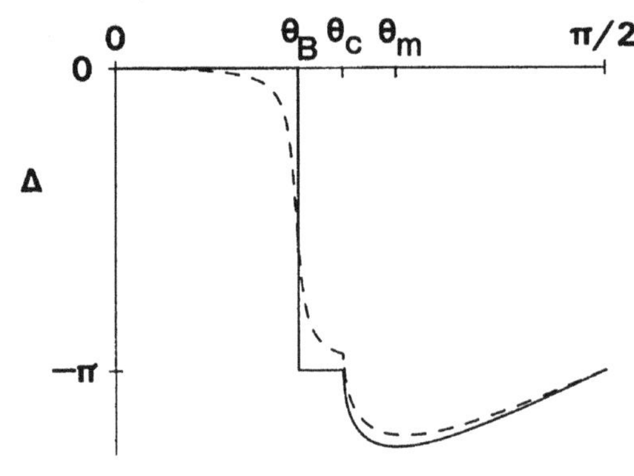

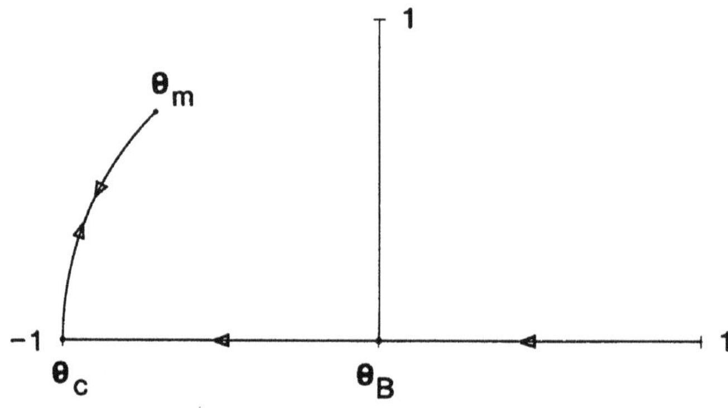

Figure 10-2. $\Delta = \delta_p - \delta_s$ and r_p/r_s, the parameters being as in Figure 10-1. The uniform film data is not shown on the r_p/r_s diagram; it was shown in Figure 2-9 on a larger scale.

for comparison we list the tangents and sines of θ_B, θ_c and θ_m:

$$\begin{aligned} \tan^2\theta_B &= \frac{\varepsilon_2}{\varepsilon_1}, & \tan^2\theta_c &= \frac{\varepsilon_2}{\varepsilon_1 - \varepsilon_2}, & \tan^2\theta_m &= \frac{2\varepsilon_2}{\varepsilon_1 - \varepsilon_2}, \\ \sin^2\theta_B &= \frac{\varepsilon_2}{\varepsilon_1 + \varepsilon_2}, & \sin^2\theta_c &= \frac{\varepsilon_2}{\varepsilon_1}, & \sin^2\theta_m &= \frac{2\varepsilon_2}{\varepsilon_1 + \varepsilon_2}. \end{aligned} \tag{31}$$

At the extremum Δ is given by

$$\Delta_m = 4\theta_B - 2\pi, \tag{32}$$

and the ratio of the reflection amplitudes takes the value

$$\left(\frac{r_p}{r_s}\right)_m = (\varepsilon_1 + \varepsilon_2)^{-2}\{\varepsilon_1^2 + \varepsilon_2^2 - 6\varepsilon_1\varepsilon_2 + i4(\varepsilon_1\varepsilon_2)^{1/2}(\varepsilon_1 - \varepsilon_2)\}. \tag{33}$$

At θ_m the trajectory of r_p/r_s in the complex plane is farthest to the right on the unit circle. The phase difference $\delta_p - \delta_s$ and the ratio r_p/r_s are shown in Figure 10-2.

Reflection phases for a uniform film. For a layer of dielectric constant ε and of thickness Δz, the s and p reflection amplitudes are given by (2.52) and (2.68):

$$r_s = \frac{q(q_1 - q_2)c + i(q^2 - q_1q_2)s}{q(q_1 + q_2)c - i(q^2 + q_1q_2)s}, \tag{34}$$

$$-r_p = \frac{Q(Q_1 - Q_2) + i(Q^2 - Q_1Q_2)s}{Q(Q_1 + Q_2)c - i(Q^2 + Q_1Q_2)s}. \tag{35}$$

Here $q^2 = \omega^2/c^2\,(\varepsilon - \varepsilon_1 \sin^2\theta_1)$, $c = \cos q\Delta z$, $s = \sin q\Delta z$, $Q = q/\varepsilon$, and the film extends from $z = 0$ to Δz. When $\varepsilon_1 > \varepsilon > \varepsilon_2$ we have to consider three ranges of θ_1: $\theta_1 \leqslant \theta_c = \arcsin(\varepsilon_2/\varepsilon_1)^{1/2}$, $\theta_c \leqslant \theta_1 \leqslant \theta_c' = \arcsin(\varepsilon/\varepsilon_1)^{1/2}$, and $\theta_1 \geqslant \theta_c'$. In the first range all the q's are real, in the second $q_2 = i|q_2|$ and q_1, q are real, and in the third $q_2 = i|q_2|$ and $q = i|q|$. Of particular interest to the beam shift to be discussed in the next section is the behaviour of the phases for θ_1 slightly above θ_c. We again find a square root singularity, with

$$\delta_s(\theta_1) = \delta_s(\theta_c) - \frac{2|q_2|/q_{1c}}{1 - \dfrac{\varepsilon_1 - \varepsilon}{\varepsilon_1 - \varepsilon_2}\sin^2 q_c\Delta z} + 0(|q_2|^2), \tag{36}$$

$$\delta_p(\theta_1) = \delta_p(\theta_c) - \frac{2|Q_2|/Q_{1c}}{1 - \left(1 - \dfrac{\varepsilon_1^2(\varepsilon - \varepsilon_2)}{\varepsilon^2(\varepsilon_1 - \varepsilon_2)}\right)\sin^2 q_c\Delta z} + 0(|q_2|^2), \tag{37}$$

where

$$q_{1c}^2 = \frac{\omega^2}{c^2}(\varepsilon_1 - \varepsilon_2), \qquad q_c^2 = \frac{\omega^2}{c^2}(\varepsilon - \varepsilon_2), \tag{38}$$

and

$$\delta_s(\theta_c) = 2\arctan\left\{\left(\frac{\varepsilon - \varepsilon_2}{\varepsilon_1 - \varepsilon_2}\right)^{1/2}\tan q_c\Delta z\right\}. \tag{39}$$

$$\delta_p(\theta_c) = -\pi + 2\arctan\left\{\frac{\varepsilon_1}{\varepsilon}\left(\frac{\varepsilon - \varepsilon_2}{\varepsilon_1 - \varepsilon_2}\right)^{1/2}\tan q_c\Delta z\right\}. \tag{40}$$

The numerators in (36) and (37) are the sharp boundary values, and have been expressed in terms of $\alpha = \theta_1 - \theta_c$ in (25) and (28). For the s wave the coefficient of $\alpha^{1/2}$ is larger for the uniform layer than for the Fresnel case; for the p wave it can be larger or smaller, depending on the dielectric constants.

Figures 10-1 and 10-2 show the s and p phase shifts for $(\omega/c)\Delta z = 1/2$. Note that there is no square root singularity in δ_s or δ_p at θ_c' where q passes through zero, the s and p phases having the variation $\delta = \delta(\theta_c') + 0(q^2)$, with

$$\delta_s(\theta_c') = -2\arctan\left\{\frac{[(\varepsilon - \varepsilon_2)/(\varepsilon_1 - \varepsilon)]^{1/2}}{1 + (\varepsilon - \varepsilon_2)^{1/2}(\omega/c)\Delta z}\right\}. \tag{41}$$

$$\delta_p(\theta_c') = -\pi - 2\arctan\left\{\frac{\dfrac{\varepsilon_1}{\varepsilon_2}[(\varepsilon - \varepsilon_2)/(\varepsilon_1 - \varepsilon)]^{1/2}}{1 + \dfrac{\varepsilon}{\varepsilon_2}(\varepsilon - \varepsilon_2)^{1/2}(\omega/c)\Delta z}\right\}. \tag{42}$$

Thus θ_c' is not a true critical angle, even though at θ_c' q makes a sharp right-angle turn in the complex plane, just as q_2 does at θ_c.

Total reflection by the tanh profile. For $\theta_1 < \theta_c$ the phase for the s polarization is given by (2.89); as θ_c is approached from below the phase tends without singularity to

$$\delta_s(\theta_c) = 2 \sum_{n=1}^{\infty} \arctan \left\{ \frac{2y_{1c}^3}{n(n^2 + 3y_{1c}^2)} \right\}, \tag{43}$$

where $y_{1c} = q_{1c}a$, a being the length characterizing the thickness of the profile, and $q_{1c} = (\omega/c)(\varepsilon_1 - \varepsilon_2)^{1/2}$. For $\theta_1 > \theta_c$ an analysis based on (2.84) and using the infinite product representation of the gamma function (2.85) gives

$$\delta_s = 2 \sum_{1}^{\infty} \arctan \left\{ \frac{\frac{2y_1}{n}(y_1^2 + |y_2|^2)}{n^2 + 3y_1^2 - |y_2|^2} \right\} - 2 \arctan \left(\frac{\tan \pi |y_2|}{\tanh \pi y_1} \right), \tag{44}$$

where $y_1 = q_1 a$, $|y_2| = |q_2|a$. Thus there is again a $|q_2|$ term in the phase just above the critical angle:

$$\delta_s = \delta_s(\theta_c) - \frac{2\pi a |q_2|}{\tanh \pi a q_{1c}} + 0(q_2^2). \tag{45}$$

As $aq_{1c} \to 0$ we regain the sharp profile result (26). For large interfacial thickness the coefficient of $|q_2|$ tends to $-2\pi a$, and the strength of the square root singularity is then proportional to the thickness. At grazing incidence $\delta_s \to -\pi$ as before.

The above examples are sufficient to make it plausible that the $(\theta_1 - \theta_c)^{1/2}$ singularity in the phase shift is a universal property. We shall give a proof for the restricted class of finite-ranged profiles, for which the s wave reflection amplitude is given by (2.25), which we write in the form

$$r_s = \frac{q_1 q_2 A + iq_1 B + iq_2 C - D}{q_1 q_2 A + iq_1 B - iq_2 C + D} \tag{46}$$

(we again set $z_1 = 0$: the reflecting inhomogeneity extends from $z = 0$ to Δz). When $\theta_1 > \theta_c$ we have $q_2 = i|q_2|$, and

$$r_s = \frac{-\alpha + i\beta}{\alpha + i\beta}, \quad \alpha = |q_2|C + D, \quad \beta = q_1(|q_2|A + B). \tag{47}$$

Thus $\delta_s = 2 \arctan(\alpha/\beta)$. The leading terms in α/β near θ_c are

$$\frac{\alpha}{\beta} = \frac{D}{q_1 B} - \frac{|q_2|}{q_1} \left(\frac{W}{B} \right)^2 + 0(|q_2|^2), \tag{48}$$

where W is the Wronskian of the solutions of the wave equation; we have used the identity $AD - BC = W^2$ (equation (2.31)). This shows that all such profiles have a term linear in $|q_2|$, with negative coefficient, leading to a square root singularity:

$$\delta_s = 2 \arctan \left(\frac{D}{q_1 B} \right)_c - \frac{2|q_2|}{q_{1c}} \frac{(W/B)_c^2}{1 + (D/q_1 B)_c^2} + 0(|q_2|^2). \tag{49}$$

For the uniform layer, with solutions $\sin qz$ and $\cos qz$ in $0 \leqslant z \leqslant \Delta z$, we have $W = q$, $B = q \cos q\Delta z$, $D = q^2 \sin q\Delta z$, and (49) gives the results contained in (36) and (39). (A similar result may be written down for δ_p, using (2.40), (2.48) and (2.49). However, since (2.40) was derived assuming continuity of the dielectric function, the resulting expression does not apply to the uniform layer.)

10-3 Reflection of beams: the lateral beam shift

We would expect a lateral beam shift in the case of total reflection from a stratified medium: the semiclassical or geometric optics picture is shown in Figure 10-3.

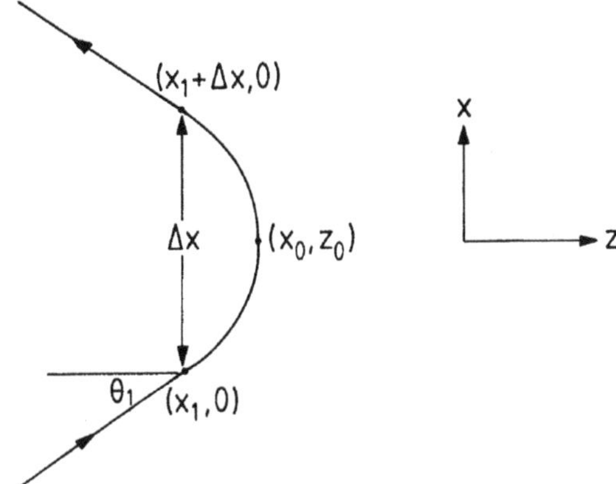

Figure 10-3. Lateral shift of a ray totally reflected from an inhomogeneous region. The inhomogeneity begins at $z = 0$, and the turning point is at z_0. The medium on the left is optically denser.

If the angle between the ray and the z axis is $\theta(z)$, geometric optics gives the lateral shift as

$$\Delta x = 2\int_{x_1}^{x_0} \mathrm{d}x = 2\int_0^{z_0} \mathrm{d}z \tan\theta(z). \tag{50}$$

For a sharp transition between media of dielectric constants ε_1 and ε_2, with $\varepsilon_1 > \varepsilon_2$ and $\theta_1 > \theta_c = \arcsin(\varepsilon_2/\varepsilon_1)^{1/2}$, the turning point z_0 and the beginning of the transition coincide, and (50) gives zero lateral shift. Goos and Hänchen (1947) however found a non-zero beam shift in this case, with a maximum lateral displacement just beyond the critical angle. This phenomenon is referred to as the Goos–Hänchen effect. It is universal for wave phenomena: a comprehensive review, with references to work in optics, acoustics, quantum mechanics and plasma physics has been given by Lotsch (1970; illustrations of acoustic beam displacement may be found in Brekhovskikh (1980). Figure 10-4 illustrates the lateral displacement of a beam at a sharp boundary.

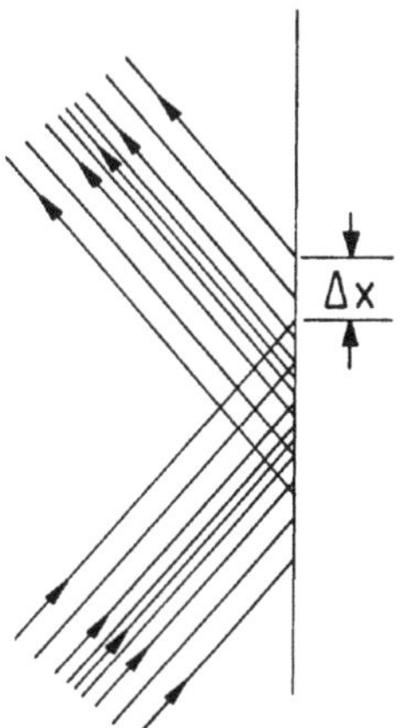

Figure 10-4. Goos–Hänchen effect: the lateral beam shift in total reflection at a discontinuous transition.

We will show that the beam shift in most cases (excluding the immediate neighbourhood of the critical angle) is well approximated by the formula

$$\Delta x = -\frac{\mathrm{d}\delta}{\mathrm{d}K}. \tag{51}$$

Here δ is the phase of the reflection amplitude, K is the lateral component of the wavevector ($K = \varepsilon^{1/2}(\omega/c)\sin\theta$), and the derivative is to be evaluated at the dominant K value of the incident beam. The two extremes illustrated in Figures 10-3 and 10-4 are encompassed by (51), except *at* the critical angle where (as we saw in the last section) δ has a square root singularity.

A finite beam may be built up by superposing plane waves with a spread of propagation directions, centred on the direction of the beam. We consider only a spread in the angle of incidence: the beam is taken to extend a large distance in the y direction. It is convenient to characterize the plane wave components by their lateral wavenumbers K, following Brekhovskikh (1980, Section 14.1). The plane waves $\exp i(Kx \pm qz - \omega t)$, with $K^2 + q^2 = \varepsilon_1\omega^2/c^2$, are solutions of the wave equation in medium 1, and the incident beam is made up of a superposition of these:

$$E_i(z, x) = \int_{-\infty}^{\infty} \mathrm{d}K\, f(K)\, \mathrm{e}^{i(Kx+qz)}. \tag{52}$$

(When $K^2 > \varepsilon_1\omega^2/c^2$, q is imaginary, implying evanescent waves. This aspect need not concern us here since we will be considering only well-collimated beams, with a narrow range of K about $K_1 = \sqrt{\varepsilon_1}(\omega/c)\sin\theta_1$.) The reflected beam is obtained by summing over the reflected component plane waves, each with its own reflection amplitude $r(K) = |r(K)| \exp(i\delta(K))$:

$$E_r(z, x) = \int_{-\infty}^{\infty} \mathrm{d}K\, f(K) r(K)\, \mathrm{e}^{i(Kx-qz)}. \tag{53}$$

If E_i at some reference plane $z = 0$ is given by $E_0(x)$, then

$$E_0(x) = \int_{-\infty}^{\infty} \mathrm{d}K\, f(K)\, \mathrm{e}^{iKx}, \qquad f(K) = \frac{1}{2\pi}\int_{-\infty}^{\infty} \mathrm{d}x\, E_0(x)\, \mathrm{e}^{-iKx}. \tag{54}$$

The reflected field at $z = 0$ is thus given by

$$E_r(0, x) = \frac{1}{2\pi}\int_{-\infty}^{\infty} \mathrm{d}K\, r(K)\, \mathrm{e}^{iKx} \int_{-\infty}^{\infty} \mathrm{d}\xi\, E_0(\xi)\, \mathrm{e}^{-iK\xi}. \tag{55}$$

We now assume that $|r(K)|$ varies slowly with K compared to the phase $\delta(K)$ (this assumption is exact for $K > K_c = \sqrt{\varepsilon_2}(\omega/c)$, when $|r| \equiv 1$) and that the variation of δ is well approximated by the linear term in its Taylor expansion about K_1:

$$|r(K)| \simeq |r(K_1)|, \qquad \delta(K) \simeq \delta(K_1) + (K - K_1)\delta'. \tag{56}$$

Here δ' stands for $\mathrm{d}\delta/\mathrm{d}K$ evaluated at K_1. The integral over K in (55) becomes, in this approximation, 2π times a delta function in ξ, which selects the value $\xi = x + \delta'$. Thus the reflected wave at $z = 0$ is

$$E_r(0, x) \simeq |r(K_1)|\, e^{i(\delta - K_1\delta')}\, E_0(x + \delta'). \tag{57}$$

This represents a beam reduced in amplitude by $|r(K_1)|$, phase shifted by $\delta - K_1\delta'$, and shifted in the x direction by the distance $-\delta'$. The last statement is equivalent to (51). Note that in this approximation the beam shape does not change: the beam is simply translated. This is analogous to the reflection of pulses considered in Section 10-1, where we saw that the linear term $(\omega - \omega_0)\, d\delta/d\omega$ leads to a time delay, with no change in shape of the reflected pulse.

We will now apply (51) to the cases illustrated in Figures 10-3 and 10-4. In the geometrical optics limit the phase is well approximated (except near grazing incidence) by the short wave formula (14). On substituting in (51) and using $q(z_0) = 0$, and $dq/dK = -K/q = -\tan\theta$, we regain (50).

In the sharp transition case, we will consider the region of total reflection. The s and p phases are given by (23) and (27), which we rewrite in terms of K:

$$\begin{aligned} \delta_s &= -2\arctan\left(\frac{K^2 - k_2^2}{k_1^2 - K^2}\right)^{1/2}, \\ \delta_p &= -\pi - 2\arctan\frac{\varepsilon_1}{\varepsilon_2}\left(\frac{K^2 - k_2^2}{k_1^2 - K^2}\right)^{1/2}. \end{aligned} \tag{58}$$

Here $k_1 = n_1\omega/c$ and $k_2 = n_2\omega/c$ are the magnitudes of the total wavevectors in media 1 and 2. The formula (51) gives the beam shifts

$$\Delta x_s = \frac{2K}{q_1|q_2|} = \frac{\lambda_1}{\pi}\frac{\tan\theta_1}{(\sin^2\theta_1 - \sin^2\theta_c)^{1/2}}, \tag{59}$$

$$\Delta x_p = \Delta x_s \frac{\sin^2\theta_c}{(1 + \sin^2\theta_c)\sin^2\theta_1 - \sin^2\theta_c}, \tag{60}$$

λ_1 being the wavelength in the first medium. Near θ_c the p wave beam shift is larger by $1/\sin^2\theta_c = \varepsilon_1/\varepsilon_2$; near grazing incidence it is smaller by the factor $\varepsilon_2/\varepsilon_1$. The formulae are not applicable *at* the critical angle or *at* grazing incidence, since they were derived by using the Taylor expansion (56), which fails at a square root singularity. The fact that δ_s and δ_p always have a term linear in $|q_2| = (K^2 - k_2^2)^{1/2}$ near θ_c was established in the last section. The behaviour near grazing incidence, with a term linear in $q_1 = (k_1^2 - K^2)^{1/2}$ is also universal, as may be seen (for example) from (47) on letting q_1 tend to zero. Equation (58) shows these singularities explicitly in the sharp boundary case.

The simple theory which gives the beam shift as $-d\delta/dK$ thus fails at the critical angle and at grazing incidence, the predicted beam shift diverging as $(K^2 - k_2^2)^{-1/2}$ and $(k_1^2 - K^2)^{-1/2}$ respectively. (Note however that the shift transverse to the reflected beam direction is $\Delta x \cos\theta_1$, which stays finite as $\theta_1 \to \pi/2$.) Horowitz and Tamir (1971) have studied the reflection of a Gaussian beam by a sharp interface, without making an approximation equivalent to (56). They find that the results given in (59) and (60) are accurate down to $\theta_1 - \theta_c \simeq 60$ millidegrees when the beam width parameter w is one thousand wavelengths, and to about 6 millidegrees when w is ten thousand wavelengths. The definition of w for a Gaussian beam is via the electric field amplitude at the beam waist:

$$A(x_t) = \frac{1}{\pi^{1/2} w} \exp\left[-(x_t/w)^2\right]. \tag{61}$$

Here x_t is the distance measured from the beam centre, transversely to the beam propagation direction. Their analysis gives a beam displacement independent of angle in the immediate neighbourhood of θ_c, with magnitude proportional to $(w\lambda)^{1/2}$:

$$\Delta x_s(\theta_c) = \frac{\Gamma(\frac{1}{4})}{2^{3/2}} \frac{(\tan\theta_c)^{1/2}}{\cos\theta_c} \frac{(w\lambda)^{1/2}}{\pi}, \tag{62}$$

with the p polarization displacement larger by $\varepsilon_1/\varepsilon_2$. Note that setting $\theta_1 = \theta_c + \lambda_1/w$ in (59) and (60) (that is, letting the angle of incidence approach the critical angle to within the diffraction-limited broadening of the beam) gives the qualitative features of (62) for small λ_1/w:

$$\Delta x_s(\theta_c + \lambda_1/w) \simeq \frac{1}{2^{1/2}} \frac{(\tan\theta_c)^{1/2}}{\cos\theta_c} \frac{(w\lambda_1)^{1/2}}{\pi}. \tag{63}$$

Experimental test of the Tamir–Horowitz prediction is difficult, since to approach the critical angle closely one must have a highly collimated beam to obtain the required angular resolution. Laser beams have the required collimation, but the wavelength is then small, and so is the beam shift. Early data of Wolter (1950) (also displayed on page 200 of the Lotsch review) are in good agreement with the simple theory.

The lateral shift on reflection is of importance in waveguides, especially in fibre optics. See for example White and Pask (1977) and Snyder and Love (1983, Chapter 10).

References

References quoted in text

V. L. Ginzburg (1964) "The propagation of electromagnetic waves in plasmas", Pergamon.

K. G. Budden (1961) "Radio waves in the ionosphere",Cambridge.

K. G. Budden (1985) "The propagation of radio waves", Cambridge.

L. Brillouin (1960) "Wave propagation and group velocity", Academic Press.

L. M. Brekhovskikh (1980) "Waves in layered media", Second Edition, Academic Press.

F. Goos and H. Hänchen (1947) "Ein neuer und fundamentaler Versuch zur Totalreflexion", Annalen der Physik **1**, 333–346.

H. K. V. Lotsch (1970) "Beam displacement at total reflection: the Goos–Hänchen effect", Parts I to IV, Optik **32**, 116–137, 189–204, 299–319 and 553–569.

B. R. Horowitz and T. Tamir (1971) "Lateral displacement of a light beam at a dielectric interface", J. Opt. Soc. Amer. **61**, 586–594.

H. Wolter (1950) "Untersuchungen zur Strahlversetzung bei Totalreflexion des Lichtes mit der Methode der Minimumstrahlkennzeichnung", Z. Naturforschung **5a**, 143–153.

I. A. White and C. Pask (1977) "Effect of Goos–Hänchen shifts on pulse widths in optical waveguides", Applied Optics **16**, 2353–2355.

A. W. Snyder and J. D. Love (1983) "Optical waveguide theory", Chapman and Hall.

Additional references on beam shifts and related topics

T. Tamir (1982) "The lateral wave" (Chapter 13), and W. P. Chen and E. Burstein (1982) "Narrow beam excitation of electromagnetic modes in prism configurations" (Chapter 14); appearing in "Electromagnetic surface modes", edited by A. D. Boardman, Wiley.

R. P. Riesz and R. Simon (1985) "Reflection of a Gaussian beam from a dielectric slab", J. Opt. Soc. Amer. **2A**, 1809–1817.

I. A. White, A. W. Snyder and C. Pask (1977) "Directional change of beams undergoing partial reflection", J. Opt. Soc. Amer. **67**, 703–705.

11

Rough surfaces

We have seen in Chapter 1 that a planar surface, or arbitrary stratification, will give specular reflection of an incident plane wave. No real surface is perfectly planar, and thus in practice there is a diffuse or scattered component, as well as a specular component of the radiation. The rougher the surface, the greater the diffuseness of the re-radiation from it. A rough surface which is planar on average (for example a liquid–vapour interface stabilized by a gravitational field) is characterized by at least two parameters: a length h giving the typical variation in the height of the surface, and another length l giving the scale of correlations between displacements at different points of the surface. The incident plane wave is characterized by its wavelength λ and angle of incidence θ (measured relative to the mean surface, assumed planar); the scattered radiation is characterized by two angles θ' and ϕ'. (We will not consider inelastic scattering by a dynamic surface here, so the wavelength of the scattered radiation is taken to be that of the incident radiation.) The characterization of scattered light is thus in terms of at least three lengths λ, h, l, and three angles θ, θ', ϕ'. In the geometrical optics limit ($\lambda \ll h, l$) the surface may be taken to be locally plane, and thus the scattered light is obtained from the statistical geometry of the surface by assuming specular reflection from each tilted element. This is a good description of the reflection of light from large bodies of water, provided that foam and spray are absent. Cox and Munk (1954), for example, measured the roughness of the sea surface from photographs of the sun's glitter. Longuet-Higgins (1960) has studied in detail the geometry of reflection and refraction at a random moving surface, of light originating a a point source. The reflection of extended objects by gently rippled water is discussed in a recent illustrated note by Lynch (1985). This chapter will concentrate on the wave theory of reflection by rough surfaces, specifically including diffraction effects which arise when the wavelength is comparable to the lengths characterizing the surface.

11-1 Reflection from rough surfaces: the Rayleigh criterion

From the wave theory point of view, a variation in height in the reflecting surface by an amount h will be significant if the resulting path difference is comparable

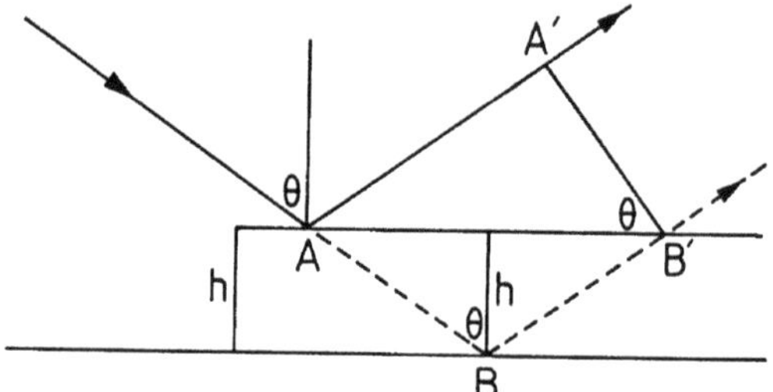

Figure 11-1. Path difference between rays reflecting specularly in the presence and absence of a step of height h. The path difference is $(AB + BB') - AA' = 2h \sec\theta - 2h \tan\theta \sin\theta = 2h \cos\theta$.

to the wavelength λ. Rayleigh (1879) noted that in the specular case the path difference is $2h \cos\theta$ for the simple geometry shown in Figure 11-1.

Since $(2\pi/\lambda)\cos\theta = q$, the normal component of the wavevector of the incident light, we may write the Rayleigh criterion for the specular reflection off a surface as

$$2\pi h \cos\theta \ll \lambda, \quad \text{or} \quad qh \ll 1. \tag{1}$$

The factor π inserted here is arbitrary: a more precise specification comes when one takes a particular model of surface roughness. For the Gaussian model considered in Section 11-4 the operative roughness parameter is

$$\mathscr{R} = (q + q')^2 \langle \zeta^2 \rangle, \tag{2}$$

where q and q' are the normal components of the incident and scattered wavectors, and $\langle \zeta^2 \rangle$ is the mean square surface height variation. The intensity of specularly reflected radiation is proportional to $\exp(-\mathscr{R})$ in this model; non-specularly scattered light depends both on $\mathscr{R}$ and on the product of the lateral correlation length l with the change in the lateral component of the wavenumber.

The Rayleigh criterion (1) in terms of path or phase differences indicates that with a given roughness long waves may be reflected specularly and short waves diffusely, or that for given roughness and wavelength there may be diffuse scattering near normal incidence and specular reflection near grazing incidence. The change from diffuse to specular reflection with angle of incidence is apparent with a plate of ground or smoked glass. Further discussion may be found in Rayleigh's lecture on "Polish" (1901) and in Wood (1934), pp. 39–41.

11-2 Corrugated surfaces: diffraction gratings

The simplest roughness to consider is that of a periodic corrugation of a sharply defined surface at $z = \zeta(x)$:

$$\begin{aligned}\zeta(x) &= \sum_1^\infty (c_n \cos npx + s_n \sin npx) \\ &= \frac{1}{2}\sum_1^\infty [(c_n - is_n)\,\mathrm{e}^{inpx} + (c_n + is_n)\,\mathrm{e}^{-inpx}].\end{aligned} \tag{3}$$

The method to be presented here was developed by Rayleigh (1896, 1907a) and rests on an assumption ("the Rayleigh hypothesis") which will be stated below and discussed again at the end of this section. Consider a plane wave incident (perpendicularly to the corrugations) at an angle of incidence θ relative to the normal to the averaged surface. If $k = 2\pi/\lambda$ is the total wavenumber in the medium of incidence, the incoming wave is

$$\psi_i = e^{ik(x \sin\theta + z \cos\theta)} = e^{i(Kx+qz)}. \tag{4}$$

The specularly reflected wave is

$$\psi_0 = A_0 \, e^{ik(x \sin\theta - z \cos\theta)} = A_0 \, e^{i(Kx-qz)}; \tag{5}$$

for spectra of the nth order the wave is represented by terms like (5) with θ_n, K_n and q_n instead of θ, K and q, where

$$k \sin\theta_n \equiv K_n = K + np, \qquad k \cos\theta_n \equiv q_n. \tag{6}$$

The period of the corrugations is $d = 2\pi/p$; the first part of (6) is the grating equation giving the direction of the nth order, which may be written as the condition that the path difference (to a distant point) between waves originating a distance d apart on the grating is an integral number of wavelengths:

$$d(\sin\theta_n - \sin\theta) = n\lambda. \tag{7}$$

The zeroth order ($n = 0$) is the specularly reflected wave ψ_0. Since the surface shape $\zeta(x)$ is expressed as a sum over positive integers n, while the grating equations (6) or (7) may have negative n, it is convenient to define the primed quantities θ'_n, K'_n and q'_n obtained by changing the sign of n in (6), and sum over positive n:

$$k \sin\theta'_n \equiv K'_n = K - np. \qquad k \cos\theta'_n = q'_n. \tag{6'}$$

The spectra of the nth order are represented by

$$\begin{aligned} \psi_n &= A_n \, e^{ik(x \sin\theta_n - z\cos\theta_n)} + A'_n \, e^{ik(x \sin\theta'_n - z \cos\theta'_n)} \\ &= A_n \, e^{i(K_n z - q_n z)} + A'_n \, e^{i(K'_n x - q'_n z)}. \end{aligned} \tag{8}$$

Since

$$K_n^2 + q_n^2 = k^2 = (K'_n)^2 + (q'_n)^2, \tag{9}$$

the wavefunction

$$\Psi = \psi_i + \psi_0 + \sum_1^\infty \psi_n \tag{10}$$

satisfies the wave equation $(\nabla^2 + k^2)\Psi = 0$ outside the reflecting surface. The Rayleigh hypothesis is that (10), which expresses the total wave as a sum of the incident wave plus all spectral orders, with arbitrary amplitudes A_0, A_n and A'_n, is *complete*. That is, it is assumed that (10) has sufficient generality to satisfy the boundary conditions on an arbitrary periodically corrugated surface. More will be said about this hypothesis later; here we note only that *evanescent waves*, that is those with

$$q_n^2 < 0 \quad \text{or} \quad (q'_n)^2 < 0 \tag{11}$$

are to be included in (10). These correspond to orders that have "passed off", which have exponentially decaying form at large $|z|$. From (6), (6′) and (9), we see that at normal incidence ($K = 0$) the maximum visible order is the integer part of d/λ, so that if $d < \lambda$ only the zeroth or specular order will be seen. At oblique incidence, the maximum visible order is the integer part of $2d/\lambda$, since the maximum value of K is k (attained at grazing incidence), and

$$(q'_n)^2 = k^2 - (K - np)^2 \tag{12}$$

will stay positive when $K = k$ for $np < 2k$ or $n < 2d/\lambda$. Thus if $d < \lambda/2$ there will be only specular reflection, at any angle of incidence, and the corrugated surface no longer acts as a diffraction grating. (There are near-field effects, but no spectra are visible in the far-field region.)

The amplitudes of the spectral orders A_0, A_n and A'_n are found from the boundary conditions to be imposed on (10) at the surface. The simplest case to consider is that of a perfect reflector with the incident wave polarized so that the electric field lies along the corrugations. The boundary condition for this case is $\Psi = 0$ on $z = \zeta$. All of the incident energy is thrown back, and is distributed between the specular beam and the reflected spectra. (The spectra represented by evanescent waves do not radiate energy.) The total wavefunction is

$$\Psi(x, z) = e^{iKx}\{e^{iqz} + A_0 e^{-iqz} + \sum_1^\infty (A_n e^{inpx - iq_n z} + A'_n e^{-inpx - iq'_n z})\}, \tag{13}$$

and is to be zero on $z = \zeta(x)$.

Two methods, both due to Rayleigh, exist for extracting the spectral amplitudes. We shall give an outline of both. In the first method, Rayleigh expands (13) in powers of ζ and equates coefficients of $e^{\pm inpx}$, $\zeta(x)$ being given by (3). If we keep the zeroth and first powers of ζ in (13), $\Psi(x, \zeta(x)) = 0$ reads

$$0 = 1 + A_0 + \sum_1^\infty (A_n e^{inpx} + A'_n e^{-inpx}) + iq\zeta(1 - A_0) + 0(\zeta^2). \tag{14}$$

To the zeroth power of ζ

$$1 + A_0 = 0, \tag{15}$$

and all the A_n and A'_n are zero. To the first power of ζ the value $A_0 = -1$ still holds, and

$$A_n = -iq(c_n - is_n), \qquad A'_n = -iq(c_n + is_n). \tag{16}$$

To this approximation, the amplitudes of $\pm n$th orders are given by the nth Fourier coefficients of the corrugation. If the corrugation is purely sinusoidal (only c_1 and s_1 non-zero) only the ± 1 diffraction orders have amplitudes which are non-zero when the calculation is taken to the first power of ζ. Rayleigh (1907a) gives the coefficients to the second power of ζ.

Rayleigh's second method uses the Jacobi expansion

$$e^{i\beta\cos\phi} = J_0(\beta) + 2\sum_1^\infty i^n \cos n\phi\, J_n(\beta), \tag{17}$$

which follows on substituting $t = i\,e^{i\phi}$ into the generating function for Bessel functions (Watson 1944, Section 2.1)

$$e^{\beta(t-1/t)/2} = \sum_{-\infty}^{\infty} t^n J_n(\beta) = J_0(\beta) + \sum_1^\infty (t^n + (-)^n t^{-n}) J_n(\beta). \tag{18}$$

We will demonstrate the method for the simplest case of normal incidence onto a pure cosine corrugation,

$$\zeta(x) = a\cos px \tag{19}$$

($c_1 = a$, all other c_n and s_n are zero in (3)). At normal incidence $K = 0$, $q = k$, $q_n = q_n' = (k^2 - n^2p^2)^{1/2}$, and for a cosine corrugation the diffraction pattern is symmetric, with $A_n = A_n'$. Thus (13) becomes

$$\Psi(x, z) = e^{ikz} + A_0\,e^{-ikz} + 2\sum_1^\infty A_n \cos npx\, e^{-iq_nz}, \tag{20}$$

and the boundary condition $\Psi(x, \zeta) = 0$ reads

$$e^{i\beta\cos px} + A_0\,e^{-i\beta\cos px} + 2\sum_1^\infty A_n\cos npx\, e^{-i\beta_n\cos px} = 0, \tag{21}$$

where $\beta = ka$, $\beta_n = q_na$. On applying (17) and setting the coefficients of $\cos npx$ equal to zero for $n = 0, 1, 2, \ldots$ one obtains an infinite set of linear equations for the coefficients A_n. The first of these, obtained by setting the coefficient of the term independent of x equal to zero, is

$$(1 + A_0)J_0(\beta) + 2\sum_1^\infty A_n(-i)^n J_n(\beta_n) = 0. \tag{22}$$

The equation resulting from setting the coefficient of $\cos npx$ to zero is, for $n \geqslant 1$,

$$[i^n + (-i)^n A_0]J_n(\beta) + A_nJ_0(\beta_n) + \sum_l\sum_m A_l(-i)^m J_m(\beta_l) = 0, \tag{23}$$

where the sum over l and m is restricted to values $\geqslant 1$ which have $|l \pm m| = n$. The infinite set of linear equations for the spectral amplitudes may be solved, approximately but without assuming β to be small, by truncating at some $n = N$. That is, all A_n for $n > N$ are set equal to zero, and the linear system is solved for the $N + 1$ unknowns $A_0, A_1, \ldots, A_N$. The solution is checked by increasing n to see if this produces an appreciable difference. Another check is to solve the modified but equally valid systems obtained by multiplying (21) by an arbitrary power of $e^{i\cos px}$. For example, the terms independent of x in (21) multiplied by $e^{i\beta\cos px}$ or $e^{-i\beta\cos px}$ are respectively

$$J_0(2\beta) + A_0 + 2\sum_1^\infty A_n i^n J_n(\beta - \beta_n) = 0, \tag{24}$$

$$1 + A_0J_0(2\beta) + 2\sum_1^\infty A_n(-i)^n J_n(\beta + \beta_n) = 0. \tag{25}$$

We now turn to consider the other polarization, with the electric field in the incident wave perpendicular to the corrugations. In this case Ψ corresponds to B_y (in the other polarization it stood for E_y), and the electric field is proportional to curl **B**, that is to $(-\partial\Psi/\partial z,\ 0,\ \partial\Psi/\partial x)$. For a perfect reflector, the component of the electric field parallel to the (local) surface is to be zero. The boundary condition is thus

$$\left(\frac{\partial\Psi}{\partial z} - \frac{\partial\Psi}{\partial x}\frac{\mathrm{d}\zeta}{\mathrm{d}x}\right)_{z=\zeta} = 0, \tag{26}$$

where Ψ is the total field, again given by (13). Expansion in powers of ζ now gives $A_0 = 1$ from the coefficients of the zeroth power, and

$$\begin{aligned} A_n &= i(c_n - is_n)(q^2 - nKp)/q_n, \\ A_n' &= i(c_n + is_n)(q^2 + nKp)/q_n', \end{aligned} \tag{27}$$

from the coefficients of the first power of ζ, A_0 being unchanged to this order. We note the fact that (in this approximation) the coefficients A_n or A_n' diverge when q_n or q_n' go through zero, that is at the passing off of the nth order. Rayleigh shows that the passing off of an order of the spectrum can have an effect on other orders. For example: in the special case of normal incidence, and a corrugation for which only c_1 and c_2 are non-zero in the Fourier expansion (3), calculation of A_1 to the second power in ζ gives

$$A_1 = \frac{ik^2c_1}{q_1} - \frac{k^2c_1c_2}{2q_1^2}(q_1^2 + 2p^2) - \frac{k^2c_1c_2}{2q_1q_2}(q_2^2 + 2p^2). \tag{28}$$

To this approximation, the coefficient of the *first* order can diverge when the *second* order is passing off. According to Rayleigh (1907a), "we may at least infer the probability of abnormalities in the brightness of any spectrum at the moment when one of a higher order is disappearing, abnormalities limited, however, to the case where the electric displacement is perpendicular to the ruling". In a subsequent paper Rayleigh (1907b) used these results to interpret and explain anomalies found by Wood (1902) under precisely these conditions. There are in fact two types of *Wood's anomalies*: a sharp anomaly, appearing as a sudden change of intensity along the spectrum at frequencies and indices corresponding to a passing off of a higher order; and a diffuse anomaly related to resonance in the production of surface waves in the grating. Grating anomalies and electromagnetic surface modes are reviewed by Maystre (1982).

The preceding analyses were based on the Rayleigh hypothesis, namely on the assumption that the wavefunction (10) is sufficiently general to satisfy the boundary conditions on an arbitrary corrugated surface. The hypothesis turns out to be true for some cases and false in others, but in a restricted least-squares sense it can always be applied. Petit and Cadilhac (1966) showed that for the pure cosine corrugation $\zeta(x) = a\cos px$, and with $\Psi = 0$ on the boundary, the Rayleigh hypothesis breaks down for $pa > 0.4477432\ldots$, that is when the amplitude a of the corrugation is greater than about 7% of its period $2\pi/p$. This number comes from solving the transcendental equation

$$e^x = \frac{x+1}{x-1}, \qquad pa = \tfrac{1}{2}(x - 1/x), \tag{29}$$

which is obtained by considering properties of the solution (10) analytically continued across the boundary $z = \zeta(x)$. Millar (1971) later demonstrated that the Rayleigh hypothesis is valid for this problem when pa is smaller than the critical value given above. These papers thus showed that there are situations in which the Rayleigh hypothesis is valid and others in which it is not. From the practical point of view, the Rayleigh hypothesis may always be used if the coefficients are determined by satisfying the boundary conditions in the least-squares sense, since it has been shown that there is a linear combination of N elements of the set of plane waves in (10) that converges on the boundary to the prescribed boundary values, in the mean-square sense, as $N \to \infty$ (Yasuura 1971, Millar 1973).

The Rayleigh method can be used to study scattering by non-periodic surfaces $z = \zeta(x)$ or $z = \zeta(x, y)$, since the function ζ can artificially be made periodic by repetition of a large section of the surface, and ζ can then be expanded in a single or double Fourier series. We noted above (see (6), (11) and the discussion following (28)) that evanescent waves, with imaginary components of the normal component of the wavenumber, and corresponding to grating orders that have "passed off", are associated with the production of surface waves. A rough surface thus enables coupling of an incoming plane wave to electromagnetic surface waves. For metal surfaces this effect is important in the study of surface roughness and of surface electromagnetic waves in metals (referred to in solid state physics as surface plasmons or surface polaritons). Reviews of this field may be found in the collection edited by Agranovich and Mills (1982): see in particular the chapters by Raether (1982) and Maradudin (1982); further references will be given at the end of this chapter. For smooth surfaces, coupling to electromagnetic surface waves is possible via attenuated total reflection, as discussed in Section 8-6.

11-3 Scattering of light by liquid surfaces

A clean and undisturbed liquid surface, such as that of mercury or water, gives an impression of perfect smoothness. The liquid surface is however roughened by thermal excitation of surface waves. Mandelstam (1913) calculated the angular distribution of the light scattered by the thermally induced fluctuations in a liquid surface, in the plane of incidence. His calculation is based on expanding the surface distortion in terms of a double Fourier series,

$$\zeta(x, y, t) = \sum_{\mathbf{k}} e^{i\mathbf{k}\cdot\mathbf{r}}\, \zeta_{\mathbf{k}}(t). \tag{30}$$

(In this section $\mathbf{k}$ is a two-dimensional wavevector in the plane of the surface, and $\mathbf{r} = (x, y)$ a two-dimensional position vector in the plane.) It follows from the equations of continuum hydrodynamics that the Fourier components $\zeta_{\mathbf{k}}$ are the normal mode coordinates for the surface vibrations, since the Hamiltonian of the excitations is, to second order in $\zeta_{\mathbf{k}}$,

$$H = \tfrac{1}{2}A \sum_{\mathbf{k}} \left\{ \frac{\varrho}{k} |\partial\zeta_{\mathbf{k}}/\partial t|^2 + (\varrho g + \sigma k^2)|\zeta_{\mathbf{k}}|^2 \right\}, \tag{31}$$

where A is the area of the surface, ϱ is the mass density of the liquid, g is the acceleration due to gravity, and σ is the surface tension. Comparison of (31) with the harmonic oscillator Hamiltonian $\frac{1}{2}m(\dot{x}^2 + \omega^2 x^2)$ shows that the effective mass m_k and angular frequency ω_k of the mode corresponding to $\zeta_{\mathbf{k}}$ are given by

$$m_k = \frac{\varrho A}{k}, \qquad \omega_k^2 = gk + \frac{\sigma k^3}{\varrho}. \tag{32}$$

It then follows from statistical mechanics that the thermal average of $|\zeta_k|^2$ is

$$\langle|\zeta_{\mathbf{k}}|^2\rangle = \frac{\hbar}{2m_k\omega_k} \coth \frac{\hbar\omega_k}{2T}. \tag{33}$$

where $\hbar$ is Planck's constant divided by 2π and T is the temperature, expressed in energy units. When $T \gg \hbar\omega_k$, which holds for most of the surface excitations of classical liquids but not for those of the quantum liquids He^3, He^4 and H_2, (33) gives

$$\langle|\zeta_{\mathbf{k}}|^2\rangle \simeq \frac{T}{m_k\omega_k^2} = \frac{T}{A(\varrho g + \sigma k^2)}. \tag{34}$$

Mandelstam's calculation is based on Rayleigh's theory of gratings, as outlined in the last section. There we assumed for simplicity that the corrugated surface was a perfect reflector. This would be a good approximation for the surfaces of liquid metals, but a very poor approximation if applied to dielectric fluids. We will therefore give a condensed treatment of Rayleigh's calculation for reflection and scattering off a corrugated interface between two media of dielectric functions ε and $\tilde{\varepsilon}$. A plane electromagnetic wave of angular frequency ω is incident at angle θ onto a one-dimensionally corrugated surface

$$\zeta(x) = \Sigma'\zeta_n \, e^{inpx}, \tag{35}$$

where the sum is over positive and negative integers n excluding zero. This Fourier series is equivalent to (3) if $\zeta_n = (c_n - is_n)/2$ for positive n and $(c_n + is_n)/2$ for negative n. The incident, specularly reflected, and nth order diffracted waves are again given by

$$\psi_i = e^{i(Kx+qz)}, \quad \psi_0 = A_0 \, e^{i(Kx-qz)}, \quad \psi_n = A_n \, e^{i(K_n x - q_n z)}, \tag{36}$$

where the tangential and normal components of the wavevectors are related by

$$K^2 + q^2 = \varepsilon\omega^2/c^2, \quad K = \varepsilon^{1/2}(\omega/c)\sin\theta, \quad q = \varepsilon^{1/2}(\omega/c)\cos\theta, \tag{37}$$

$$K_n^2 + q_n^2 = \varepsilon\omega^2/c^2, \quad K_n = \varepsilon^{1/2}(\omega/c)\sin\theta_n, \quad q_n = \varepsilon^{1/2}(\omega/c)\cos\theta_n, \tag{38}$$

with K_n given by the grating equation (6) or (6′), $K_n = K + np$. The total wave on the incidence side of the surface is

$$\Psi = \psi_i + \psi_0 + \Sigma'\psi_n. \tag{39}$$

The total wave on the other side is

$$\tilde{\Psi} = \tilde{\psi}_0 + \Sigma'\tilde{\psi}_n = B_0\,\mathrm{e}^{i(Kx+\tilde{q}z)} + \Sigma'\,B_n\,\mathrm{e}^{i(K_nx+\tilde{q}_nz)}, \tag{40}$$

where

$$K^2 + \tilde{q}^2 = \tilde{\varepsilon}\omega^2/c^2, \quad K = \tilde{\varepsilon}^{1/2}(\omega/c)\sin\tilde{\theta}, \quad \tilde{q} = \tilde{\varepsilon}^{1/2}(\omega/c)\cos\tilde{\theta}, \tag{41}$$

$$K_n^2 + \tilde{q}_n^2 = \tilde{\varepsilon}\omega^2/c^2, \quad K_n = K + np, \quad \tilde{q}_n = \tilde{\varepsilon}^{1/2}(\omega/c)\cos\tilde{\theta}_n. \tag{42}$$

We will consider the case where the electric field vector is along the corrugations. Then Ψ and $\tilde{\Psi}$ represent the electric field, and the boundary conditions are the equality of Ψ and $\tilde{\Psi}$ and of their normal derivatives on $z = \zeta(x)$. The latter condition implies that

$$\frac{\partial\Psi}{\partial z} - \frac{\partial\Psi}{\partial x}\frac{\mathrm{d}\zeta}{\mathrm{d}x} = \frac{\partial\tilde{\Psi}}{\partial x} - \frac{\partial\tilde{\Psi}}{\partial x}\frac{\mathrm{d}\zeta}{\mathrm{d}x} \tag{43}$$

on the surface (compare (26)). The equality of Ψ and $\tilde{\Psi}$ on $z = \zeta(x)$ gives, after removal of the common factor e^{iKx},

$$\mathrm{e}^{iq\zeta} + A_0\,\mathrm{e}^{-iq\zeta} + \Sigma'A_n\,\mathrm{e}^{i(npx-q\zeta)} = B_0\,\mathrm{e}^{i\tilde{q}\zeta} + \Sigma'B_n\,\mathrm{e}^{i(npx+\tilde{q}\zeta)}. \tag{44}$$

To first order in ζ this reads

$$1 + A_0 - B_0 + \Sigma'\,\mathrm{e}^{inpx}(A_n - B_n) + i\zeta(q - qA_0 - \tilde{q}B_0) = 0. \tag{45}$$

The zeroth order part gives $B_0 = 1 + A_0$, and the first order part (on using (35) and equating coefficients of e^{inpx}) gives

$$B_n - A_n = i\zeta_n\{q(1 - A_0) - \tilde{q}B_0\}. \tag{46}$$

The boundary condition (43) may be written as

$$\frac{\partial}{\partial z}(\Psi - \tilde{\Psi}) = \frac{\mathrm{d}\zeta}{\mathrm{d}x}\frac{\partial}{\partial x}(\Psi - \tilde{\Psi}) \quad \text{on} \quad z = \zeta(x), \tag{47}$$

in which we use

$$\Psi - \tilde{\Psi} = \mathrm{e}^{iKx}\{\mathrm{e}^{iqz} + A_0\,\mathrm{e}^{-iqz} - B_0\,\mathrm{e}^{i\tilde{q}z} + \Sigma'\,\mathrm{e}^{inpx}(A_n\,\mathrm{e}^{-iq_nz} - B_n\,\mathrm{e}^{i\tilde{q}_nz})\}. \tag{48}$$

Thus (47) reads, to the first order in ζ,

$$\begin{aligned} q(1 - A_0) - \tilde{q}B_0 + i\zeta[q^2(1 + A_0) - \tilde{q}^2B_0] - \Sigma'\,\mathrm{e}^{inpx}(q_nA_n + \tilde{q}_nB_n) \\ = pK(1 + A_0 - B_0)\Sigma'n\zeta_n\,\mathrm{e}^{inpx}, \end{aligned} \tag{49}$$

of which the zeroth order part is $q(1 - A_0) = \tilde{q}B_0$, and the first order part gives, on equating coefficients of e^{inpx} and using the fact that $1 + A_0 = B_0$,

$$q_nA_n + \tilde{q}_nB_n = i\zeta_n[q^2(1 + A_0) - \tilde{q}^2B_0]. \tag{50}$$

The zeroth order parts $1 + A_0 = B_0$ and $q(1 - A_0) = \tilde{q}B_0$ give

$$A_0 = \frac{q - \tilde{q}}{q + \tilde{q}}, \qquad B_0 = \frac{2q}{q + \tilde{q}}. \tag{51}$$

Thus A_0 and B_0 are the Fresnel reflection and transmission amplitudes for a flat surface (Section 1-1). Also (46) implies $B_n = A_n$ (to this order in ζ) and (50) gives

$$\begin{aligned} A_n = B_n &= 2i\zeta_n q(q - \tilde{q})/(q_n + \tilde{q}_n) \\ &= 2i\zeta_n \varepsilon^{1/2} \frac{\omega}{c} \frac{\cos\theta \sin(\tilde{\theta} - \theta)}{\sin\tilde{\theta}\cos\theta_n + \sin\theta\cos\tilde{\theta}_n}. \end{aligned} \tag{52}$$

Equation (52) gives the angular dependence of the amplitude of the non-specularly reflected or transmitted radiation due to the Fourier component ζ_n of the corrugation. The average intensity is proportional to $\langle |A_n|^2 \rangle$ and thus to $\langle |\zeta_n|^2 \rangle$; the latter is found from (35) on identifying the wavenumber k of the surface waves with the change in the lateral component of the radiation wavevector, namely $np = K_n - K = \varepsilon^{1/2}(\omega/c)(\sin\theta_n - \sin\theta)$. The intensity (in the plane of incidence) scattered into the direction θ_n by a thermally roughened surface is thus proportional to

$$\langle |A_n|^2 \rangle = 4\varepsilon \frac{\omega^2}{c^2} \frac{T/A}{\varrho g + \sigma(K_n - K)^2} \left\{ \frac{q(q - \tilde{q})}{q_n + \tilde{q}_n} \right\}^2. \tag{53}$$

This, in essence, is the result of Mandelstam (1913). Andronov and Leontovich (1926) and Gans (1926) later obtained formulae for the scattered intensity which are not restricted to the plane of incidence. Bouchiat and Langevin (1978) have extended these results by including scattering by surface structural and orientational fluctuations, but still keeping the assumption of a discontinuous interface. Earlier, but in an unpublished thesis, Triezenberg (1973) had in fact constructed a perturbation theory for light scattering from thermal fluctuations in a thin diffuse interface. Triezenberg's results are however restricted to the s polarization (electric field perpendicular to the plane of incidence), and thus are not applicable to ellipsometric data. Beaglehole (1980) first gave a theoretical expression for the ellipsometric ratio r_p/r_s to first order in the distortions of an interface. It takes the form (3.46), namely that obtained for a planar diffuse interface, with

$$\mathscr{I}_1 = \frac{(\varepsilon - \tilde{\varepsilon})^2}{\varepsilon + \tilde{\varepsilon}} \frac{Tk_{\max}}{\pi\sigma}, \tag{54}$$

where $k_{\max}$ is a wavenumber cut-off for the thermally excited surface waves. This result was deduced using the work of Bedeaux and Vlieger (1973) and Kretschman and Kröger (1975). A more complete theory leading to (54) has been given by Zielinska, Bedeaux and Vlieger (1981, 1983). Further discussion may be found in Beaglehole (1982) and (1983).

11-4 The surface integral formulation of scattering by rough surfaces

The preceding two sections have introduced and applied the Rayleigh method of calculating scattering by rough surfaces. An alternative method is based on the Kirchhoff formulation of diffraction theory. It is sometimes called the Kirchhoff method, although the approach was developed by M. L. Antokolskii, L. M. Brekhovskikh, M. A. Isakovich, H. Davis, P. Beckmann and others. References to the original work may be found in Beckmann and Spizzichino (1963), Shmelev (1972) and Bass and Fuks (1979); the outline given below is based on Beckmann's part of the book by Beckmann and Spizzichino.

The method to be described is based on the following theorem of Helmholtz (Baker and Copson 1950, Section 4.2): if E is a solution of $(\nabla^2 + k^2)E = 0$ whose first and second partial derivatives are continuous within and on a closed surface S, R is the distance from a fixed point P and $\partial/\partial n$ denotes differentiation along the inward normal to S, then if P lies inside S,

$$E(P) = \frac{1}{4\pi}\int\int \mathrm{d}S\left\{E\frac{\partial}{\partial n}\left(\frac{\mathrm{e}^{ikR}}{R}\right) - \frac{\mathrm{e}^{ikR}}{R}\frac{\partial E}{\partial n}\right\}. \tag{55}$$

Thus the solution of the wave equation $(\nabla^2 + k^2)E = 0$ at an interior point of a region can be found in terms of the values of E and $\partial E/\partial n$ on the boundary of the region.

Consider now the evaluation of the integral in (55) in terms of the (presumed known) values of E on a rough surface, when the point P is taken to be far from the surface. If R is the distance to P from a radiating element on the surface, which is at $\mathbf{r}$ relative to an origin near the surface, and $\mathbf{k}$ and $\mathbf{k}'$ are the wavevectors of the incoming and scattered radiation, then $R \simeq R_0 - \mathbf{k}'\cdot\mathbf{r}/k$, where R_0 is the distance from the origin to P, and k is the magnitude of $\mathbf{k}$ and $\mathbf{k}'$. In the surface integral (55), the spherical wave radiating from the element at $\mathbf{r}$ may be approximated by

$$\frac{\mathrm{e}^{ikR}}{R} \simeq \frac{\mathrm{e}^{ikR_0}}{R_0}\,\mathrm{e}^{-i\mathbf{k}'\cdot\mathbf{r}}. \tag{56}$$

The unknowns in (55), namely the field E and its normal derivatives $\partial E/\partial n$, are approximated from the field that would be present at a given point on the surface if the surface were replaced by its tangent plane at that point. If the radiation is excited by an incident plane wave $\exp(i\mathbf{k}\cdot\mathbf{r})$, and is assumed to have $\mathbf{E}$ locally along the surface (the latter can only be true in an average sense for a surface with two-dimensional roughness), and if further the surface is assumed to reflect perfectly, the scattered intensity which follows from the squared modulus of (55) is shown to be proportional to

$$F^2\int_{-\infty}^{\infty}\mathrm{d}x\int_{-\infty}^{\infty}\mathrm{d}y\,\mathrm{e}^{ix\Delta K_x + iy\Delta K_y}\,\chi(\Delta q, \varrho). \tag{57}$$

Here ΔK_x and ΔK_y are the x and y components of $\mathbf{k} - \mathbf{k}'$, and $\Delta q = q + q'$ is the z component of $\mathbf{k} - \mathbf{k}'$. χ is the expectation value of $\exp(i\Delta q(\zeta_1 - \zeta_2))$, where ζ_1 and ζ_2 are the displacements of the surface (from a mean value of zero) at two points separated on the surface by the distance $\varrho = (x^2 + y^2)^{1/2}$. The factor F depends on the angle of incidence θ, and on the direction of the scattered radiation (θ', ϕ'), with θ and θ' being measured relative to the normal to the averaged surface, and ϕ' being the azimuthal deviation of the scattered radiation from the specular plane:

$$F(\theta, \theta', \phi') = \frac{1 - \cos\theta\cos\theta' - \sin\theta\sin\theta'\cos\phi'}{\cos\theta(\cos\theta + \cos\theta')}. \tag{58}$$

On transforming the integration variables from x, y to ϱ, ϕ via $x = \varrho\cos\phi$, $y = \varrho\sin\phi$, (57) becomes

$$F^2 \int_0^\infty \mathrm{d}\varrho\, \varrho\, \chi(\Delta q, \varrho) \int_0^{2\pi} \mathrm{d}\phi\, \mathrm{e}^{i(\alpha\cos\phi + \beta\sin\phi)}, \tag{59}$$

where $\alpha = \varrho\Delta K_x$ and $\beta = \varrho\Delta K_y$. The integral over ϕ is equal to $2\pi J_0\{(\alpha^2 + \beta^2)^{1/2}\}$; this follows from the Jacobi expansion (17). Thus the scattered intensity is proportional to

$$2\pi F^2 \int_0^\infty \mathrm{d}\varrho\, \varrho\, \chi(\Delta q, \varrho) J_0(\varrho\Delta K), \tag{60}$$

where $\Delta K = (\Delta K_x^2 + \Delta K_y^2)^{1/2}$. The scattered intensity is seen to depend on both Δq and ΔK, the magnitudes of the normal and lateral components of the change in wavevector due to the scattering, $\mathbf{k} - \mathbf{k}'$.

An important special case is that of scattering by a normally distributed surface (Beckmann, Section 5.3) for which

$$\chi(\Delta q, \varrho) = \exp\{-\mathscr{R}(1 - C(\varrho))\}. \tag{61}$$

In (61) $\mathscr{R}$ is the roughness parameter defined in Section 11-1,

$$\mathscr{R} = \langle\zeta^2\rangle(\Delta q)^2, \tag{62}$$

and $C(\varrho)$ is the correlation function for vertical displacements of the surface, characterized by a lateral correlation length l:

$$C(\varrho) = \frac{\langle\zeta(0)\zeta(\varrho)\rangle}{\langle\zeta^2\rangle} = \exp(-\varrho^2/l^2). \tag{63}$$

In this case (60) becomes

$$2\pi F^2\, \mathrm{e}^{-\mathscr{R}} \sum_0^\infty \frac{\mathscr{R}^n}{n!} \int_0^\infty \mathrm{d}\varrho\, \varrho\, \mathrm{e}^{-n\varrho^2/l^2} J_0(\varrho\Delta K). \tag{64}$$

The integral in (64) is known as Weber's first exponential integral (Watson 1944, Section 13.3), and is equal to $(l^2/2n)\exp[-(l\Delta K)^2/4n]$. The $n = 0$ term is zero, except for exactly specular reflection ($\Delta K = 0$), in which case it diverges. This divergence is a consequence of having calculated the intensity for an infinite and perfectly reflecting surface (see also the discussion following (3.34) in Rice, 1951). We omit the $n = 0$ term; the non-specular intensity is thus proportional to

$$\pi F^2 l^2\, \mathrm{e}^{-\mathscr{R}} \sum_1^\infty \frac{\mathscr{R}^n}{n!n} \mathrm{e}^{-(l\Delta K)^2/4n}. \tag{65}$$

The series in (65) has the upper and lower bounds

$$\mathscr{R}\,\mathrm{e}^{-(l\Delta K)^2/4} < \sum_1^\infty \frac{\mathscr{R}^n}{n!n}\,\mathrm{e}^{-(l\Delta K)^2/4n} \leqslant \int_0^{\mathscr{R}} \frac{\mathrm{d}r}{r}(\mathrm{e}^r - 1). \tag{66}$$

The lower bound comes from taking the first term in the series, the upper bound from setting $l\Delta K = 0$. For a slightly rough surface (small $\mathscr{R}$), (65) is approximately equal to

$$\pi F^2 l^2 \mathscr{R} \exp\{-[\mathscr{R} + (l\Delta K/2)^2]\}. \tag{67}$$

For other than specular reflection the intensity depends on both the mean-square height variation $\langle\zeta^2\rangle$, and on the lateral correlation length l.

The specular intensity has to be obtained separately. Beckmann shows that it depends on $(\Delta q)^2\langle\zeta^2\rangle$ only, being proportional to $\mathrm{e}^{-\mathscr{R}}$. Details may be found in Beckmann (Section 5.3) and Chandley and Welford (1975); see also Bennett and Porteus (1961) for a discussion and experimental test of the $\mathrm{e}^{-\mathscr{R}}$ law at normal incidence, and Nieto-Vesperinas and García (1981) for an analysis of the validity of the surface integral method.

References

References quoted in the text

C. Cox and W. Munk (1954) "Measurement of the roughness of the sea surface from photographs of the sun's glitter", J. Opt. Soc. Amer. **44**, 838–850.

M. S. Longuet-Higgins (1960) "Reflection and refraction at a random moving surface: I – Pattern and paths of specular points; II – Number of specular points in a Gaussian surface; III – Frequency of twinkling in a Gaussian surface", J. Opt. Soc. Amer. **50**, 838–856.

D. K. Lynch (1985) "Reflections on closed loops", Nature **316**, 216–217.

J. W. S. Rayleigh (1879) "On the accuracy required in optical surfaces", Section 5 of Article 62, Scientific Papers, Vol. I (Cambridge 1899, Dover 1964).

J. W. S. Rayleigh (1901) "Polish", Article 268 in Scientific Papers, Vol. IV.

R. W. Wood (1934) "Physical Optics", Macmillan.

J. W. S. Rayleigh (1896) "Theory of sound", Section 272a (Dover 1954).

J. W. S. Rayleigh (1907a) "On the dynamical theory of gratings", Article 322 in Scientific Papers, Vol. V.

G. N. Watson (1944) "Theory of Bessel functions", Cambridge.

J. W. S. Rayleigh (1907b) "Note on the remarkable case of diffraction spectra described by Prof. Wood", Article 323 in Scientific Papers, Vol. V.

R. W. Wood (1902) "On a remarkable case of uneven distribution of light in a diffraction grating spectrum", Phil Mag. **4**, 396–402.

D. Maystre (1982) "General study of grating anomalies from electromagnetic surface modes", Chapter 17 in "Electromagnetic surface modes", A. D. Boardman (ed.), Wiley.

R. Petit and M. Cadilhac (1966) "Sur la diffraction d'une onde plane par un réseau infiniment conducteur", C.R. Acad Sc. Paris **B263**, 468–471.

R. F. Millar (1971) "On the Rayleigh assumption in scattering by a periodic surface, 2", Proc. Camb. Phil. Soc. **69**, 217–225.

K. Yasuura (1971) "A view of numerical methods in diffraction problems", Progress in Radio Science 1966–1969, Vol. 3, 257–270, (International Union of Radio Science, Brussels).

R. F. Millar (1973) "The Rayleigh hypothesis and a related least-squares solution to scattering problems for periodic surfaces and other scatterers", Radio Science **8**, 785–796.
V. M. Agranovich and D. L. Mills (eds) (1982) "Surface polaritons: electromagnetic waves at surfaces and interfaces", North-Holland.
H. Raether (1982) "Surface plasmons and roughness", Chapter 9 in the volume edited by Agranovich and Mills.
A. A. Maradudin (1982) "Interaction of surface polaritons and plasmons with surface roughness", Chapter 10 in the volume edited by Agranovich and Mills.
L. Mandelstam (1913) "Über die Rauhigkeit freier Flüssigkeitsoberflächen", Annalen der Physik **41**, 609–624.
A. A. Andronov and M. A. Leontovich (1926) "Zur Theorie der molekularen Lichtzerstreuung an Flüssigkeitsoberflächen", Z. Physik **38**, 485–501.
R. Gans (1926) "Lichtzerstreuung infolge der molekularen Rauhigkeit der Trennungsfläche zweier durchsichtiger Medien", Annalen der Physik **79**, 204–226.
M. A. Bouchiat and D. Langevin (1978) "Relation between molecular properties and the intensity scattered by a liquid interface", J. of Colloid and Interface Science, **63**, 193–211.
D. G. Triezenberg (1973) "Capillary surface waves in a diffuse liquid–gas interface", Ph.D. Thesis, University of Maryland.
D. Beaglehole (1980) "Ellipsometric stud- of the surface of simple liquids", Physica **100B**, 163–174.
D. Bedeaux and J. Vlieger (1973) "A phenomenological theory of the dielectric properties of thin films", Physica **67**, 55–73.
E. Kretschmann and E. Kröger (1975) "Reflection and transmission of light by a rough surface, including results for surface-plasmon effects", J. Opt. Soc. Amer. **65**, 150–154.
B. J. A. Zielinska, D. Bedeaux and J. Vlieger (1981, 1983) "Electric and magnetic susceptibilities for a fluid–fluid interface": I "The ellipsometric coefficient", Physica **107A**, 81–108; II "Critical behaviour", Physica **117A**, 28–46.
D. Beaglehole (1982) "Short-ranged roughness on the near-critical liquid interface", Physica **112B**, 320–330.
D. Beaglehole (1983) "Ellipsometry of liquid surfaces", J. de Physique, **C10**, 147–154.
P. Beckmann and A. Spizzichino (1963) "The scattering of electromagnetic waves from rough surfaces", Pergamon.
A. B. Shmelev (1972) "Wave scattering by statistically uneven surfaces", Sov. Phys. Uspekhi **15**, 173–183.
F. G. Bass and I. M. Fuks (1979) "Wave scattering by statistically rough surfaces", Pergamon.
B. B. Baker and E. T. Copson (1950) "The mathematical theory of Huygen's principle", Oxford.
S. O. Rice (1951) "Reflection of electromagnetic waves from slightly rough surfaces", Common. Pure Appl. Math. **4**, 351–378.
P. J. Chandley and W. T. Welford (1975) "A re-formulation of some results of P. Beckmann for scattering from rough surfaces", Optical and Quantum Electronics **7**, 393–397.
H. E. Bennett and J. O. Porteus (1961) "Relation between surface roughness and specular reflectance at normal incidence", J. Opt. Soc. Amer. **51**, 123–129.
M. Nieto-Vesperinas and N. García (1981) "A detailed study of the scattering of scalar waves from random rough surfaces", Optica Acta **289**, 1651–1672.

Additional references on gratings

G. W. Stroke (1967) "Diffraction gratings", Handbuch der Physik **29**, 426–754, Springer.
R. Petit (ed) (1980) "Electromagnetic theory of gratings", Springer.

Metallic rough surfaces and surface plasmons

E. Kröger and E. Kretschmann (1970) "Scattering of light by slightly rough surfaces on thin films including plasmon resonance emission", Z. Physik **237**, 1–15.
D. Beaglehole and O. Hunderi (1970) "Study of the interaction of light with rough metal surfaces. I – Experiment, II-Theory", Phys. Rev. **B2**, 309–321, 321–329.

J. M. Elson and R. H. Ritchie (1971) "Photon interactions at a rough metal surface", Phys. Rev. **B4**, 4129–4138.

V. Celli, A. Marvin and F. Toigo (1975) "Light scattering from rough surfaces", Phys. Rev. **B11**, 1779–1786.

A. Marvin, F. Toigo and V. Celli (1975) "Light scattering from rough surfaces: general incidence angle and polarization", Phys. Rev. **B11**, 2777–2782.

J. M. Keller, R. Fuchs, and K. L. Kliever (1975) "*p*-Polarized optical properties of a metal with a diffusely scattering surface", Phys. Rev. **B12**, 2012–2029.

O. Hunderi (1980) "Optics of rough surfaces, discontinuous films and heterogeneous materials", Surface Science **96**, 1–31.

D. L. Hornauer and H. Raether (1981) "Determination of roughness of LiF films using guided light modes", Opt. Commun. **40**, 105–110.

Scattering from liquid surfaces

A. Vrij, J. G. H. Joosten and H. M. Fijnaut (1981) "Light scattering from thin liquid films", Adv. Chem. Phys. **48**, 329–396.

J. Peternelj (1981) "Scattering of light due to the density fluctuations in the interface region of a liquid-vapour system", Can. J. Phys. **59**, 1009–1019.

J. Meunier and D. Langevin (1982) "Optical reflectivity of a diffuse interface", J. de Physique Lett. **43**, 185–191.

Experimental studies of surface roughness

J. M. Elson, H. E. Bennett and J. M. Bennett (1979) 'Scattering from optical surfaces", Ch. 7 of Vol. 7, Applied Optics and Optical Engineering, Academic Press.

C. L. Rufenach and W. R. Alpers (1981) "Imaging ocean waves by synthetic aperture radars with long integration times", IEEE Trans. **AP-29**, 422–428.

Y. Wang and W. L. Wolfe (1983) "Scattering from microrough surfaces: comparison of theory and experiment", J. Opt. Soc. Amer. **73**, 1596–1602.

J. M. Bennett (1985) "Comparison of techniques for measuring the roughness of optical surfaces", Opt. Eng. **24**, 380–387.

Laser speckle and surface roughness

J. C. Dainty (ed) (1975) "Laser speckle and related phenomena", Springer.

W. T. Welford (1980) "Laser speckle and surface roughness", Contemp. Phys. **21**, 401–412.

12

Matrix methods

The idea of representing an arbitrary stratification by a series of uniform layers goes back at least to Rayleigh (1912). The problem of wave propagation through the transition is solved by matching the wave amplitude and derivative at the boundaries of the uniform layers, and letting the number of layers increase and their thickness decrease. (In the case of finite number of uniform layers, such as optical coatings, this limiting process is not required.) Rayleigh carried through the necessary algebra without reference to matrices; Weinstein (1947), Herpin (1947) and Abelès (1950, 1967) have shown how matrix algebra simplifies and systematizes this approach. We will give three versions of the matrix method, of which the last (given in Section 12-2) is the closest to that currently in use, but differs from it in having all matrix elements real in the absence of absorption.

12-1 Matrices relating the coefficients of linearly independent solutions

We consider the reflection problem for waves satisfying

$$\frac{d^2\psi}{dz^2} + q^2\psi = 0, \qquad e^{iq_a z} + r\,e^{-iq_a z} \leftarrow \psi \rightarrow t\,e^{iq_b z}. \tag{1}$$

(A change of notation from our usual q_1, q_2 designation of the limiting values of the normal components of the wavevector is required here, since the subscripts 1, 2, . . . , N, $N + 1$ will be needed for quantities belonging to the N layers and $N + 1$ boundaries.) The function $q^2(z) = \omega^2/c^2(\varepsilon(z) - \varepsilon_a \sin^2\theta_a)$ is either given by or approximated by a series of steps. Figure 12-1 shows the corresponding $\varepsilon(z)$.

Let q_n be the value of q in the interval (z_n, z_{n+1}). The general solution of $d^2\psi/dz^2 + q^2\psi = 0$ in this interval may be written as

$$\psi_n = \alpha_n\,e^{iq_n z} + \beta_n\,e^{-iq_n z}. \tag{2}$$

In the interval (z_{n-1}, z_n) the solution is

$$\psi_{n-1} = \alpha_{n-1}\,e^{iq_{n-1} z} + \beta_{n-1}\,e^{-iq_{n-1} z}. \tag{3}$$

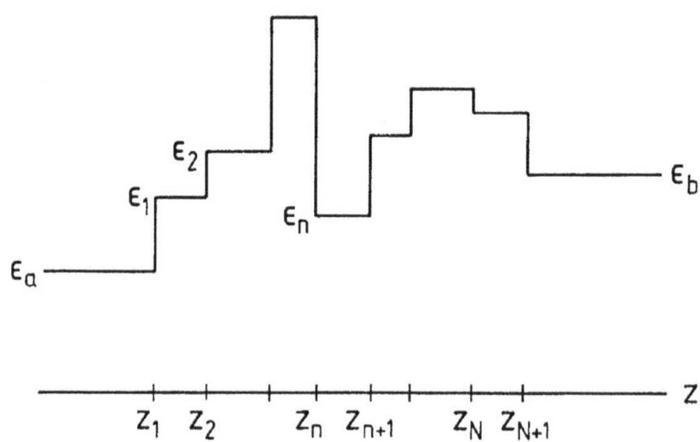

Figure 12-1. A stack of N uniform layers, bounded by media with dielectric constants ε_a and ε_b. The nth layer extends from z_n to z_{n+1} and has dielectric constant ε_n.

At $z = z_n$ we have $\psi_{n-1} = \psi_n$ and $d\psi_{n-1}/dz = d\psi_n/dz$; these continuity conditions imply

$$\alpha_{n-1}\, e^{iq_{n-1}z_n} + \beta_{n-1}\, e^{-iq_{n-1}z_n} = \alpha_n\, e^{iq_nz_n} + \beta_n\, e^{-iq_nz_n} \tag{4}$$

$$q_{n-1}(\alpha_{n-1}\, e^{iq_{n-1}z_n} - \beta_{n-1}\, e^{-iq_{n-1}z_n}) = q_n(\alpha_n\, e^{iq_nz_n} - \beta_n\, e^{-iq_nz_n}) \tag{5}$$

Solving for α_n and β_n in terms of α_{n-1} and β_{n-1}, we find

$$\begin{pmatrix}\alpha_n\\ \\ \beta_n\end{pmatrix} = \begin{pmatrix}\frac{1}{2}\left(1+\frac{q_{n-1}}{q_n}\right)e^{i(q_{n-1}-q_n)z_n} & \frac{1}{2}\left(1-\frac{q_{n-1}}{q_n}\right)e^{-i(q_{n-1}+q_n)z_n}\\ \frac{1}{2}\left(1-\frac{q_{n-1}}{q_n}\right)e^{-i(q_{n-1}+q_n)z_n} & \frac{1}{2}\left(1+\frac{q_{n-1}}{q_n}\right)e^{-i(q_{n-1}-q_n)z_n}\end{pmatrix}\begin{pmatrix}\alpha_{n-1}\\ \\ \beta_{n-1}\end{pmatrix}. \tag{6}$$

The two-by-two matrix in (6) will be written as M_n; it gives us the coefficients of the nth layer in terms of the coefficients of the $(n - 1)$th layer. Note that the determinant of M_n is

$$\det M_n = \frac{q_{n-1}}{q_n}; \tag{7}$$

this value is relevant to the conservation law of Section 2-1,

$$q_a(1 - |r|^2) = q_b|t|^2, \tag{8}$$

as we shall see shortly. To find the reflection and transmission amplitudes r and t we note from (1) that

$$\begin{pmatrix}\alpha_1\\ \beta_1\end{pmatrix} = M_1\begin{pmatrix}1\\ r\end{pmatrix}, \qquad M_{N+1}\begin{pmatrix}\alpha_N\\ \beta_N\end{pmatrix} = \begin{pmatrix}\alpha_{N+1}\\ \beta_{N+1}\end{pmatrix} = \begin{pmatrix}t\\ 0\end{pmatrix}. \tag{9}$$

Thus

$$\begin{pmatrix}t\\ 0\end{pmatrix} = M\begin{pmatrix}1\\ r\end{pmatrix} = \begin{pmatrix}m_{11} & m_{12}\\ m_{21} & m_{22}\end{pmatrix}\begin{pmatrix}1\\ r\end{pmatrix}, \tag{10}$$

where the characteristic or profile matrix M is the product of $N+1$ two-by-two matrices,

$$M = M_{N+1}M_N \ldots M_n \ldots M_2 M_1. \tag{11}$$

From (7) and the fact that the determinant of a product of matrices is the product of their determinants,

$$\det M = q_a/q_b. \tag{12}$$

From (10) and (12) the reflection and transmission amplitudes are given by

$$r = -\frac{m_{21}}{m_{22}}, \quad t = m_{11} - \frac{m_{12}m_{21}}{m_{22}} = \frac{\det M}{m_{22}} = \frac{q_a/q_b}{m_{22}}. \tag{13}$$

When q is real everywhere the matrix M_n defined by (6) has the form

$$M_n = \begin{pmatrix} \mu_{11} & \mu_{12} \\ \mu_{12}^* & \mu_{11}^* \end{pmatrix}, \tag{14}$$

that is, its elements are related by $\mu_{22} = \mu_{11}^*$, $\mu_{21} = \mu_{12}^*$. The product of two such matrices will also have this property (as may be verified by direct multiplication), and thus M has this property. Therefore

$$|m_{22}|^2 - |m_{21}|^2 = m_{11}m_{22} - m_{12}m_{21} = \det M, \tag{15}$$

and the conservation law (8) is satisfied by the reflection and transmission amplitude given by (13).

The above is for waves originating in medium a and transmitted into medium b. The reflection and transmission amplitudes for this case are

$$r_{ab} = -\frac{m_{21}}{m_{22}}, \qquad t_{ab} = \frac{q_a/q_b}{m_{22}}. \tag{16}$$

For waves incident from medium b the wavefunction has the limiting forms

$$t_{ba}\,\mathrm{e}^{-iq_a z} \leftarrow \psi \rightarrow \mathrm{e}^{-iq_b z} + r_{ba}\,\mathrm{e}^{iq_b z}, \tag{17}$$

and thus, for the same profile matrix M as before,

$$\begin{pmatrix} r_{ba} \\ 1 \end{pmatrix} = M \begin{pmatrix} 0 \\ t_{ba} \end{pmatrix} = \begin{pmatrix} m_{11} & m_{12} \\ m_{21} & m_{22} \end{pmatrix} \begin{pmatrix} 0 \\ t_{ba} \end{pmatrix}. \tag{18}$$

Thus

$$r_{ba} = \frac{m_{12}}{m_{22}}, \qquad t_{ba} = 1/m_{22}. \tag{19}$$

Comparison with (16) gives the reciprocity relations

$$q_b t_{ab} = q_a t_{ba}, \tag{20}$$

and

$$\frac{r_{ba}}{r_{ab}^*} = -\frac{m_{12}\,m_{22}^*}{m_{21}^*\,m_{22}} = -\frac{m_{12}}{m_{21}^*}\frac{t_{ba}}{t_{ba}^*}. \tag{21}$$

The first of these is the same as (2.14), and the second reduces to (2.18) when q is real everywhere.

We now return to the case of waves incident from medium a, and consider a slightly different formulation which has the advantage of having real matrix elements when q is real. This is obtained by writing ψ_n in the interval (z_n, z_{n+1}) as

$$\psi_n = \alpha_n \cos q_n z + \beta_n \sin q_n z. \tag{22}$$

In the interval (z_{n-1}, z_n) the solution is

$$\psi_{n-1} = \alpha_{n-1} \cos q_{n-1} z + \beta_{n-1} \sin q_{n-1} z, \tag{23}$$

and continuity of ψ and $d\psi/dz$ at z_n now gives

$$\alpha_n c_n + \beta_n s_n = \alpha_{n-1} c_{n-1} + \beta_{n-1} s_{n-1}, \tag{24}$$

$$q_n(-\alpha_n s_n + \beta_n c_n) = q_{n-1}(-\alpha_{n-1} s_{n-1} + \beta_{n-1} c_{n-1}), \tag{25}$$

where c_n, s_n and c_{n-1}, s_{n-1} stand for the cosines and sines of $q_n z_n$ and $q_{n-1} z_n$. On solving for α_n and β_n in terms of α_{n-1} and β_{n-1} we now have

$$\begin{pmatrix} \alpha_n \\ \beta_n \end{pmatrix} = \begin{pmatrix} c_n c_{n-1} + \dfrac{q_{n-1}}{q_n} s_n s_{n-1} & c_n s_{n-1} - \dfrac{q_{n-1}}{q_n} s_n c_{n-1} \\ s_n c_{n-1} - \dfrac{q_{n-1}}{q_n} c_n s_{n-1} & s_n s_{n-1} + \dfrac{q_{n-1}}{q_n} c_n c_{n-1} \end{pmatrix} \begin{pmatrix} \alpha_{n-1} \\ \beta_{n-1} \end{pmatrix}. \tag{26}$$

The determinant of the two-by-two matrix M_n in (26) is again q_{n-1}/q_n, and so the determinant of the profile matrix $M = M_{N+1} M_N \ldots M_n \ldots M_2 M_1$ is q_a/q_b as before. The reflection and transmission amplitudes are now given by

$$\begin{pmatrix} t \\ it \end{pmatrix} = \begin{pmatrix} m_{11} & m_{12} \\ m_{21} & m_{22} \end{pmatrix} \begin{pmatrix} 1 + r \\ i - ir \end{pmatrix}; \tag{27}$$

and thus

$$r = -\frac{m_{11} - m_{22} + im_{12} + im_{21}}{m_{11} + m_{22} - im_{12} + im_{21}}. \tag{28}$$

$$t = \frac{2(m_{11}m_{22} - m_{12}m_{21})}{m_{11} + m_{22} - im_{12} + im_{21}} = \frac{2q_a/q_b}{m_{11} + m_{22} - im_{12} + im_{21}}. \tag{29}$$

These formulae are not as simple as (13), but the advantage of the cosine and sine representation of ψ is that the matrix elements m_{ij} are real when q is real. This advantage is shared by the method introduced in the next section.

12-2 Matrices relating fields and their derivatives

The matrix methods developed above were based on recurrence relations for the coefficients of two linearly independent solutions within a given layer (chosen to be $e^{\pm iq_n z}$, or $\cos q_n z$ and $\sin q_n z$). Here we give the matrices which relate the field and its derivative, layer to neighbouring layer. This version of the matrix formulation will be used in the remainder of this chapter, and also in the next chapter (on numerical methods). We will consider the s polarization first. The electric field is $\mathbf{E} = (0, e^{iKx} E(z), 0)$, with

$$\frac{d^2E}{dz^2} + q^2E = 0, \qquad e^{iq_az} + r_s e^{-iq_az} \leftarrow E \rightarrow t_s e^{iq_bz}. \tag{30}$$

The second order differential equation for E may be written as a pair of coupled first order differential equations (as in Section 5-1):

$$\frac{dE}{dz} = D, \qquad \frac{dD}{dz} = -q^2E. \tag{31}$$

If $q(z)$ takes the value q_n in (z_n, z_{n+1}) as before, and E_n and D_n are the values of the field and of its derivative at z_n, the solution of (31) in $z_n \leqslant z \leqslant z_{n+1}$ is

$$E = E_n \cos q_n(z - z_n) + \frac{D_n}{q_n} \sin q_n(z - z_n), \tag{32}$$

$$D = D_n \cos q_n(z - z_n) - E_n q_n \sin q_n(z - z_n). \tag{33}$$

Since E and D are continuous at z_{n+1} (continuity follows from the differential equation for E, provided q^2 has no delta function singularities), it follows that

$$E_{n+1} = E_n \cos \delta_n + \frac{D_n}{q_n} \sin \delta_n, \tag{34}$$

$$D_{n+1} = D_n \cos \delta_n - E_n q_n \sin \delta_n. \tag{35}$$

Here

$$\delta_n = q_n(z_{n+1} - z_n) \tag{36}$$

is the phase increment in propagating from z_n to z_{n+1}. The relation between the coefficients at z_{n+1} and those at z_n is thus

$$\begin{pmatrix} E_{n+1} \\ D_{n+1} \end{pmatrix} = \begin{pmatrix} \cos \delta_n & \dfrac{\sin \delta_n}{q_n} \\ -q_n \sin \delta_n & \cos \delta_n \end{pmatrix} \begin{pmatrix} E_n \\ D_n \end{pmatrix}. \tag{37}$$

The matrix in (37) is unimodular (has unit determinant). Note that when $\delta_n = l\pi$, l an integer, the matrix is $(-)^l$ times the unit matrix. This corresponds to the layer thickness $z_{n+1} - z_n$ being an integer (for even l) or half-integer (for odd l) multiple of the effective wavelength for motion in the z direction, $\lambda_z = 2\pi/q_n$.

Before deriving expressions for r_s and t_s in terms of the elements of the profile matrix $M = M_N \ldots M_1$, we will give the corresponding matrix formulation for the p polarization. This has $\mathbf{B} = (0, e^{iKx} B(z), 0)$, with $B(z)$ satisfying

$$\frac{d}{dz}\left(\frac{1}{\varepsilon}\frac{dB}{dz}\right) + \frac{q^2}{\varepsilon} B = 0, \qquad e^{iq_az} - r_p e^{-iq_az} \leftarrow B \rightarrow \left(\frac{\varepsilon_b}{\varepsilon_a}\right)^{1/2} t_p e^{iq_bz}. \tag{38}$$

We again write the second order differential equation for B as a pair of coupled first order equations (as in Section 5-3):

$$\frac{1}{\varepsilon}\frac{\mathrm{d}B}{\mathrm{d}z} = C, \qquad \frac{\mathrm{d}C}{\mathrm{d}z} = -\frac{q^2}{\varepsilon}B. \tag{39}$$

If B_n and C_n are the values at z_n, within $z_n \leqslant z \leqslant z_{n+1}$ we have

$$B = B_n \cos q_n(z - z_n) + \frac{\varepsilon_n C_n}{q_n} \sin q_n(z - z_n), \tag{40}$$

$$C = C_n \cos q_n(z - z_n) - \frac{B_n q_n}{\varepsilon_n} \sin q_n(z - z_n). \tag{41}$$

B and C are continuous at a discontinuity in ε; their continuity at z_{n+1} leads to the matrix relation

$$\begin{pmatrix} B_{n+1} \\ C_{n+1} \end{pmatrix} = \begin{pmatrix} \cos \delta_n & \dfrac{\sin \delta_n}{Q_n} \\ -Q_n \sin \delta_n & \cos \delta_n \end{pmatrix} \begin{pmatrix} B_n \\ C_n \end{pmatrix}, \tag{42}$$

where $Q_n = q_n/\varepsilon_n$. The p wave matrix is also unimodular, and equal to $(-)^l$ times a unit matrix when $\delta_n = l\pi$ with integer l.

Note that the *layer* matrices in (37) and (42) depend on the thickness of the nth layer and on its dielectric properties. This is in contrast to the *boundary* matrices of the last section, where M_n was a function of the boundary position z_n, and of the wavenumber components q_{n-1} and q_n on either side of the boundary. The N-layer system of Figure 12-1 can be characterized by N layer matrices or by $N + 1$ boundary matrices.

The profile matrix in the present case is

$$M = \begin{pmatrix} m_{11} & m_{22} \\ m_{21} & m_{22} \end{pmatrix} = M_N M_{N-1} \ldots M_n \ldots M_2 M_1. \tag{43}$$

The reflection and transmission amplitudes for the s wave are obtained by matching the limiting form (3) to (32) at z_1 and z_{N+1}. We will use the notation $\alpha = q_a z_1$, $\beta = q_b z_{N+1}$. Then

$$E_1 = \mathrm{e}^{i\alpha} + r_s\,\mathrm{e}^{-i\alpha}, \qquad D_1 = iq_a(\mathrm{e}^{i\alpha} - r_s\,\mathrm{e}^{-i\alpha}), \tag{44}$$

$$E_{N+1} = t_s\,\mathrm{e}^{i\beta}, \qquad D_{N+1} = iq_b t_s\,\mathrm{e}^{i\beta}. \tag{45}$$

Since

$$\begin{pmatrix} E_{n+1} \\ D_{n+1} \end{pmatrix} = M_n \begin{pmatrix} E_n \\ D_n \end{pmatrix}, \tag{46}$$

we have

$$\begin{pmatrix} t_s\,\mathrm{e}^{i\beta} \\ iq_b t_s\,\mathrm{e}^{i\beta} \end{pmatrix} = \begin{pmatrix} m_{11} & m_{12} \\ m_{21} & m_{22} \end{pmatrix} \begin{pmatrix} \mathrm{e}^{i\alpha} + r_s\,\mathrm{e}^{-i\alpha} \\ iq_a(\mathrm{e}^{i\alpha} - r_s\,\mathrm{e}^{-i\alpha}) \end{pmatrix}. \tag{47}$$

Thus

$$r_s = e^{2i\alpha} \frac{q_a q_b m_{12} + m_{21} - iq_b m_{11} + iq_a m_{22}}{q_a q_b m_{12} - m_{21} + iq_b m_{11} + iq_a m_{22}}, \tag{48}$$

$$t_s = e^{i(\alpha-\beta)} \frac{2iq_a}{q_a q_b m_{12} - m_{21} + iq_b m_{11} + iq_a m_{22}}. \tag{49}$$

(In the numerator of (49) we have replaced $m_{11}m_{22} - m_{12}m_{21}$ by unity, since the matrix M is a product of unimodular matrices.) These formulae have the same form as (2.25) and (2.26), if a unit Wronskian is assumed in the latter. The matrix elements m_{ij} are real if ε is real, just as in Section 2-2 the functions F and G may be taken to be real in the same circumstances. When q_a, q_b and the matrix elements are real, the reflectance and transmittance are given by

$$R_s = |r_s|^2 = \frac{(q_a q_b m_{12} + m_{21})^2 + (q_b m_{11} - q_a m_{22})^2}{(q_a q_b m_{12} - m_{21})^2 + (q_b m_{11} + q_a m_{22})^2}, \tag{50}$$

$$T_s = \frac{q_b}{q_a}|t_s|^2 = \frac{4q_a q_b}{(q_a q_b m_{12} - m_{21})^2 + (q_b m_{11} + q_a m_{22})^2}. \tag{51}$$

The corresponding formulae for the p wave may be obtained from (38) and (42). We find

$$-r_p = e^{2i\alpha} \frac{Q_a Q_b m_{12} + m_{21} - iQ_b m_{11} + iQ_a m_{22}}{Q_a Q_b m_{12} - m_{21} + iQ_b m_{11} + iQ_a m_{22}}, \tag{52}$$

$$\left(\frac{\varepsilon_b}{\varepsilon_a}\right)^{1/2} t_p = e^{i(\alpha-\beta)} \frac{2iQ_a}{Q_a Q_b m_{12} - m_{21} + iQ_b m_{11} + iQ_a m_{22}}. \tag{53}$$

The meaning of α and β is $q_a z_1$ and $q_b z_{N+1}$ as before; $Q_a = q_a/\varepsilon_a$ and $Q_b = q_b/\varepsilon_b$. The matrix elements here are not the same as in (48) and (49): they are found by taking the product of the N matrices given by (42) instead of (37). The analogous formulae (2.40) and (2.41) are not of precisely the same form, since they apply to the case of continuous $\varepsilon(z)$ only. For real ε the matrix elements are real, and in the absence of total reflection the p wave reflectance and transmittance are given by

$$R_p = |r_p|^2 = \frac{(Q_a Q_b m_{12} + m_{21})^2 + (Q_b m_{11} - Q_a m_{22})^2}{(Q_a Q_b m_{12} - m_{21})^2 + (Q_b m_{11} + Q_a m_{22})^2}, \tag{54}$$

$$T_p = \frac{q_b}{q_a}|t_p|^2 = \frac{4Q_a Q_b}{(Q_a Q_b m_{12} - m_{21})^2 + (Q_b m_{11} + Q_a m_{22})^2}. \tag{55}$$

An interesting special configuration is a stack of half-wavelength layers. We noted above that when a layer has phase increment equal to an integer times π, the layer matrix is equal to plus or minus times the unit matrix, depending on whether the integer is even or odd. We can define an effective wavelength for propagation in the z direction as $2\pi/q$; for propagation in the zx plane the two component wavelengths are $\lambda_x = 2\pi/K$ and $\lambda_z = 2\pi/q$, with the total wavelength λ given by $\lambda^{-2} = \lambda_x^{-2} + \lambda_z^{-2}$. The condition $q\delta z = l\pi$ is equivalent to

$$\delta z = l\lambda_z/2; \tag{56}$$

the layer thickness is an integer times one half of the effective z component wavelength within the layer. When this condition holds all the multiply reflected waves (see Section 2-4, and especially Figure 2-5b) are in phase. For thick layers this may happen at several angles of incidence: the condition $q\delta z = l\pi$ is satisfied for angles of incidence

$$\theta_a(l) = \arcsin\left\{\frac{\varepsilon - [l\pi/(\omega/c)\delta z]^2}{\varepsilon_a}\right\}^{1/2}. \tag{57}$$

Possible values of l lie in the range

$$\frac{1}{\pi}\frac{\omega}{c}\delta z(\varepsilon - \varepsilon_a)^{1/2} \leqslant l \leqslant \frac{1}{\pi}\frac{\omega}{c}\delta z\varepsilon^{1/2}. \tag{58}$$

For example, when $\varepsilon_a = 1$, $\varepsilon = (4/3)^2$, $\varepsilon_b = (3/2)^2$ and $(\omega/c)\delta z = 27$ (as in Figure 5-3), the possible values of l are 8 to 11. At the corresponding angles $\theta_a(l)$ the layer is invisible: the reflectivity is the same as if the layer were absent.

A stack of N uniform layers, each with thickness equal to an integer l_n times half of the effective wavelength in the layer, namely with $q_n\delta z_n = l_n\pi$, will have a profile matrix which is equal to $(-)^L$ times the unit matrix, where $L = \Sigma_1^N l_n$. This stack will be invisible as far as reflectivity measurement is concerned (at the given angle of incidence and frequency). When L is even, the reflection *amplitudes* (given by (48) and (52)), are also identical to that of a step in the dielectric function from ε_a to ε_b, *located at* z_1. This is remarkable, since no restriction has been placed on the thickness $\Delta z = \Sigma_1^N \delta z_n$ of the stack.

12-3 Periodically stratified media

The problem of reflection and transmission of light by a pile of plates possessing periodic properties in the direction perpendicular to the plates was considered by Stokes (1862/1904) and Rayleigh (1917). The modern theory based on the properties of unimodular matrices was given by Abelès (1950, 1967). Tunnelling through a periodic system of barriers has been discussed by Pshenichnov (1962) and by Heading (1973, 1982).

Consider a stratification of period Δz, consisting of a structure repeated p times. If

$$M(q, \Delta z) = \begin{pmatrix} \mu_{11} & \mu_{12} \\ \mu_{21} & \mu_{22} \end{pmatrix} \tag{59}$$

is the matrix for one period, then M^p is the matrix for p periods. For matrices such as those defined in (37) and (42) which are unimodular (having a determinant equal to one), the pth power can be computed in terms of Chebyshev polynomials of the second kind:

$$\begin{pmatrix} \mu_{11} & \mu_{12} \\ \mu_{21} & \mu_{22} \end{pmatrix}^p = \begin{pmatrix} \mu_{11}U_{p-1}(x) - U_{p-2}(x) & \mu_{12}U_{p-1}(x) \\ \mu_{21}U_{p-1}(x) & \mu_{22}U_{p-1}(x) - U_{p-2}(x) \end{pmatrix}. \tag{60}$$

Here x is one half of the trace of the original matrix,

$$x = \tfrac{1}{2}(\mu_{11} + \mu_{22}). \tag{61}$$

The first few polynomials are

$$U_0(x) = 1, \quad U_1(x) = 2x, \quad U_2(x) = 4x^2 - 1, \quad U_3(x) = 8x^3 - 4x. \tag{62}$$

A general formula for U_p is

$$U_p(x) = (1 - x^2)^{-1/2} \sin [(p + 1) \arccos x], \tag{63}$$

from which we may verify the recurrence relation

$$U_p(x) = 2xU_{p-1}(x) - U_{p-2}(x). \tag{64}$$

The expression (60) holds for $p = 2$, and if true for p then from (64) it follows that it is true for $p + 1$. Thus it holds for all p. The result is due to Abelès (1950); the proof outlined here follows Born and Wolf (1970), Section 1.6.5.

The definition (63) takes a simpler form in terms of a variable θ, where

$$x = \cos \theta. \tag{65}$$

Then

$$U_{p-1}(x) = \frac{\sin p\theta}{\sin \theta}. \tag{66}$$

When θ takes one of the values θ_l,

$$\theta_l = l\pi/p, \qquad l = 1, 2, \ldots, p - 1, \tag{67}$$

$U_{p-1} = 0$ and $U_{p-2} = (-)^{l+1}$. Thus when

$$x = \tfrac{1}{2}(\mu_{11} + \mu_{22}) = \cos l\pi/p \tag{68}$$

the periodically stratified medium has a profile matrix M^p which is equal to the unit matrix times $(-)^{l+1}$; the stratification then has no effect on the reflectivity.

When $|x| > 1$, θ is complex. In the next section we will need expressions for the case when $x < -1$. We then define a variable ξ such that

$$\pm(x^2 - 1)^{1/2} - x = e^{\pm\xi}, \qquad -x = \cosh \xi \tag{69}$$

(compare Ninham and Parsegian, 1970); in this case $\theta = \pi + i\xi$ and

$$U_{p-1}(x) = (-)^{p-1} \frac{\sinh p\xi}{\sinh \xi}. \tag{70}$$

The reflection amplitudes (48) and (52) are unchanged when all the matrix elements m_{ij} of the profile matrix are multiplied by the same factor. We set

$$\sigma_p(x) = \frac{U_{p-2}(x)}{U_{p-1}(x)}; \tag{71}$$

this function takes the forms

$$\sigma_p(x) = \frac{\sin (p-1)\theta}{\sin p\theta} \qquad (|x| < 1), \tag{72}$$

$$\sigma_p(x) = -\frac{\sinh (p-1)\xi}{\sinh p\xi} \qquad (x < -1). \tag{73}$$

The profile matrix is equal to

$$M^p = U_{p-1}(x)\begin{pmatrix} \mu_{11}-\sigma_p(x) & \mu_{12} \\ \mu_{21} & \mu_{22}-\sigma_p(x) \end{pmatrix}, \tag{74}$$

and the s reflectivity is (when q is real everywhere, and $U_{p-1} \neq 0$)

$$R_s = \frac{(q_a q_b \mu_{12} + \mu_{21})^2 + (q_b\mu_{11} - q_a\mu_{22} + (q_a - q_b)\sigma_p)^2}{(q_a q_b \mu_{12} - \mu_{21})^2 + (q_b\mu_{11} + q_a\mu_{22} + (q_a - q_b)\sigma_p)^2}. \tag{75}$$

The p reflectivity has the same form with Q_a and Q_b replacing q_a and q_b, and with different definitions of the matrix elements μ_{ij}.

12-4 Multilayer dielectric mirrors

High reflectivity mirrors (used, for example, to form the optical cavity of lasers) are made by depositing alternating layers of high and low dielectric constant materials on a substrate, as shown in Figure 12-2.

These mirrors are wavelength-selective, high reflectivity at a particular frequency being obtained by constructive interference of the waves reflected at each discontinuity in refractive index. Each layer is made a quarter of a wavelength thick (at the design frequency) so as to make all the reflected waves in phase. For example: if in Figure 12-2 the front of the mirror (on the left) is at $z = 0$, the reflection amplitude off the first face is $(q_a - q_h)/(q_a + q_h)$, which is real and negative. The

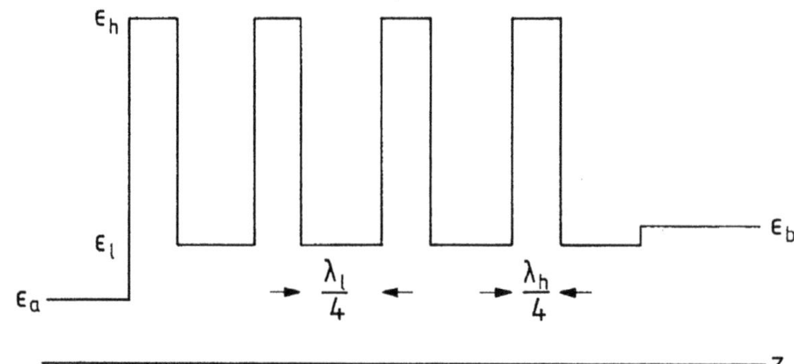

Figure 12-2. Dielectric function profile of a multilayer dielectric mirror, drawn to scale for an $(HL)^4$ configuration, with the refractive indices for the high and low index materials $n_h = 2.35$ and $n_l = 1.38$. These correspond to ZnS and MgF_2 at 633 nm (data from Table 1.1 of Yariv and Yeh, 1984). The substrate is glass, with $n_b = 1.5$.

reflection amplitude off the second (high/low) interface is a positive fraction times $e^{2i\delta_h}(q_h - q_l)/(q_h + q_l)$ (from (1.15)), where $\delta_h = q_h \delta z_h$. When $\delta_h = \pi/2$, which at normal incidence amounts to $\delta z_h = \lambda_h/4$ where λ_h is the wavelength in the high-index material, this second reflection amplitude is also real and negative. Similarly, the contribution from the next (low/high) interface will be in phase with the preceding if $\delta z_l = \lambda_l/4$. Thus constructive interface is obtained by making each layer a quarter of a wavelength thick (or in general an odd integer times a quarter wavelength).

The theory for periodically strafied media developed in the last section applies directly to multilayer configurations of the kind illustrated in Figure 12-2. One high/low combination forms a period of this structure, and has the matrix

$$\begin{pmatrix} \mu_{11} & \mu_{12} \\ \mu_{21} & \mu_{22} \end{pmatrix} = \begin{pmatrix} c_l & s_l/q_l \\ -q_l s_l & c_l \end{pmatrix} \begin{pmatrix} c_h & s_h/q_h \\ -q_h s_h & c_h \end{pmatrix} = \begin{pmatrix} c_l c_h - \dfrac{q_h}{q_l} s_l s_h & \dfrac{c_l s_h}{q_h} + \dfrac{s_l c_h}{q_l} \\ -q_l s_l c_h - q_h s_h c_l & c_l c_h - \dfrac{q_l}{q_h} s_l s_h \end{pmatrix}. \tag{76}$$

Here

$$c = \cos\delta, \quad s = \sin\delta, \quad \delta = q\delta z, \tag{77}$$

and the subscripts h and l denote the values for the high and low refractive index materials. The matrix elements are given for the s wave; the p wave elements are obtained by replacing q_h by Q_h and q_l by Q_l in (76). The reflectivity for an $(\mathrm{HL})^p$ mirror, with light incident from a medium of dielectric constant ε_a, resting on a substrate of dielectric constant ε_b, is obtained by substituting the matrix elements μ_{ij} into (75).

For a perfect $\lambda/4$ stack at normal incidence and at the design angular frequency ω_0, $\delta_h = \pi/2 = \delta_l$, and from (76)

$$\mu_{11} = -\frac{n_h}{n_l}, \quad \mu_{22} = -\frac{n_l}{n_h}, \quad \mu_{12} = 0 = \mu_{21}, \tag{77}$$

where $n_h = \sqrt{\varepsilon_h}$ and $n_l = \sqrt{\varepsilon_l}$ are the refractive indices of the alternating layers. In this case the μ-matrix is diagonal, and the profile matrix is equal to

$$M^p = \begin{pmatrix} \left(-\dfrac{n_h}{n_l}\right)^p & 0 \\ 0 & \left(-\dfrac{n_l}{n_h}\right)^p \end{pmatrix}. \tag{78}$$

The normal reflectivity at the design frequency is thus

$$R_n(\omega_0) = \left\{ \frac{\dfrac{n_b}{n_a}\left(\dfrac{n_h}{n_l}\right)^{2p} - 1}{\dfrac{n_b}{n_a}\left(\dfrac{n_h}{n_l}\right)^{2p} + 1} \right\}^2, \tag{79}$$

This increases rapidly with p, the number of HL strata. For example: when $n_a = 1$, $n_b = 1.5$, $n_h = 2.35$ and $n_l = 1.38$ (as in Figure 12-2) the $p = 1, 2, 3, 4, 5,$ and 6 stacks give $R_n(\omega_0) \simeq 0.392, 0.728, 0.896, 0.963, 0.987$ and 0.996.

At normal incidence, but away from the design wavelength, the phase changes $\delta_h = n_h(\omega/c)\delta z_h$ and $\delta_l = n_l(\omega/c)\delta z_l$ remain equal (again for the $\lambda/4$ stack) but are no longer $\pi/2$. Let $\delta = (\pi/2)(\omega/\omega_0)$ denote the common value of δ_h and δ_l. The μ-matrix is now

$$\begin{pmatrix} \cos^2\delta - \dfrac{n_h}{n_l}\sin^2\delta & \dfrac{c}{\omega}\left(\dfrac{1}{n_h} + \dfrac{1}{n_l}\right)\cos\delta\sin\delta \\ -\dfrac{\omega}{c}(n_h + n_l)\cos\delta\sin\delta & \cos^2\delta - \dfrac{n_l}{n_h}\sin^2\delta \end{pmatrix}. \tag{80}$$

Figure 12-3 shows the frequency dependence of the reflectivity of an $(HL)^4$ stack at normal incidence.

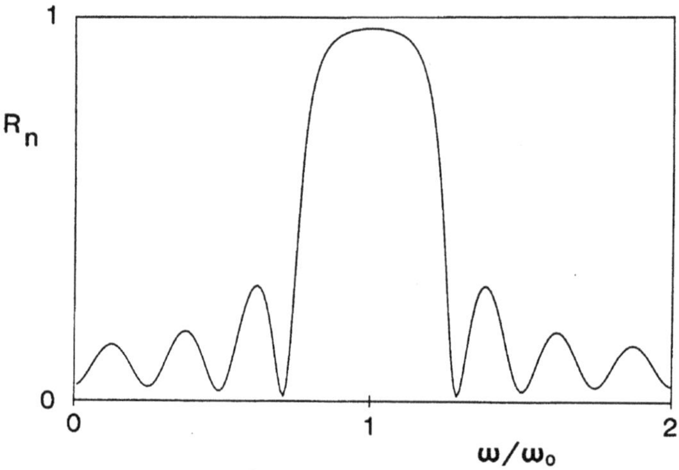

Figure 12-3. Normal incidence reflectivity of an $(HL)^4$ stack of dielectric layers, as a function of the frequency. The reflectivity is maximum at the design frequency ω_0, where the wavelength in both the high and low index layers is four times the layer thickness. The reflectivity is periodic in ω, with period $2\omega_0$. The refractive indices are as in Figure 12-2.

From (75) we see that the function $\sigma_p(x)$ changes the reflectivity from that of an HL bilayer to that of an $(HL)^p$ multilayer. In particular, the minima in the reflectivity are approximately at the poles of $\sigma_p(x)$ (and hence at the zeros at $U_{p-1}(x)$). As we shall see, there are $p - 1$ of these on either side of the central peak (in one frequency period). Equation (75) suggests that the reflectivity at these minima is roughly $(q_a - q_b)^2/(q_a - q_b)^2$; this is true for the outlying minima. The argument of the Chebyshev polynomials is

$$x = \tfrac{1}{2}(\mu_{11} + \mu_{22}) = \cos^2\delta - \frac{1}{2}\left(\frac{n_h}{n_l} + \frac{n_l}{n_h}\right)\sin^2\delta. \tag{81}$$

The cross-over from high to low reflectivity takes place as x increases through -1 from its minimum value of $-\frac{1}{2}[(n_h/n_l) + (n_l/n_h)]$, the latter being attained at the design frequency. The value $x = -1$ occurs at the frequencies $\omega = \omega_0 \pm \Delta\omega$, where from (81) and $\delta = (\pi/2)(\omega/\omega_0)$,

$$\frac{\Delta\omega}{\omega_0} = \frac{2}{\pi}\arcsin\left(\frac{n_h - n_l}{n_h + n_l}\right). \tag{82}$$

Within the band $\omega = \omega_0 \pm \Delta\omega$ the value of x is below -1, and the Chebyshev polynomials are non-oscillatory (see (70) for example). From zero frequency to $\omega_0 - \Delta\omega$, and from $\omega_0 + \Delta\omega$ to $2\omega_0$ (we stay within one frequency period in this characterization), $|x| < 1$ and the Chebyshev polynomials are oscillatory, with $U_{p-1}(x)$ having $p - 1$ zeros as x goes from -1 at $\omega_0 \pm \Delta\omega$ to $+1$ at $\omega = 0$ or $2\omega_0$. The last statement follows from (66).

At oblique incidence the s and p reflectivities behave differently. The increment in phase on passing through a high refractive index layer is $\delta_h = q_h\delta z_h$, with

$$q_h^2 = \varepsilon_h^2\frac{\omega^2}{c^2} - K^2 = \frac{\omega^2}{c^2}(\varepsilon_h - \varepsilon_a\sin^2\theta_a), \tag{83}$$

where θ_a is the angle of incidence onto the front face of the mirror. Thus

$$\delta_h = n_h\frac{\omega}{c}\delta z_h\left(1 - \frac{\varepsilon_a}{\varepsilon_h}\sin^2\theta_a\right)^{1/2}, \tag{84}$$

with similar formula for δ_l. If, as before, the mirror consists of a stack of layers each a quarter of a wavelength thick at normal incidence, we have

$$\delta_h = \frac{\pi}{2}\frac{\omega}{\omega_0}\left(1 - \frac{\varepsilon_a}{\varepsilon_h}\sin^2\theta_a\right)^{1/2}, \qquad \delta_l = \frac{\pi}{2}\frac{\omega}{\omega_0}\left(1 - \frac{\varepsilon_a}{\varepsilon_l}\sin^2\theta_a\right)^{1/2}. \tag{85}$$

The s wave μ-matrix is given by (76), and the reflectivity by (75), with

$$x = c_lc_h - \frac{1}{2}\left(\frac{q_h}{q_l} + \frac{q_l}{q_h}\right)s_ls_h. \tag{86}$$

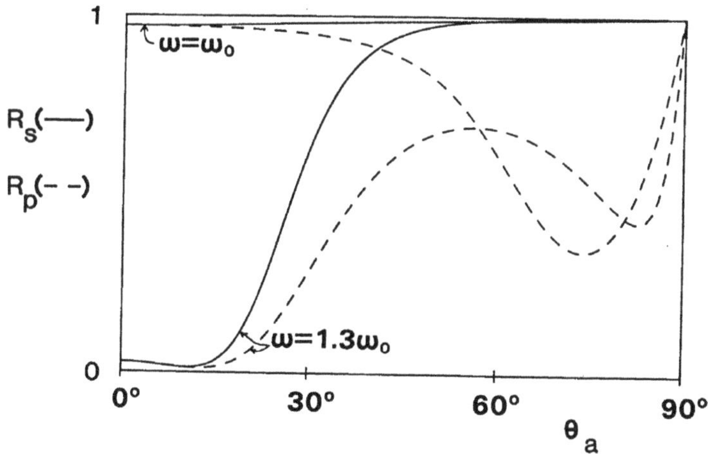

Figure 12-4. Reflectivity of an (HL)^4 stack on glass, as a function of the angle of incidence. The solid and dashed curves are s and p reflectivities. Two values of the frequency are shown: the design frequency ω_0 at which the normal incidence reflectivity is maximum, and $\omega = 1.3\omega_0$, which is near the first reflectivity minimum in Figure 12-3. The refractive indices are as in Figures 12-2 and 12-3.

The results for the p wave are similar, with Q replacing q except in the arguments of the trigonometric functions. Figure 12-4 shows the angular dependence of R_s and R_p at the design frequency, and near the first reflectivity minimum at normal incidence ($\omega \simeq 1.3\omega_0$).

12-5 Reflection of long waves

We return now to the problem of reflection by an arbitrary profle, as treated by the matrix method of Section 12-2, and consider the case where the total profile thickness is small compared to the wavelength of the radiation. The interface is represented by N uniform layers. As N increases to infinity the phase increments $\delta_n = q_n \delta z_n$ become infinitesimal. At the start we will keep first and second order terms in δ_n in the s wave matrix of equation (37),

$$M_n = \begin{pmatrix} 1 - \delta_n^2/2 & \dfrac{\delta_n}{q_n} \\ -q_n\delta_n & 1 - \delta_n^2/2 \end{pmatrix} + \ldots, \tag{87}$$

but it will turn out that only the first order terms of M_n play a role as $N \to \infty$. We write (87) as

$$M_n = (1 - \delta_n^2/2)I + \frac{\delta_n}{q_n} J - q_n \delta_n \tilde{J} + \ldots, \tag{88}$$

where I is the identity (or unit) matrix, and

$$J = \begin{pmatrix} 0 & 1 \\ 0 & 0 \end{pmatrix}, \quad \tilde{J} = \begin{pmatrix} 0 & 0 \\ 1 & 0 \end{pmatrix}. \tag{89}$$

The profile matrix

$$M = M_N M_{N-1} \ldots M_n \ldots M_2 M_1 \tag{90}$$

may be expanded in powers of δ_n. The fact that

$$J^2 = 0 = \tilde{J}^2 \tag{91}$$

simplifies the result, which reads

$$\begin{aligned} M = {} & \left(1 - \frac{1}{2}\sum_1^N \delta_n^2\right) I + \left(\sum_1^N \delta z_n\right) J - \left(\sum_1^N q_n^2\, \delta z_n\right) \tilde{J} \\ & - \sum_{n=1}^{N-1} \sum_{l=n+1}^{N} (q_n^2 \delta z_n \delta z_l J\tilde{J} + \delta z_n q_l^2 \delta z_l \tilde{J}J) + \ldots . \end{aligned} \tag{92}$$

In the limit as $N \to \infty$ and $\delta z_n \to 0$, $\Sigma_1^N \delta_n^2 \to 0$ and

$$\sum_1^N \delta z_n f_n \to \int_a^b \mathrm{d}z\, f(z), \tag{93}$$

$$\sum_{n=1}^{N-1} \sum_{l=n+1}^{N} \delta z_n f_n \delta z_l g_l \to \int_a^b \mathrm{d}z\, f(z) \int_z^b \mathrm{d}\zeta\, g(\zeta), \tag{94}$$

where $a = z_1$ and $b = z_{N+1}$ denote the left and right boundaries of the stack (see Figure 12-1). Since

$$J\tilde{J} = \begin{pmatrix} 1 & 0 \\ 0 & 0 \end{pmatrix}, \quad \tilde{J}J = \begin{pmatrix} 0 & 0 \\ 0 & 1 \end{pmatrix}, \tag{95}$$

the limit as $N \to \infty$ of (92) becomes

$$M = \begin{pmatrix} 1 - \int_a^b \mathrm{d}z\, q^2(z)(b - z) & b - a \\ -\int_a^b \mathrm{d}z\, q^2(z) & 1 - \int_a^b \mathrm{d}z\, q^2(z)(z - a) \end{pmatrix} + \ldots. \tag{96}$$

The s wave reflectivity is given by (50). After some reduction, the result to second order in the thickness $b - a$ is found to be

$$R_s = \left(\frac{q_a - q_b}{q_a + q_b}\right)^2 + \frac{4q_a q_b}{(q_a + q_b)^4}\left[(q_b^2 - q_a^2)\int_a^b \mathrm{d}z\,(2z - a - b)q^2 \right.$$
$$\left. + \int_a^b \mathrm{d}z(q^2 - q_a^2)\int_a^b \mathrm{d}\zeta\,(q^2 - q_b^2)\right] + \ldots \tag{97}$$

The substitution $q^2 = \varepsilon\omega^2/c^2 - K^2$ reduces the square bracket in (97) to ω^4/c^4 times the angle-independent term

$$(\varepsilon_b - \varepsilon_a)\int_a^b \mathrm{d}z\,(2z - a - b)\varepsilon + \int_a^b \mathrm{d}z\,(\varepsilon - \varepsilon_a)\int_a^b \mathrm{d}\zeta\,(\varepsilon - \varepsilon_b). \tag{98}$$

At this stage the s reflectivity has been reduced to the same form as (5.67). The subsequent analysis of Section 5-5 shows that this is equivalent to the second order result of Chapter 3, equation (3.51).

For the p wave, the profile matrix is the product of matrices of the form (42), which to second order in the phase increment δ_n is

$$M_n \begin{pmatrix} 1 - \delta_n^2/2 & \delta_n/Q_n \\ -Q_n\delta_n & 1 - \delta_n^2/2 \end{pmatrix} + \ldots, \tag{99}$$

where $Q_n = q_n/\varepsilon_n$. We again write M_n in terms of the matrices I, J and $\tilde{J}$, and form the product (90) to obtain the profile matrix M. The resulting elements of m are, after taking the limit of $N \to \infty$ as before,

$$\begin{aligned} m_{11} &= 1 - \int_a^b \mathrm{d}z\, q^2(z)/\varepsilon(z)\int_z^b \mathrm{d}\zeta\, \varepsilon(\zeta) + \ldots \\ m_{12} &= \int_a^b \mathrm{d}z\, \varepsilon(z) + \ldots \\ m_{21} &= -\int_a^b \mathrm{d}z\, q^2(z)/\varepsilon(z) + \ldots \\ m_{22} &= 1 - \int_a^b \mathrm{d}z\, \varepsilon(z)\int_z^b \mathrm{d}\zeta\, q^2(\zeta)/\varepsilon(\zeta) + \ldots. \end{aligned} \tag{100}$$

The p wave reflectivity is given by (54); to second order in the total thickness of the inhomogeneity this takes the form

$$R_p = \left(\frac{Q_a - Q_b}{Q_a + Q_b}\right)^2 + \frac{4Q_aQ_b}{(Q_a + Q_b)^4}[(Q_b^2 - Q_a^2)(m_{11} - m_{22}) + Q_a^2Q_b^2m_{12}^2 + m_{12}m_{21}(Q_a^2 + Q_b^2) + m_{21}^2] + \ldots. \tag{101}$$

A substantial reduction of (101) is required in order to regain the invariant form (3.50).

12-6 Absorbing stratified media: some general results

General theorems for arbitrary stratifications have already been given in Sections 2-1 to 2-3; Section 8-2 briefly discussed results for absorbing media. Here we give three theorems which follow from the matrix analysis of wave propagation through layered media.

(i) The transmittance of a stratified medium is independent of the direction of propagation. The transmittance T is defined as the ratio of the energy leaving a unit area of the interface in unit time to the energy incident on a unit area in unit time. The transmittance for propagation from a non-absorbing medium a, through an arbitrary stratification (which may be absorbing), to a non-absorbing medium b, is

$$T_{ab} = \frac{q_b}{q_a}|t_{ab}|^2 \tag{102}$$

(see the discussion following (2.8) and Figure 2-1). The equality of T_{ab} and T_{ab} follows from (16) and (19), of which the relevant parts are

$$t_{ab} = \frac{q_a/q_b}{m_{22}}, \qquad t_{ba} = 1/m_{22}. \tag{103}$$

(ii) An arbitrary stratified medium is equivalent to two suitably chosen adjacent uniform layers (Herpin 1947). Equivalence here means that the profile matrix elements m_{ij} of the two systems are the same. A general profile matrix

$$M = \begin{pmatrix} m_{11} & m_{12} \\ m_{21} & m_{22} \end{pmatrix} \tag{104}$$

has four elements (in general complex), which are linked by one constraint, namely the value of the determinant $m_{11}m_{22} - m_{12}m_{21}$. The latter is equal to q_a/q_b for the boundary matrices defined in Section 12-1, and to unity for the layer matrices used from Section 12-2 onward. A single uniform layer has the layer matrix

$$M_1 = \begin{pmatrix} \cos\delta_1 & \dfrac{\sin\delta_1}{q_1} \\ -q_1\sin\delta_1 & \cos\delta_1 \end{pmatrix}.$$

This cannot represent (104) since it has its diagonal elements equal, and has only two free parameters (δ_1 and q_1). The profile matrix for two homogeneous layers 1 and 2 is

$$M = M_2 M_1 = \begin{pmatrix} c_1 c_2 - \dfrac{q_1}{q_2} s_1 s_2 & \dfrac{s_1 c_2}{q_1} + \dfrac{c_1 s_2}{q_2} \\ -c_1 q_2 s_2 - c_2 q_1 s_1 & c_1 c_2 - \dfrac{q_2}{q_1} s_1 s_2 \end{pmatrix}, \tag{105}$$

where $c_1 = \cos\delta_1, \ldots, s_2 = \sin\delta_2$. This has unit determinant and four parameters (δ_1, q_1 and δ_2, q_2); it is thus sufficiently general to represent (104). But note that the equivalence, established by making the elements of (104) and (105) equal, will hold at a given angle of incidence and a given frequency only: as either changes, so do the parameters of the two-layer system.

(iii) For non-absorbing media the reflectance R and transmittance T are related by $R + T = 1$. For an absorbing stratification between two non-absorbing media the conservation law becomes $R + T + A = 1$, where A is the absorptance, a positive quantity for passive media. Thus the ratio $(1 - R)/T = 1 + A/T$ is in general greater than unity. Abelès (1950) has shown that *if an arbitrary non-absorbing layer is inserted in front of the absorbing layer, causing the reflectance to change to R' and the transmittance to T', the ratio of* $1 - R$ *to* T *is unaltered*:

$$\frac{1 - R}{T} = \frac{1 - R'}{T'}. \tag{106}$$

The unprimed and primed configurations are illustrated in Figure 12-5.

Let m_{ij} be the complex elements of the matrix representing the left-hand configuration in Figure 12-5, and m'_{ij} be those representing the right-hand configuration. Then from (48) and (49) we have

$$R = \left| \frac{q_a q_b m_{12} + m_{21} - i q_b m_{11} + i q_a m_{22}}{q_a q_b m_{12} - m_{21} + i q_b m_{11} - i q_a m_{22}} \right|^2, \tag{107}$$

$$T = \frac{4 q_a q_b}{|q_a q_b m_{12} - m_{21} + i q_b m_{11} - i q_a m_{22}|^2}, \tag{108}$$

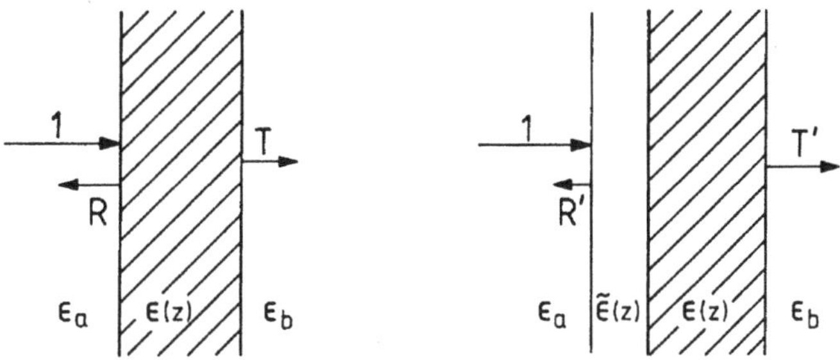

Figure 12-5. Two configurations which have equal value of the ratio of $1 - R$ to T. The dielectric function $\varepsilon(z)$ may be complex, while $\tilde{\varepsilon}(z)$ is real; ε_a and ε_b are real constants.

so that

$$\frac{1 - R}{T} = \operatorname{Re}(m_{22}^* m_{11} - m_{12}^* m_{21}) - \operatorname{Im}(q_b m_{12}^* m_{11} + q_b^{-1} m_{22}^* m_{21}). \quad (109)$$

A similar expression holds for $(1 - R')/T'$, with m_{ij} replaced by m'_{ij}, the elements of $M' = M\tilde{M}$, where $\tilde{M}$ is the layer matrix for the profile with dielectric function $\tilde{\varepsilon}(z)$. The matrix elements of M' are

$$\begin{aligned} m'_{11} &= m_{11}\tilde{m}_{11} + m_{12}\tilde{m}_{21} \qquad & m'_{12} &= m_{11}\tilde{m}_{12} + m_{12}\tilde{m}_{22} \\ m'_{21} &= m_{21}\tilde{m}_{11} + m_{22}\tilde{m}_{21} \qquad & m'_{22} &= m_{21}\tilde{m}_{12} + m_{22}\tilde{m}_{22} \end{aligned} \quad (110)$$

On using the fact that the elements of $\tilde{M}$ are real, and that $\tilde{M}$ has unit determinant, the expression for $(1 - R')/T'$ reduces to that for $(1 - R)/T$. Since $(1 - R)/T = 1 + A/T$, the Abelès result is equivalent to $A/T = A'/T'$: the absorptance to transmittance ratio is unchanged by the insertion of a non-absorbing layer in front of the absorber.

12-7 High transparency of an absorbing film in a frustrated total reflection configuration

In Section 8-6 we discussed attenuated total reflection, the phenomenon where a metallic layer or substrate converts a total internal reflection situation into one of low (or even zero) reflection of the p polarized wave. The physical basis of the phenomenon is the excitation of surface waves at the metal|dielectric boundaries, as explained in Section 8-6. There we considered two out of the three configurations illustrated in Figure 8-4. Here we consider the remaining one, the high|low|complex|low|high dielectric function configuration on the right of Figure 8-4. The dielectric function profile corresponding to this configuration is shown in Figure 12-6.

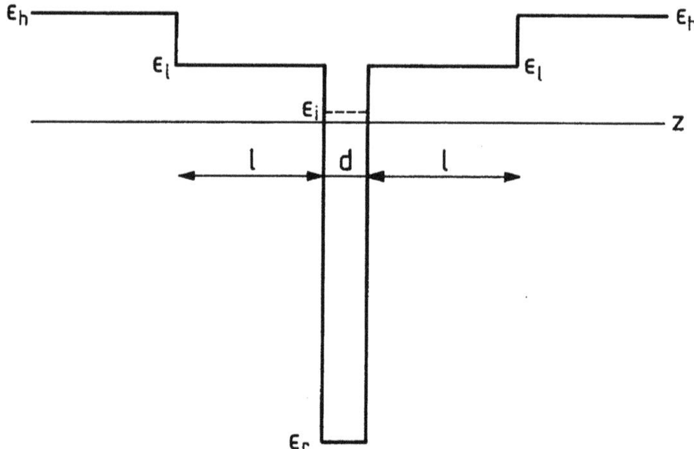

Figure 12-6. The dielectric function profile for a metal film, with complex dielectric constant $\varepsilon = \varepsilon_r + i\varepsilon_i$, sandwiched between two layers of low dielectric constant ε_l, which in turn are bounded by a material of high dielectric constant ε_h. The profile is drawn to scale for high refractive index glass ($\varepsilon_h = 3.617$), lithium fluoride ($\varepsilon_l = 1.938$) and silver ($\varepsilon = -10.755 + 0.361i$) at 546 nm, as in Figures 8-5 to 8-9.

We will calculate R_p and T_p for the symmetric case where the low-index dielectric layers have the same thickness l. The profile matrix for this case is

$$M = M_l M_m M_l, \tag{111}$$

where, from (42),

$$M_l = \begin{pmatrix} \cos\delta_l & Q_l^{-1}\sin\delta_l \\ -Q_l\sin\delta_l & \cos\delta_l \end{pmatrix}, \qquad M_m = \begin{pmatrix} \cos\delta & Q^{-1}\sin\delta \\ -Q\sin\delta & \cos\delta \end{pmatrix}. \tag{112}$$

Here $Q_l = q_l/\varepsilon_l$, $\delta_l = q_l l$, $Q = q/\varepsilon$ and $\delta = qd$. (The quantities for the metal film are complex, with $q = q_r + iq_i$ and so on.) Thus the elements of the profile matrix M are

$$\begin{aligned} m_{11} &= (c_l^2 - s_l^2)c - c_l s_l\left(\frac{Q_l}{Q} + \frac{Q}{Q_l}\right)s = m_{22} \\ m_{12} &= c_l^2 Q^{-1}s + 2Q_l^{-1}c_l s_l c - Q_l^{-2}s_l^2 Qs \\ m_{21} &= Q_l^2 s_l^2 Q^{-1}s - 2Q_l c_l s_l c - c_l^2 Qs, \end{aligned} \tag{113}$$

where c, s and c_l, s_l stand for the cosines and sines of δ and δ_l. From (52) and (53) the p reflectance and transmittance are given by

$$R_p = \left|\frac{Q_h^2 m_{12} + m_{21}}{Q_h^2 m_{12} - m_{21} + 2iQ_h m_{11}}\right|^2, \tag{114}$$

$$T_p = \frac{4Q_h^2}{|Q_h^2 m_{12} - m_{21} + 2iQ_h m_{11}|^2}. \tag{115}$$

We are most interested in the attenuated total reflection case, for which the angle of incidence exceeds the critical angle at the h, l interface. Then q_l, Q_l and δ_l are positive imaginary, $c_l = \cosh|\delta_l|$ and $s_l = i\sinh|\delta_l|$. The reflectivity can be zero

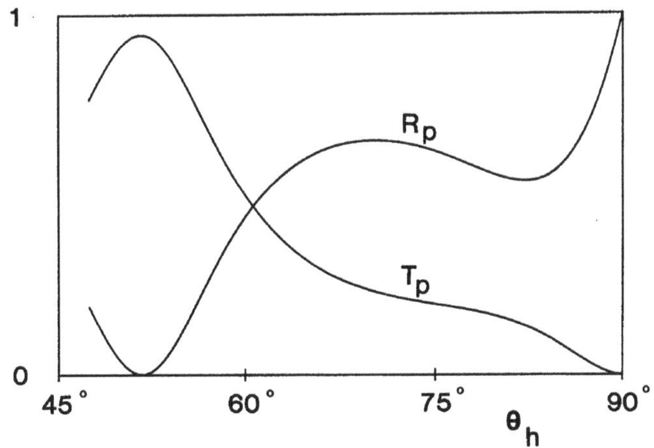

Figure 12-7. Reflectance and transmittance for the p wave, for the configuration of Figure 12-6, with $l = 100\,\text{nm}$, $d = 30\,\text{nm}$. The reflectance minimum is near 51.8°, with $R_p \simeq 1.3 \times 10^{-4}$. Only angles of incidence greater than the critical angle for the h–l interface are shown.

if the real and imaginary parts of $Q_h^2 m_{12} + m_{21}$ can be made zero simultaneously. For a given set of materials the variables are the angle of incidence, and the thicknesses of the low refractive index dielectric and of the metal film. Figure 12-7 gives an example of the p wave reflectance and transmittance for the configuration shown in Figure 12-6. Further examples can be found in Otto (1976) (see especially his Figure 14(b) and (c)), and in Dragila, Luther-Davies and Vukovic (1985). The physical interpretation of the reflectance minimum in terms of the excitation of surface waves is discussed in these references and in Section 8-6.

References

References quoted in text

J. W. S. Rayleigh (1912) "On the propagation of waves through a stratified medium, with special reference to the question of reflection", Proc. Roy. Soc. **A86**, 207–266.

W. Weinstein (1947) "The reflectivity and transmissivity of multiple thin coatings", J. Opt. Soc. Amer. **37**, 576–581.

A. Herpin (1947) "Calcul du pouvoir réflecteur d'un système stratifié quelconque", Comptes Rendus **225**, 182–183.

F. Abelès (1950) "Recherches sur la propagation des ondes électromagnétiques sinusoidales dans les milieux stratifiés. Application aux couches minces", Annales de Physique **5**, 596–640, 706–782.

F. Abelès (1967) "Optics of thin films", Chapter 5 of "Advanced Optical Techniques", edited by A. C. S. van Heel, North-Holland.

G. G. Stokes (1862/1904) "On the intensity of the light reflected from or transmitted through a pile of plates", Mathematical and Physical Papers **4**, 145–156.

J. W. S. Rayleigh (1917) "On the reflection of light from a regularly stratified medium", Proc. Roy. Soc. **A93**, 565–577.

E. A. Pshenichnov (1962) "The tunnel effect through a system of identical potential barriers", Sov. Phys. – Solid State **4**, 819–821.

J. Heading (1973) "Exact and approximate methods for the investigation of the propagation of waves through a system of barriers", Proc. Camb. Phil. Soc. **74**, 161–178.

J. Heading (1982) "Four-parameter formulae for wave propagation through a periodic system of inhomogeneous slabs", Wave Motion **4**, 127–139.

M. Born and E. Wolf (1970) "Principles of optics", (4th ed.) Pergamon.

B. W. Ninham and V. A. Parsegian (1970) "van der Waals interaction in multilayer systems", J. Chem. Phys. **53**, 3398–3402.

A. Yariv and P. Yeh (1984) "Optical waves in crystals", Wiley.

A. Otto (1976) "Spectroscopy of surface polaritons by attenuated total reflection", Chapter 13 in "Optical properties of solids: new developments", edited by B. O. Seraphin, North-Holland.

R. Dragila, B. Luther-Davies and S. Vukovic (1985) "High transparency of classically opaque metallic films", Phys. Rev. Lett. **55**, 1117–1120.

Additional references on optical properties of multilayers and on dielectric mirrors

O. S. Heavens (1955) "Optical properties of thin solid films", Butterworths.

O. S. Heavens (1970) "Thin film physics", Methuen.

E. Ritter (1972) "Optical coatings and thin film techniques", Section C9 of the "Laser Physics Handbook", Vol. 1, edited by F. T. Arecchi and E. O. Schulz-Dubois, North-Holland.

13

Numerical methods

Approximate analytical results for the reflection amplitudes have been given in the long wave and short wave cases (Chapters 3 and 6). The long wave region of validity is extended by the perturbation and variational theories in Chapter 4, and the Rayleigh approximation of Chapter 5 is good at all wavelengths provided the reflection is weak. All these analytical methods share the drawback that higher-order approximations rapidly become cumbersome and thus of little practical value. For accurate results at intermediate wavelengths, and for a profile which is not among the few exactly soluble, numerical methods are needed.

This chapter describes numerical methods based on the matrix theory of the previous chapter. We do not give details of the direct solution of the wave equation, because the complications of that approach are greater, as the following outline shows. Let $\psi(z)$ satisfy

$$\frac{d^2\psi}{dz^2} + q^2\psi = 0, \qquad e^{iq_a z} + r\,e^{-iq_a z} \leftarrow \psi \rightarrow t\,e^{iq_b z}. \tag{1}$$

Since r and t are unknown, to integrate the differential equation we change the boundary conditions to

$$\alpha\,e^{iq_a z} + \beta\,e^{-iq_a z} \leftarrow \psi \rightarrow e^{iq_b z}, \tag{1'}$$

integrate backward from some z_b at which $q(z)$ is close enough to q_b, and extract α and β in the region where $q(z)$ is close enough to q_a. Then the reflection and transmission amplitudes are found from $r = \beta/\alpha$, $t = 1/\alpha$. There are two complications in this method, both avoided by the matrix methods to be given later. The first is that r, t and α, β are in general complex, and thus solutions for both Re ψ and Im ψ are required. The second is that the extraction of the real and imaginary parts of α and β requires matching the real part of ψ to

$$(\alpha_r + \beta_r)\cos q_a z - (\alpha_i - \beta_i)\sin q_a z, \tag{2}$$

and the imaginary part of ψ to

$$(\alpha_i + \beta_i)\cos q_a z + (a_r - \beta_r)\sin q_a z. \tag{3}$$

By matching at points where $q_a z$ is an even and an odd multiple of $\pi/2$ (for example) one obtains the four quantities $\alpha_r + \beta_r$, $\alpha_r - \beta_r$, $\alpha_i + \beta_i$ and $\alpha_i - \beta_i$, and thus the real and imaginary parts of α and β.

The complications in the direct solution of the differential equation outweigh (in our view) the advantage of ready access to a very large literature on the numerical solution of ordinary differential equations (see, for example, Hildebrand 1956, Hartree 1958, Ralston 1965, Kelly 1967, Goldstine 1977). The matrix methods we will use, in contrast, evaluate only real quantities (in the absence of absorption), and the matching is done automatically: see for example the derivation of the expression for r and t in terms of the elements of the profile matrix in Section 12-2. The calculation of the profile matrix involves merely the computation of a product of two-by-two matrices, which is easily programmed.

13-1 Numerical methods based on the layer matrices

Two kinds of matrices were introduced in Chapter 12: the boundary matrices of Section 12-1, and the layer matrices of Section 12-2 onward. The latter are more convenient for numerical work and will be used here. Figure 13-1 shows how a given profile is approximated by N uniform layers.

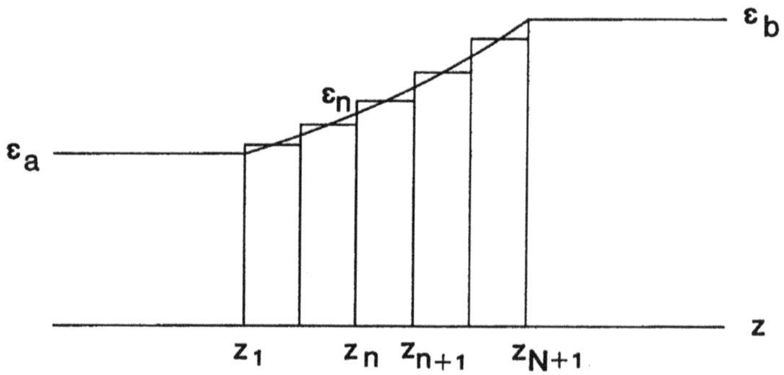

Figure 13-1. The Rayleigh profile

$$\varepsilon(z) = [n_a^{-1} + (n_b^{-1} - n_a^{-1})(z - z_1)/\Delta z]^{-2} \quad (z_1 \leqslant z \leqslant z_{N+1},\ \Delta z = z_{N+1} - z_1)$$

approximated by a set of uniform layers. The figure is drawn for $n_a = 1$, $n_b = 4/3$.

The nth layer extends from z_n to z_{n+1}, and in the case illustrated is uniform, with dielectric constant ε_n. The corresponding layer matrices for the s and p waves are given by (12.37) and (12.42):

$$M_n = \begin{pmatrix} \cos\delta_n & \dfrac{\sin\delta_n}{q_n} \\ -q_n \sin\delta_n & \cos\delta_n \end{pmatrix} \quad \text{or} \quad \begin{pmatrix} \cos\delta_n & \dfrac{\sin\delta_n}{Q_n} \\ -Q_n \sin\delta_n & \cos\delta_n \end{pmatrix}, \tag{4}$$

where $\delta_n = q_n(z_{n+1} - z_n) \equiv q_n \delta z_n$, $q_n^2 = \varepsilon_n \omega^2/c^2 - K^2 = \omega^2/c^2(\varepsilon_n - \varepsilon_a \sin^2\theta_a)$, and $Q_n = q_n/\varepsilon_n$. To first order in the layer thickness δz_n these matrices are

$$\begin{pmatrix} 1 & \delta z_n \\ -q_n^2 \delta z_n & 1 \end{pmatrix} \quad \text{or} \quad \begin{pmatrix} 1 & \varepsilon_n \delta z_n \\ -q_n^2 \delta z_n/\varepsilon_n & 1 \end{pmatrix}. \tag{5}$$

As N gets large, the layer thicknesses δz_n become small, and the matrices in (4) are well approximated by (5). This approximation for the layer matrices is in fact equivalent to the first order Euler method of solving the differential equation (1). To see this, let u be the real part of ψ, and $v = du/dz$. The second order equation for u, $d^2u/dz^2 + q^2u = 0$, can be replaced by the pair of coupled first order equations, $du/dz = v$ and $dv/dz = -q^2u$. The discretized version of this pair is

$$\frac{u_{n+1} - u_n}{\delta z_n} = v_n, \qquad \frac{v_{n+1} - v_n}{\delta z_n} = -q_n^2 u_n. \tag{6}$$

In matrix form this reads

$$\begin{pmatrix} u_{n+1} \\ v_{n+1} \end{pmatrix} = \begin{pmatrix} 1 & \delta z_n \\ -q_n^2\delta z_n & 1 \end{pmatrix} \begin{pmatrix} u_n \\ v_n \end{pmatrix}. \tag{7}$$

The matrix method with the profile replaced by a stack of uniform layers, and with the layer matrices calculated to first order in the layer thickness, is thus equivalent in accuracy to the Euler method.

We will not use this simplest approach, since it is easy to improve on the uniform layer approximation without much complication in the matrices and the consequent programming. The improvement consists in approximating the profile by a set of layers in which the dielectric function varies linearly within each layer (Law and Beaglehole, 1981). This is illustrated in Figure 13-2.

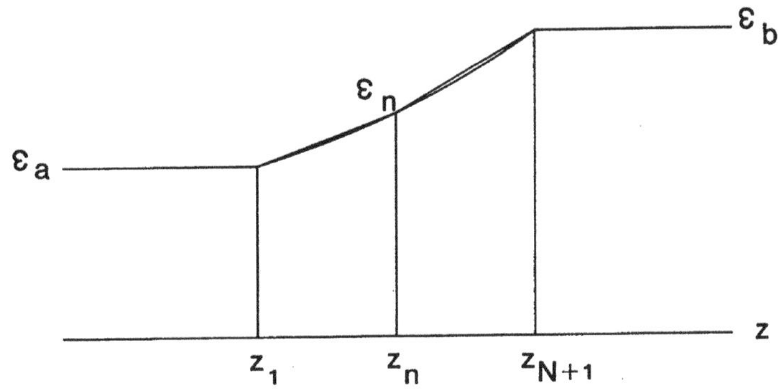

Figure 13-2. Approximation of a profile by layers with linear variation in ε. The diagram is drawn for the Rayleigh profile, with parameters as in Figure 13-1.

The variation of $\varepsilon(z)$ in $[z_n, z_{n+1}]$ is approximated by

$$\varepsilon(z) = \varepsilon_n + (z - z_n)\,\delta\varepsilon_n/\delta z_n, \tag{8}$$

where $\delta z_n = z_{n+1} - z_n$ as before, and $\delta\varepsilon_n = \varepsilon_{n+1} - \varepsilon_n$. Let $\Delta z = z_{N+1} - z_1$ be the total thickness of the profile. (When the values ε_a and ε_b are attained at minus and plus infinity, the profile must be truncated at some points z_1 and z_{N+1} as discussed later.) Then if at a given angular frequency ω a large enough number N of the layers is taken so that $(\omega/c)\delta z_n \ll 1$ (or $(\omega/c)\Delta z \ll N$ assuming the δz_n are roughly equal), each layer matrix will be well approximated by its long-wave form as given in Section 12-5 to second order in the layer thickness.

For the s wave we find from (8) and (12.96) that the elements s_{ij} of the matrix M_n representing the nth layer are given by

$$\begin{aligned} s_{11} &= 1 + (\delta z_n)^2\left[K^2/2 - \frac{\omega^2}{c^2}(2\varepsilon_n + \varepsilon_{n+1})/6\right], \\ s_{12} &= \delta z_n, \qquad s_{21} = \delta z_n\left[K^2 - \frac{\omega^2}{c^2}(\varepsilon_n + \varepsilon_{n+1})/2\right], \\ s_{22} &= 1 + (\delta z_n)^2\left[K^2/2 - \frac{\omega^2}{c^2}(\varepsilon_n + 2\varepsilon_{n+1})/6\right]. \end{aligned} \tag{9}$$

The p wave matrix elements p_{ij} to second order in δz_n are found from (8) and (12.100): they are

$$\begin{aligned} p_{11} &= 1 + (\delta z_n)^2\left[K^2\left\{\frac{2\varepsilon_{n+1}^2}{\delta\varepsilon_n}\log\frac{\varepsilon_{n+1}}{\varepsilon_n} - \varepsilon_{n+1} - \varepsilon_n\right\}\Big/(4\delta\varepsilon_n) - \frac{\omega^2}{c^2}(\varepsilon_n + 2\varepsilon_{n+1})/6\right], \\ p_{12} &= \delta z_n(\varepsilon_n + \varepsilon_{n+1})/2, \qquad p_{21} = \delta z_n\left[\frac{K^2}{\delta\varepsilon_n}\log\frac{\varepsilon_{n+1}}{\varepsilon_n} - \frac{\omega^2}{c^2}\right], \\ p_{22} &= 1 + (\delta z_n)^2\left[K^2\left\{\varepsilon_n + \varepsilon_{n+1} - \frac{2\varepsilon_n^2}{\delta\varepsilon_n}\log\frac{\varepsilon_{n+1}}{\varepsilon_n}\right\}\Big/(4\delta\varepsilon_n) - \frac{\omega^2}{c^2}(2\varepsilon_n + \varepsilon_{n+1})/6\right]. \end{aligned} \tag{10}$$

In computation it is better to replace the expressions involving $\log(\varepsilon_{n+1}/\varepsilon_n)$ by the leading terms in their $\delta\varepsilon_n/\varepsilon_n$ expansion. The resulting matrix elements are

$$\begin{aligned} p_{11} &\simeq 1 + (\delta z_n)^2\left[K^2(2\varepsilon_n + \varepsilon_{n+1})/6\varepsilon_n - \frac{\omega^2}{c^2}(\varepsilon_n + 2\varepsilon_{n+1})/6\right], \\ p_{21} &\simeq \delta z_n\left[K^2(1/\varepsilon_n + 1/\varepsilon_{n+1})/2 - \frac{\omega^2}{c^2}\right], \\ p_{22} &\simeq 1 + (\delta z_n)^2\left[K^2(\varepsilon_n + 2\varepsilon_{n+1})/6\varepsilon_{n+1} - \frac{\omega^2}{c^2}(2\varepsilon_n + \varepsilon_{n+1})/6\right], \end{aligned} \tag{11}$$

with p_{12} unchanged. For comparison with (9) and (11) we write down the uniform layer matrices to second order in δz_n. These are, from (4),

$$\begin{pmatrix} 1 - (\delta z_n q_n)^2/2 & \delta z_n \\ -\delta z_n q_n^2 & 1 - (\delta z_n q_n)^2/2 \end{pmatrix}, \quad \begin{pmatrix} 1 - (\delta z_n q_n)^2/2 & \varepsilon_n \delta z_n \\ -\delta z_n q_n^2/\varepsilon_n & 1 - (\delta z_n q_n)^2/2 \end{pmatrix}, \tag{12}$$

and are seen to be the degenerate forms of (9) and (11), obtained by setting $\varepsilon_{n+1} = \varepsilon_n$ in the linear layer formulae.

The linear layer formulae taken to first order in δz_n will be referred to as L1, and those retaining the terms second order in δz_n as L2. The results of Law and Beaglehole (1981) are equivalent to L1, with the difference (for real ε) that their off-diagonal matrix elements are imaginary whereas the ones used here are real. The results obtained by the L1 and L2 methods, for the Rayleigh profile with parameters as in Figures 13-1 and 2, are compared in Figures 13-3 and 4.

From these results and similar ones for other profiles we draw the conclusion that the second order method is preferable to the first order method. There is a

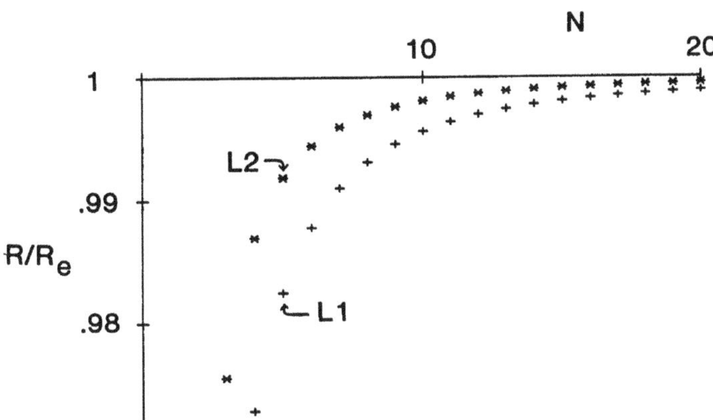

Figure 13-3. Ratios of the calculated to the exact reflectivities, for the Rayleigh profile at normal incidence, with $(\omega/c)\Delta z = 1$. The letter L denotes that a linear variation of dielectric function within each layer is used; 1 or 2 denote that first or second order terms in δz_n are retained in each layer matrix. N is the number of layers; a constant step size was used, $\delta z_n = \Delta z/N$.

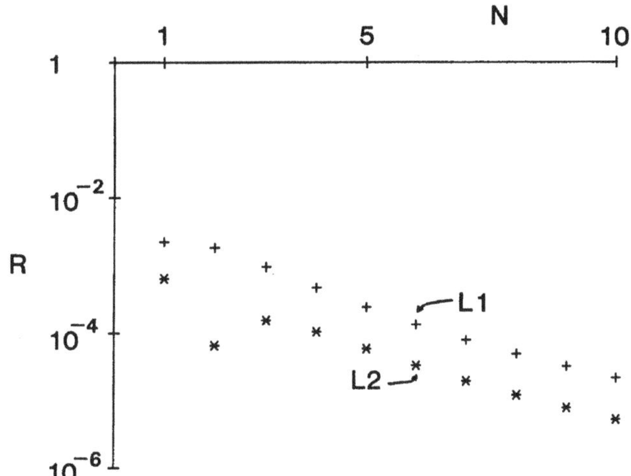

Figure 13-4. Calculated normal incidence reflectivities at the first reflectivity zero for the Rayleigh profile $[(\omega/c)\Delta z = 2.73295\ldots$, from (2.110)]. The notation is as in Figure 13-3. The second order method is better by a factor of four or more for $2 \leqslant N \leqslant 10$.

small increase in programming complexity and execution time in going from L1 and L2. For example: in the case shown in Figure 13-3, for $N = 10$ the L1 and L2 methods gave 4 and 2 parts-per-thousand accuracy, and ran for 2.4 and 3.1 seconds on a personal computer programmed in BASIC.

Further improvements are possible, by better than linear approximations to $\varepsilon(z)$ within each layer, and by going to higher order in δz_n. For example, one may approximate $\varepsilon(z)$ by a cubic in $[z_n, z_{n+1}]$ by using the derivatives ε_n' and ε_{n+1}' at the end-points. The formula resulting from matching to ε and ε' at z_n and z_{n+1} is

$$\varepsilon(z) \simeq \varepsilon_n + (z - z_n)\varepsilon_n' + \left(\frac{z - z_n}{\delta z_n}\right)^2 \{3\delta\varepsilon_n - \delta z_n(2\varepsilon_n' + \varepsilon_{n+1}')\}$$
$$+ \left(\frac{z - z_n}{\delta z_n}\right)^3 \{\delta z_n(\varepsilon_n' + \varepsilon_{n+1}') - 2\delta\varepsilon_n\}. \tag{13}$$

The method C2, obtained by using (13) and calculating the matrix elements to second order in δz_n, was found to be not much better than L2. The cubic third order formulae, C3, promise to be better but are not available since the matrix elements for a general profile are not yet known to third order in the layer thickness.

13-2 Variable step size, profile truncation, total reflection and tunnelling, absorption, and calculation of wavefunctions

We now briefly discuss some further aspects of the numerical application of these matrix methods.

A uniform step size $\delta z_n = \Delta z/N$ was chosen in the calculations discussed above. This is convenient, but not necessary; the matrix formule given here are valid for *variable step size*. However, a constant step size is normally the simplest to program, and in most cases is just as accurate as (for example) a variable step size chosen to make $\delta\varepsilon_n = \varepsilon_{n+1} - \varepsilon_n$ a constant.

The Rayleigh profile shown in Figures 13-1 and 2, and for which the results of Figures 13-3 and 4 were calculated, is an example of a profile of strictly finite range. For dielectric functions in which the inhomogeneity extends to infinity, such as the hyperbolic tangent profile

$$\varepsilon(z) = \tfrac{1}{2}(\varepsilon_a + \varepsilon_b) - \tfrac{1}{2}(\varepsilon_a - \varepsilon_b)\tanh\frac{z}{2\Delta z} = \frac{\varepsilon_a + \varepsilon_b\, e^{z/\Delta z}}{1 + e^{z/\Delta z}}, \tag{14}$$

profile truncation is necessary for the application of numerical methods. By truncation is meant that ε is set equal to ε_a for $z < a$ and to ε_b for $z > b$, where a and b are chosen so that $\varepsilon(a) - \varepsilon_a$ and $\varepsilon(b) - \varepsilon_b$ are sufficiently small to cause negligible error. For example, suppose we take $a = -7\Delta z$ and $b = 7\Delta z$ for the hyperbolic tangent profile. Since $e^7 \approx 10^3$, truncation at a and b can be expected to introduce an error of the order of one part per thousand. Larger values of $|a|$ and b will introduce smaller errors, but correspondingly larger numbers of layer matrices will be required to attain convergence to the truncated profile matrix elements. This is illustrated in Figure 14 of Law and Beaglehole (1981), and in the following table.

Table 13-1. The ratio of the calculated to the exact reflectivity, for the tanh profile truncated at a and b, as a function of the number of layer matrices. The values given are for $(\omega/c)\Delta z = 0.2$, $\varepsilon_a = 1$, $\varepsilon_b = (4/3)^2$, at normal incidence.

N	$-a, b = 7\Delta z$	$-a, b = 9\Delta z$
10	0.91	0.85
20	0.976	0.962
40	0.993	0.990
80	0.9977	0.9974
160	0.9988	0.9992
320	0.99908	0.99962
640	0.99915	0.99974

We see from the table that truncation at the larger value of $|a|$ and b ultimately gives a more accurate reflectivity, but in the case illustrated the smaller cut-off gives a better reflectivity up to about 95 layers. This is because the smaller effective thickness of the profile is better approximated by a given number of layers.

The formulae given in this chapter remain valid when $q(z)$ is imaginary (and $q^2(z) < 0$), as is the case for a range of z values in *total internal reflection*, and in *tunnelling*. No change is required in the calculation of the elements of the profile matrix, which remain real. The reflection and transmission amplitudes are still given by (12.48) and (12.49) in the s wave case, and by (12.52) and (12.53) in the p wave case. In total reflection q_b and Q_b are positive imaginary, and both R_s and R_p are unity. The quantity of interest is the phase of the reflected wave, given by

$$\delta_s = 2q_a z_1 - 2 \operatorname{atn} [s_{21} + |q_b|s_{11}, q_a(s_{12}|q_b| + s_{22})], \tag{15}$$

$$\delta_p = 2q_a z_1 + 2 \operatorname{atn} [Q_a(p_{12}|Q_b| + p_{22}), p_{21} + |Q_b|p_{11}], \tag{16}$$

where atn (y, x) is the arctangent of y/x placed in the correct quadrant according to the signs of x and y.

In the presence of *absorption* the dielectric function becomes complex. If only the substrate (characterized by ε_b, q_b and Q_b) is absorbing, the matrix elements remain real, and only the calculation of the reflectivity from the reflection amplitude is modified. (The expressions (12.50) and (12.54) for the s and p reflectances no longer apply.) When however the stratification is itself absorbing, the matrix elements are complex, and four matrix multiplications are needed in place of one performed in the non-absorbing case: if $R + iS$ and $U + iV$ represent the real and imaginary parts of two matrices, their product is

$$(U + iV)(R + iS) = UR - VS + i(VR + US). \tag{17}$$

Thus calculations involving absorption within the interface are roughly four times longer than those which do not.

We turn finally to the problem of the *calculation of wavefunctions* within the stratification. These are obtained, if required, as a by-product of the calculation of the elements of the profile matrix. In the s wave case, for example, we have

$$\begin{pmatrix} E_{n+1} \\ D_{n+1} \end{pmatrix} = M_n \begin{pmatrix} E_n \\ D_n \end{pmatrix} = M_n M_{n-1} \ldots M_2 M_1 \begin{pmatrix} E_1 \\ D_1 \end{pmatrix}. \tag{18}$$

Let v_{ij} be the elements of the product of n matrixes in (18). Then

$$E_{n+1} = v_{11} E_1 + v_{12} D_1 \tag{19}$$

gives the wavefunction at z_{n+1} in terms of the wavefunction and its derivative at z_1. These are given by

$$E_1 = \mathrm{e}^{i\alpha} + r_s \mathrm{e}^{-i\alpha}, \qquad D_1 = iq_a(\mathrm{e}^{i\alpha} - r_s \mathrm{e}^{-i\alpha}), \tag{20}$$

where $\alpha = q_a z_1$, and $r_s = r_r + ir_i$ is the reflection amplitude. The latter is found first, by calculating the product up to $n = N$. If the elements v_{11} and v_{12} are stored for all intermediate n, the wavefunction may then be plotted at the completion of the calculation of r_s. From (19) and (20), we have in the absence of absorption (real v_{ij}) that

$$\text{Re}\,(E_{n+1}) = v_{11}\{(1 + r_r)c + r_i s\} + v_{12}q_a\{r_i c - (1 + r_r)s\}. \tag{21}$$

$$\text{Im}\,(E_{n+1}) = v_{11}\{r_i c + (1 - r_r)s\} + v_{12}q_a\{(1 - r_r)c - r_i s\}, \tag{22}$$

where $c = \cos\alpha$ and $s = \sin\alpha$.

References

F. B. Hildebrand (1956) "Introduction to numerical analysis", McGraw-Hill.

D. R. Hartree (1958) "Numerical analysis" (second edition), Oxford.

A. Ralston (1965) "A first course in numerical analysis", McGraw-Hill.

L. G. Kelly (1967) "Handbook of numerical methods and applications", Addison-Wesley.

H. H. Goldstine (1977) "A history of numerical analysis from the 16th to the 19th century", Springer.

B. M. Law and D. Beaglehole (1981) "Model calculations of the ellipsometric properties of inhomogeneous dielectric surfaces", J. Phys. D **14**, 115–126.

Appendix. Reflection of particle waves

In Section 1-3 we saw that there is a one-to-one correspondence between the propagation in planar-stratified media of the s-polarized electromagnetic wave, and of the non-relativistic particle wave satisfying the Schrödinger equation

$$-\frac{\hbar^2}{2m}\nabla^2\Psi + V\Psi = \mathscr{E}\Psi. \tag{1}$$

($\mathscr{E}$ and m are particle energy and mass, V is the potential energy, and $\hbar$ is Planck's constant divided by 2π). In this Appendix we give a representative selection from the main results derived in the book for the electromagnetic s wave, translated into quantum mechanical language and notation.

A-1 General results

Equation (1) may be written as $\nabla^2\Psi + k^2\Psi = 0$, where

$$k^2 = \frac{2m}{\hbar^2}(\mathscr{E} - V) \tag{2}$$

is the square of wavevector. We assume that V is a function of one spatial coordinate only, $V = V(z)$, and takes the limiting values V_1 and V_2 at $z = -\infty$ and $z = \infty$. Then

$$\frac{2m}{\hbar^2}(\mathscr{E} - V_1) = k_1^2 \leftarrow k^2(z) \rightarrow k_2^2 = \frac{2m}{\hbar^2}(\mathscr{E} - V_2). \tag{3}$$

For planar stratification the wave equation separates. For propagation in the zx plane the wavefunction is $\Psi = e^{iKx}\psi(z)$, where K is the x-component of the wave vector,

$$K = k_1 \sin\theta_1 = k_1 \sin\theta_2, \tag{4}$$

θ_1 and θ_2 being the angles of incidence and refraction. Equation (4) gives Snell's Law for particle waves, and shows that the refractive index is proportional to $(\mathscr{E} - V)^{1/2}$, that is, to the square root of the kinetic energy of the particle.

The function ψ satisfies

$$\frac{d^2\psi}{dz^2} + q^2\psi = 0, \qquad q^2(z) = k^2(z) - K^2. \tag{5}$$

The normal component of the wavevector, $q(z)$, has limiting values $q_1 = (k_1^2 - K^2)^{1/2}$ and $q_2 = (k_2^2 - K^2)^{1/2}$. The reflection and transmission amplitudes r and t are defined by

$$e^{iq_1z} + r\,e^{-iq_1z} \leftarrow \psi(z) \rightarrow t\,e^{iq_2z}. \tag{6}$$

The left side of (6) represents an incident plane wave of unit amplitude, and a reflected plane wave of amplitude r. With x and time dependence included, the incident plane wave is $\exp i(Kx + qz - \mathscr{E}t/\hbar)$. The factor $\exp i(Kx - \mathscr{E}t/\hbar)$ is common to all parts of the wavefunction in the propagation of plane waves through stratified media, and will usually be omitted. The t and x dependence is needed only in the treatment of wavepackets (Section A-9), and of finite beams.

The reflection and transmission amplitudes are found by solving the wave equation (5). In the simplest case of a potential step at (say) z_1, where $V(z)$ changes from V_1 to V_2, ψ is given by

$$\psi = \begin{cases} e^{iq_1z} + r_0\,e^{-iq_1z} & (z \leqslant z_1) \\ t_0\,e^{iq_2z} & (z \geqslant z_1), \end{cases} \tag{7}$$

and continuity of ψ and $d\psi/dz$ at z_1, gives

$$r_0 = e^{2iq_1z_1}\frac{q_1 - q_2}{q_1 + q_2}, \qquad t_0 = e^{i(q_1-q_2)z_1}\frac{2q_1}{q_1 + q_2}. \tag{8}$$

Other exactly solvable potential energy profiles will be considered in the next section. Here we give some results valid for arbitrary profiles. Let $V(z)$ and $\tilde{V}(z)$ be two potential functions with the same limiting values, V_1 and V_2, at $-\infty$ and $+\infty$. Consider the reflection of plane waves off the two profiles, at the same angle of incidence (the limiting values q_1 and q_2 are thus also common to the two profiles). In Section 2-1 we showed that the corresponding reflection amplitudes, r and $\tilde{r}$, are related by the *comparison identities*

$$r = \tilde{r} - \frac{im}{\hbar^2 q_1}\int_{-\infty}^{\infty}(V - \tilde{V})\psi\tilde{\psi}, \tag{9}$$

$$q_1(1 - r\tilde{r}^*) - q_2 t\tilde{t}^* = \frac{im}{\hbar^2}\int_{-\infty}^{\infty} dz\,(V - \tilde{V})\psi\tilde{\psi}^*. \tag{10}$$

The second identity holds for real $\tilde{V}$ only. For $\tilde{V} = V$ it gives

$$q_1(1 - |r|^2) = q_2|t|^2, \tag{11}$$

which expresses the conservation of particle flux at a non-absorbing barrier, since the probability current density

$$\mathbf{J} = \frac{\hbar}{2im}(\Psi^*\nabla\Psi - \Psi\nabla\Psi^*) = \frac{\hbar}{m}\,\mathrm{Im}\,(\Psi^*\nabla\Psi) \tag{12}$$

has x-component $\hbar K/m$, and z-component limiting values

$$\frac{\hbar q_1}{m}(1 - |r|^2) \leftarrow J_z \rightarrow \frac{\hbar q_2}{m}|t|^2. \tag{13}$$

Other identities can be obtained (as in Section 2-1) by comparing the wave incident from medium 1 with that incident from medium 2. If these are denoted by the subscripts 12 and 21 respectively, an identity relating the corresponding transmission amplitudes is

$$q_2\tilde{t}_{12} - q_1 t_{21} = \frac{im}{\hbar^2}\int_{-\infty}^{\infty} \mathrm{d}z\,(V - \tilde{V})\psi_{21}\tilde{\psi}_{12}. \tag{14}$$

Setting $\tilde{V} = V$ in (14) gives

$$q_2 t_{12} = q_1 t_{21}, \tag{15}$$

so that (14) may be rewritten as

$$t_{12} = \tilde{t}_{12} - \frac{im}{\hbar^2 q_2}\int_{-\infty}^{\infty} \mathrm{d}z\,(V - \tilde{V})\psi_{21}\tilde{\psi}_{12}. \tag{16}$$

Comparison of ψ_{21} with $\tilde{\psi}^*_{12}$ gives, for real $\tilde{V}$,

$$q_2 r_{21}\tilde{t}^*_{12} + q_1 t_{21}\tilde{r}^*_{12} = -\frac{im}{\hbar^2}\int_{-\infty}^{\infty} \mathrm{d}z\,(V - \tilde{V})\psi_{21}\tilde{\psi}^*_{12}. \tag{17}$$

For $\tilde{V} = V$ this implies

$$q_2 r_{21} t^*_{12} + q_1 t_{21} r^*_{12} = 0, \tag{18}$$

which gives, on using (15),

$$r_{21} = -\frac{t_{12}}{t^*_{12}}\, r^*_{12}. \tag{19}$$

The last relation shows that $|r_{12}|^2 = |r_{21}|^2$: the reflectance is the same in either direction for a non-absorbing barrier of arbitrary profile.

For a potential energy $V(z)$ which varies only in the interval $z_1 \leqslant z \leqslant z_2$, with $V = V_1$ for $z \leqslant z_1$ and $V = V_2$ for $z \geqslant z_2$, general expressions for r and t may be written down in terms of two independent solutions $F(z)$ and $G(z)$ of (5). In the interval (z_1, z_2), $\psi = \alpha F + \beta G$. Also ψ and ψ' are continuous at z_1 and z_2, assuming the $V(z)$ has no delta function singularities (or worse) at the end-points. The continuity conditions give four equations for r, t, α and β, the solution of which gives

$$\begin{aligned}
r &= \mathrm{e}^{2iq_1z_1}\{q_1q_2(F_1G_2 - G_1F_2) + iq_1(F_1G_2' - G_1F_2') + iq_2(F_1'G_2 - G_1'F_2) \\
&\quad - (F_1'G_2' - G_1'F_2')\}/D \\
t &= \mathrm{e}^{i(q_1z_1 - q_2z_2)}\, 2iq_1 W/D \\
\alpha &= \mathrm{e}^{iq_1z_1}\, 2iq_1(G_2' - iq_2G_2)/D \\
\beta &= -\,\mathrm{e}^{iq_1z_1}\, 2iq_1(F_2' - iq_2F_2)/D,
\end{aligned} \tag{20}$$

where W is the Wronskian $FG' - GF'$ (a constant), and the common denominator is given by

$$D = q_1 q_2 (F_1 G_2 - G_1 F_2) + iq_1 (F_1 G_2' - G_1 F_2') - iq_2 (F_1' G_2 - G_1' F_2) + (F_1' G_2' - G_1' F_2'). \quad (21)$$

Some general properties of r and t follow directly from (20), as shown in Sections 2-2 and 2-3. For non-absorbing interfaces, $|r| = 1$ when q_2 is imaginary: total reflection occurs for

$$\theta_1 \geqslant \theta_c = \arcsin \left\{ \frac{\mathscr{E} - V_2}{\mathscr{E} - V_1} \right\}^{1/2}. \quad (22)$$

As the thickness $\Delta z = z_2 - z_1$ of the profile tends to zero, r and t tend to the step (Fresnel) values given in (8). The probability density current conservation law (11) follows from (20), for non-absorbing interfaces. And finally, at grazing incidence the reflection amplitude tends to -1, for profiles of arbitrary shape, with or without absorption, and even for internally reflected waves. Thus Lloyd's mirror experiment should produce diffraction fringes with destructive interference at the mirror's edge, for particle as well as for electromagnetic waves.

Potential energy profiles of the form

$$V(z) = \tfrac{1}{2}(V_1 + V_2) - \tfrac{1}{2}(V_1 - V_2) f(z, a), \quad (23)$$

where the function f depends on parameters (collectively denoted by a) which are *independent of* V_1, V_2 *and* $\mathscr{E}$, have following property. If the reflection amplitude at *normal incidence* is known as a function of the limiting magnitudes k_1 and k_2 of the wavevector, the same functional form gives the reflection amplitude at *oblique incidence*, with the normal components of the wavevector q_1 and q_2 replacing k_1 and k_2 (see Section 2-5). The uniform, linear, and hyperbolic tangent profiles have the scaling property (23); the exponential and Rayleigh profiles do not. (These exactly solvable profiles are among those discussed in the next section.)

If $V(z)$ changes monotonically between V_1 and V_2, the reflectivity of this profile cannot be more than that of the potential step between V_1 and V_2 (at the same energy and angle of incidence):

$$R \leqslant \left(\frac{q_1 - q_2}{q_1 + q_2} \right)^2. \quad (24)$$

This upper bound was obtained in Section 5-4.

A-2 Some exactly solvable profiles

Uniform layer. For a potential barrier or well with V constant in (z_1, z_2), we have from Section 2-4 that

$$r = e^{2iq_1 z_1} \frac{q(q_1 - q_2)c + i(q^2 - q_1 q_2)s}{q(q_1 + q_2)c - i(q^2 + q_1 q_2)s}, \quad (25)$$

$$t = e^{i(q_1 z_1 - q_2 z_2)} \frac{2q_1 q}{q(q_1 + q_2)c - i(q^2 + q_1 q_2)s}, \quad (26)$$

where $q^2 = k^2 - K^2$, $\hbar^2 k^2/2m = \mathscr{E} - V$, $K = k_1 \sin\theta_1$, $c = \cos q\Delta z$ and $s = \sin \Delta z$. Equivalent formulae may be written down in terms of the reflection amplitudes at the discontinuities in potential,

$$r_1 = \frac{q_1 - q}{q_1 + q}, \qquad r_2 = \frac{q - q_2}{q + q_2}. \tag{27}$$

These are

$$r = e^{2iq_1 z_1} \frac{r_1 + r_2 e^{2iq\Delta z}}{1 + r_1 r_2 e^{2iq\Delta z}}, \tag{28}$$

$$t = e^{i(q_1 z_1 - q_2 z_2)} \frac{(1 + r_1)(1 + r_2) e^{iq\Delta z}}{1 + r_1 r_2 e^{2iq\Delta z}}. \tag{29}$$

Zero reflection (and thus perfect transparency) is possible if (i) $r_1 = r_2$ and $e^{2iq\Delta z} = -1$, or if (ii) $r_1 = -r_2$ and $e^{2iq\Delta z} = 1$. Condition (i) is satisfied if $q^2 = q_1 q_2$ and $2q\Delta z$ is an odd multiple of π. At normal incidence $q^2 = q_1 q_2$ is satisfied if

$$\mathscr{E} = U \equiv \frac{V^2 - V_1 V_2}{2V - V_1 - V_2}. \tag{30}$$

At oblique incidence $q^2 = q_1 q_2$ can be satisfied only if $\mathscr{E} > U$; it then holds at one angle,

$$\theta_1 = \arcsin\left\{\frac{\mathscr{E} - U}{\mathscr{E} - V_1}\right\}^{1/2}. \tag{31}$$

The second condition for zero reflection can be satisfied only when $V_1 = V_2$, and can hold at more than one angle, given by

$$\theta_1 = \arcsin\left\{\frac{\mathscr{E} - V - \dfrac{\hbar^2}{2m}\left(\dfrac{n\pi}{\Delta z}\right)^2}{\mathscr{E} - V_1}\right\}^{1/2}, \tag{32}$$

where n is an integer. Conditions (i) and (ii) hold when Δz is respectively equal to an odd or an even multiple of $\lambda/4$, where $\lambda = 2\pi/q$ is the effective wavelength, within the layer, for propagation in the z direction. Zero reflection is not possible in the tunnelling case, $\mathscr{E} < V$.

Provided V lies between V_1 and V_2, the reflectance $R = |r|^2$ of a uniform layer does not exceed the reflectance $(q_1 - q_2)^2/(q_1 + q_2)^2$ of a single potential step between the same values V_1 and V_2 (at the same energy and angle of incidence). This is a special case of a general theorem proved in Section 5-4 that a monotonic profile cannot reflect more than the corresponding step profile.

From (28) and (29) we see that for real q, that is when

$$\mathscr{E} - V > (\mathscr{E} - V_1)\sin^2\theta_1, \tag{33}$$

the reflectance R and transmittance $T = (q_2/q_1)|t|^2$ are periodic functions of $q\Delta z$, with period π. The reflectance

$$R = \frac{r_1^2 + 2r_1 r_2 \cos 2q\Delta z + r_2^2}{1 + 2r_1 r_2 \cos 2q\Delta z + (r_1 r_2)^2} \tag{34}$$

has extrema with respect to Δz when $\cos 2q\Delta z = \pm 1$; these are

$$R^{+} = \left(\frac{q_1 - q_2}{q_1 + q_2}\right)^2, \qquad R^{-} = \left(\frac{q^2 - q_1 q_2}{q^2 + q_1 q_2}\right)^2. \tag{35}$$

R^- is less than R^+ when V lies between V_1 and V_2; when V is outside this range, R^+ becomes the minimum value.

Linear profile. A transition region where the classical force $-\mathrm{d}V/\mathrm{d}z$ is constant has the potential energy

$$V(z) = \begin{cases} V_1 & z \leqslant z_1 \\ V_1 + \dfrac{\Delta V}{\Delta z}(z - z_1) & z_1 < z < z_2 \\ V_2 & z \geqslant z_2, \end{cases} \tag{36}$$

where $\Delta z = z_2 - z_1$ and $\Delta V = V_2 - V_1$. Within (z_1, z_2) the wave equation (5) for motion normal to the interface can be written in terms of a dimensionless variable $x = l^2 q^2$ as

$$\frac{\mathrm{d}^2\psi}{\mathrm{d}x^2} + x\psi = 0, \tag{37}$$

where the length l is given by

$$l = \left(\frac{\hbar^2}{2m}\left|\frac{\Delta z}{\Delta V}\right|\right)^{1/3}. \tag{38}$$

The standard pair of independent solutions of (37) are the Airy functions $Ai(-x)$, $Bi(-x)$ (definitions and elementary properties of the Airy functions are given in Section 5-2). The reflection and transmission amplitudes may be obtained from (20), care being taken to convert between derivatives with respect to z and x via

$$\frac{\mathrm{d}\psi}{\mathrm{d}z} = l^2 \frac{\Delta q^2}{\Delta z}\frac{\mathrm{d}\psi}{\mathrm{d}x} = l^2 \frac{2m}{\hbar^2}\left(-\frac{\Delta V}{\Delta z}\right)\frac{\mathrm{d}\psi}{\mathrm{d}x}. \tag{39}$$

The linear profile has the scaling property (23), the function f being given by

$$f(z, z_1, z_2) = \begin{cases} -1 & z \leqslant z_1 \\ (2z - z_1 - z_2)/(z_2 - z_1) & z_1 < z < z_2 \\ 1 & z \geqslant z_2. \end{cases} \tag{40}$$

Hyperbolic tangent profile. This may be written in several equivalent ways, the first of which explicitly shows the scaling property (23):

$$\begin{aligned} V(z) &= \tfrac{1}{2}(V_1 + V_2) - \tfrac{1}{2}(V_1 - V_2)\tanh z/2a \\ &= \frac{V_1 + V_2\,\mathrm{e}^{z/a}}{1 + \mathrm{e}^{z/a}} = \frac{V_1}{1 + \mathrm{e}^{z/a}} + \frac{V_2}{1 + \mathrm{e}^{-z/a}}. \end{aligned} \tag{41}$$

The solution for this profile in terms of hypergeometric functions is discussed in detail in Section 2-5. There is no need to translate the formulae given there into quantum mechanical notation, since they are given in terms of the variables $y_1 = q_1 a$ and $y_2 = q_2 a$ and thus can be applied directly to either electromagnetic or particle waves.

sech² profile. The potential energy is given by

$$V(z) = V_0 + \Delta V \operatorname{sech}^2 z/a. \tag{42}$$

The equation for the probability amplitude ψ reads

$$\frac{d^2\psi}{dz^2} + \left[q_0^2 - \frac{2m\Delta V}{\hbar^2}\operatorname{sech}^2 z/a\right]\psi = 0. \tag{43}$$

Here q_0 is the limiting value of the normal component of the wavevector, given by

$$q_0^2 = \frac{2m}{\hbar^2}(\mathscr{E} - V_0) - K^2 = k_0^2\cos^2\theta, \tag{44}$$

where

$$k_0^2 = \frac{2m}{\hbar^2}(\mathscr{E} - V_0), \qquad K = k_0 \sin\theta. \tag{45}$$

A solution of (43) can be found in terms of the hypergeometric function, as discussed for the electromagnetic case in Section 4-3. The solution is characterized by two dimensionless parameters,

$$\alpha = -2ma^2\Delta V/\hbar^2, \qquad \beta = q_0 a. \tag{46}$$

Formulae for the reflection and transmission amplitudes in terms of α and β are given in Section 4-3. Tunnelling can occur for positive ΔV, namely when

$$(\mathscr{E} - V_0)\cos^2\theta < \Delta V, \quad \text{or} \quad \beta^2 < -\alpha. \tag{47}$$

Exponential profile. The potential energy is given by

$$V(z) = \begin{cases} V_1 & z \leqslant z_1 \\ \mathscr{E} - (\mathscr{E} - V_1)\exp(z - z_1)/a & z_1 < z < z_2 \\ V_2 & z \geqslant z_2 \end{cases} \tag{48}$$

where the length a depends on V_1, V_2 and $\mathscr{E}$, and can be positive or negative:

$$a = (z_2 - z_1)/\log\left(\frac{\mathscr{E} - V_2}{\mathscr{E} - V_1}\right). \tag{49}$$

Transformation to a dimensionless independent variable proportional to the local magnitude of the wavevector,

$$u = 2ka = 2a\left\{\frac{2m}{\hbar^2}(\mathscr{E} - V)\right\}^{1/2}, \tag{50}$$

converts (5) to Bessel's equation

$$\frac{d^2\psi}{du^2} + \frac{1}{u}\frac{d\psi}{du} + \left[1 - \frac{(2Ka)^2}{u^2}\right]\psi = 0. \tag{51}$$

The general solution within (z_1, z_2) is thus $\alpha J_s(u) + \beta Y_s(u)$, where $s = 2Ka$. Note that both the order s and the argument u of the Bessel functions are proportional to the thickness of the inhomogeneity. The order also depends on the angle of incidence, increasing from zero at normal incidence to $2k_1 a$ at grazing incidence. Reflection and transmission amplitudes may be obtained from (20), on converting between derivatives with respect to z and u via

$$\frac{d\psi}{dz} = k\frac{d\psi}{du} \tag{52}$$

Some reduction of the formulae is possible by using the properties of Bessel functions. Details are given in Section 2-5.

Rayleigh profile. In this transition the wavelength $2\pi/k$ is linear in z. It is useful to work in terms of a dimensionless variable which is linear in z,

$$\eta(z) \equiv (k\Delta z)^{-1} = \eta_1 + (z - z_1)\Delta\eta/\Delta z. \tag{53}$$

The interface extends from z_1 to $z_2 = z_1 + \Delta z$, and

$$\Delta\eta = \eta_2 - \eta_1 = (k_2^{-1} - k_1^{-1})/\Delta z. \tag{54}$$

The potential energy is given by

$$\mathscr{E} - V(z) = \frac{\hbar^2}{2m\eta^2(z)(\Delta z)^2}. \tag{55}$$

At normal incidence the wave equation has a simple power-law solution: on changing to the variable η the equation $d^2\psi/dz^2 + k^2\psi = 0$ becomes

$$\frac{d^2\psi}{d\eta^2} + \frac{\psi}{\eta^2(\Delta\eta)^2} = 0, \tag{56}$$

and has the solutions $\psi_\pm = \eta^{1/2\pm\nu}$, where

$$\nu^2 = \tfrac{1}{4} - (\Delta\eta)^{-2}. \tag{57}$$

(The parameter ν is introduced for mathematical convenience, and to increase commonality with the results of Section 2-5.) On converting between z and η derivatives via

$$\frac{d\psi}{dz} = \frac{\Delta\eta}{\Delta z}\frac{d\psi}{d\eta}, \tag{58}$$

the reflection and transmission amplitudes at normal incidence are found from (20) to be

$$r_n = e^{2ik_1z_1}\frac{\tfrac{1}{2}i\Delta\eta[\varrho^\nu - 1]}{\varrho^\nu - 1 - i\nu\Delta\eta[\varrho^\nu + 1]}, \tag{59}$$

$$t_n = -e^{i(k_1z_1 - k_2z_2)}\frac{2i\nu\Delta\eta\,\varrho^{\nu/2-1/4}}{\varrho^\nu - 1 - i\nu\Delta\eta[\varrho^\nu + 1]}, \tag{60}$$

where ϱ is the square of the ratio of the refractive indices,

$$\varrho = \left(\frac{\eta_1}{\eta_2}\right)^2 = \left(\frac{k_2}{k_1}\right)^2 = \frac{\mathscr{E} - V_2}{\mathscr{E} - V_1}. \tag{61}$$

The reflectivity at normal incidence takes different forms, depending on whether $(\Delta\eta)^2$ is greater than or less than four. When $(\Delta\eta)^2 > 4$, ν is real, and

$$R_n = \frac{\frac{1}{4}(\Delta\eta)^2(\varrho^\nu - 1)^2}{(\varrho^\nu - 1)^2 + (\nu\Delta\eta)^2(\varrho^\nu + 1)^2}. \tag{62}$$

When $(\Delta\eta)^2 < 4$, $\nu = i|\nu|$ and

$$R_n = \frac{\sin^2(\frac{1}{2}|\nu| \log \varrho)}{4|\nu|^2 + \sin^2(\frac{1}{2}|\nu| \log \varrho)}. \tag{63}$$

At $\nu = 0$ these two forms take the common value

$$R_n(\nu = 0) = \frac{\log^2 \varrho}{16 + \log^2 \varrho}. \tag{64}$$

From (63) we see that the normal incidence reflectivity is zero when $|\nu| \log \varrho$ is an integer multiple of 2π. This happens when the interface thickness Δz takes one of the values

$$\Delta z = \frac{|k_1 - k_2|}{k_1 k_2}\left\{\frac{1}{4} + \left(\frac{n\pi}{\log (k_2/k_1)}\right)^2\right\}^{1/2}, \quad n = 1, 2, \ldots . \tag{65}$$

At oblique incidence the wave equation in the η variable reads

$$\frac{d^2\psi}{d\eta^2} + \left[\frac{\frac{1}{4} - \nu^2}{\eta^2} - \left(\frac{K\Delta z}{\Delta\eta}\right)^2\right]\psi = 0, \tag{66}$$

and has solutions proportional to $\eta^{1/2}$ times Bessel functions of order ν and imaginary argument $\pm i(K\Delta z/\Delta\eta)\eta$. The reflection and transmission amplitudes may thus be obtained from (20). A reference to explicit expressions obtained in this way is given in Section 2-5.

References to listings and systematic generation of solvable profiles are given at the end of Chapter 2.

A-3 Perturbation and variational theories

Perturbation theory gives expressions for the unknown function ψ satisfying (5) and (6) (and thus also for its reflection and transmission amplitudes r and t) in terms of some known function ψ_0, satisfying

$$\frac{d^2\psi_0}{dz^2} + q^2\psi_0 = 0, \quad e^{iq_1 z} + r_0\, e^{-iq_1 z} \leftarrow \psi_0 \rightarrow t_0\, e^{iq_2 z}. \tag{67}$$

The solution is given in terms of $\psi_0(z)$ and a Green's function $G(z, \zeta)$, which satisfies

$$\frac{\partial^2 G}{\partial z^2} + q_0^2(z)G = \delta(z - \zeta), \tag{68}$$

and has the appropriate limiting forms to make ψ, given by the integral equation

$$\psi(z) = \psi_0(z) - \int_{-\infty}^{\infty} \mathrm{d}\zeta \, \Delta q^2(\zeta) G(z, \zeta)\psi(\zeta), \tag{69}$$

take the limiting forms (6). In the above, q and q_0 share the limiting values q_1 at $-\infty$ and q_2 at $+\infty$, and

$$\Delta q^2 = q^2 - q_0^2 = -\frac{2m}{\hbar^2}(V - V_0). \tag{70}$$

(V and V_0 also share the limiting values V_1 and V_2, and the equations for ψ and ψ_0 are to be solved at the same particle energy.) Note that the perturbation Δq^2 is independent of energy and of the angle of incidence.

In the long wave case the perturbation is built up from the step potential energy profile

$$V_0(z) = \tfrac{1}{2}(V_1 + V_2) - \tfrac{1}{2}(V_1 - V_2)\,\mathrm{sgn}\,(z), \tag{71}$$

for which the wavefunction ψ_0 and Green's function $G(z, \zeta)$ are given in Section 3-1. The perturbation theory expression for r_n (the contribution of order $(V - V_0)^n$ to the reflection amplitude) is given in (3.12). When $V_0(z)$ is constant ($V_1 = V_2 = V_0$), $r_0 = (q_1 - q_2)/(q_1 + q_2)$ is zero, and r_1 is the Fourier transform of the deviation of the potential from V_0:

$$r_1 = -\frac{im}{q_0\hbar^2}\int_{-\infty}^{\infty} \mathrm{d}z\,(V - V_0)\,\mathrm{e}^{2iq_0 z}. \tag{72}$$

This is divergent at grazing incidence, when $q_0 \to 0$.

In the short wave case the perturbation theory is constructed from the Liouville–Green functions of Section 6-2:

$$\psi^+ = (q_1/q)^{1/2} \exp(i\phi), \quad \psi^- = (q_2/q)^{1/2} \exp(-i\phi), \quad \phi(z) = \int^z \mathrm{d}\zeta\, q(\zeta), \tag{73}$$

the Green's function being given in Section 6-5. The corresponding first order perturbation result for the reflection amplitude is

$$r^{(1)} = \frac{1}{4i}\int_{-\infty}^{\infty} \mathrm{d}z \left[\frac{\mathrm{d}\gamma}{\mathrm{d}z} + \tfrac{1}{2}q\gamma^2\right] \mathrm{e}^{2i\phi}, \tag{74}$$

where the dimensionless function $\gamma = \mathrm{d}q/q^2\,\mathrm{d}z$ must be small everywhere for the perturbation result to be accurate.

Variational expressions for the reflection amplitude may be derived from the perturbation theories. The general variational principle is $\delta(F^2/S) = 0$, where

F and S are first and second order in the unknown ψ:

$$F = \int_{-\infty}^{\infty} dz\, \Delta q^2(z)\psi(z)\psi_0(z), \tag{75}$$

$$S = \int_{-\infty}^{\infty} dz\, \Delta q^2(z)\psi^2(z) + \int_{-\infty}^{\infty} dz\, \Delta q^2(z)\psi(z) \int_{-\infty}^{\infty} d\zeta\, \Delta q^2(\zeta)\psi(\zeta)G(z, \zeta). \tag{76}$$

The variational estimate for the reflection amplitude is

$$r^{\text{var}} = r_0 - F^2/2iq_1S. \tag{77}$$

Further details are given in Chapter 4. Here we note only that the variational theory based on the long-wave perturbation theory removes the grazing incidence divergence which troubles all orders of the perturbation theory when $V_1 = V_2$.

A-4 Long waves, integral invariants

In the long wave limit a given potential energy profile reflects predominantly as a step profile, with a small correction which depends on the deviation of the profile from the step $V_0(z)$ given by (71). This correction can be expressed as a geometric series in the ratio of the interface thickness to the wavelength of the wave. The reflection amplitude, written as $r_0 + r_1 + r_2 \ldots$ where the subscript n here refers to the order (power) in the interface thickness, is found from the long-wave perturbation theory of the previous section to be given by

$$r_0 = \frac{q_1 - q_2}{q_1 + q_2}, \qquad r_1 = \left(-\frac{2m}{\hbar^2}\right)\frac{2iq_1\mu_1}{(q_1 + q_2)^2},$$
$$r_2 = \left(\frac{2m}{\hbar^2}\right)\frac{2q_1}{(q_1 + q_2)^3}\left\{2(q_1 + q_2)q_2\mu_2 - \frac{2m}{\hbar^2}\mu_1^2\right\}, \tag{78}$$

where

$$\mu_n = \int_{-\infty}^{\infty} dz\, [V(z) - V_0(z)]z^{n-1}. \tag{79}$$

The integrals μ_n depend on the relative positioning of the profiles V and V_0, with the exception of μ_1 in the case where V_0 is a constant ($V_1 = V_2$).

The reflectivity $R = |r|^2$ must be independent of the relative positioning of the actual and step profiles. When $q(z)$ is real everywhere, R differs from $R_0 = r_0^2$ by a term which is of second order in the interface thickness:

$$\begin{aligned} R &= R_0 + |r_1|^2 + 2r_0r_2 + \ldots \\ &= R_0 - \frac{4q_1q_2}{(q_1 + q_2)^4}\left(\frac{2m}{\hbar^2}\right)^2 \{2(V_1 - V_2)\mu_2 - \mu_1^2\} + \ldots . \end{aligned} \tag{80}$$

The expression in braces, which we call i_2, is invariant to the relative positioning of V and V_0. It is the first in an infinite set of integral invariants which may be constructed from the μ_n (see Section 3-3).

The result (80) shows that the reflectivity takes a *universal form* in the long wave limit:

$$R = \left(\frac{q_1 - q_2}{q_1 + q_2}\right)^2 - \frac{4q_1q_2}{(q_1 + q_2)^4}\left(\frac{2m}{\hbar^2}\right)^2 i_2 + \dots . \tag{81}$$

This holds for all non-absorbing profiles $V(z)$ which do not contain delta function singularities or worse. When $V_1 = V_2$ the μ_2 integral is not needed (to this order) and μ_1 is separately invariant. The reflectivity is then proportional to μ_1^2, and is independent of the sign of $V - V_0$, to this order. When $V_1 \neq V_2$ and V is real everywhere it is possible to position the profiles V and V_0 to make $\mu_1 = 0$. The second order invariant i_2 may then be put in the form

$$i_2 = (V_2 - V_1)\int_{-\infty}^{\infty} dz \frac{dV}{dz} z^2 \qquad (\text{when } \mu_1 = 0), \tag{82}$$

which shows that $R - R_0$ is proportional to $(V_2 - V_1)$ times the second moment of the force $-dV/dz$. When V increases or decreases monotonically, i_2 is positive and the second order term decreases the reflectivity from that of a step profile, as may be expected. The last result may be strengthened by writing the second order invariant as

$$i_2 = -\int_{-\infty}^{\infty} dz_1 \int_{-\infty}^{\infty} dz_2 \, [V(z_1) - V_0(z_1 - z_2)][V(z_2) - V_0(z_2 - z_1)]. \tag{83}$$

This form shows that i_2 is positive if $V(z)$ lies between V_1 and V_2 for all z, as can be seen by considering the sign of the integrand for $z_1 < z_2$ and for $z_1 > z_2$. Thus the second order term decreases the reflectivity if

$$\min(V_1, V_2) \leqslant V(z) \leqslant \max(V_1, V_2). \tag{84}$$

The functional form of i_2 for the profiles considered in Section 3-6 (all but one of which, the double exponential profile, were transcribed to the quantum mechanical case in Section A-3) can be obtained from Table 3-1 by the substitution $\varepsilon \to \mathscr{E} - V$. For example, the uniform layer of thickness Δz and potential V has

$$i_2/(\Delta z)^2 = (V_1 - V)(V - V_2), \tag{85}$$

and the exponential profile of thickness Δz has

$$i_2/(\Delta z)^2 = \left\{\frac{V_1 - V_2}{\log \dfrac{\mathscr{E} - V_2}{\mathscr{E} - V_1}}\right\}^2 - (\mathscr{E} - V_1)(\mathscr{E} - V_2). \tag{86}$$

(The energy dependence in (86) comes from the energy dependence of the potential for the exponential profile, given by equation (48).)

Profiles of the form (23), for example the linear, tanh, error function and double exponential profiles (the last being defined in (3.69)) all have i_2 proportional to $(V_1 - V_2)^2$, provided the function f is continuous.

A-5 Riccati-type equations; the Rayleigh approximation

The second order differential equation (5) is equivalent to a pair of coupled first order equations in ψ and $\psi' = \mathrm{d}\psi/\mathrm{d}z$. On setting

$$\psi = F + G, \qquad \psi' = iq(F - G), \tag{87}$$

we find that F and G satisfy

$$F' = iqF - \frac{q'}{2q}(F - G), \tag{88}$$

$$G' = -iqG + \frac{q'}{2q}(F - G). \tag{89}$$

From (6) we see that when incidence is from medium 1, $F \to \mathrm{e}^{iq_1 z}$ and $G \to r\,\mathrm{e}^{-iq_1 z}$ as $z \to -\infty$. Thus the ratio $\varrho = G/F$ tends to $\mathrm{e}^{-2iq_1 z}$ times the reflection amplitude r as $z \to -\infty$, and to zero as $z \to \infty$. The equation satisfied by ϱ is of the generalized Riccati type:

$$\varrho' + 2iq\varrho - \frac{q'}{2q}(1 - \varrho^2) = 0. \tag{90}$$

On writing $\varrho = |\varrho|\,\mathrm{e}^{i\theta}$, separating the real and imaginary parts, and integrating the equation for $|\varrho|'$ over all z, one finds

$$\log \frac{1 + |r|}{1 - |r|} = -\int_{-\infty}^{\infty} \mathrm{d}z \frac{q'}{q} \cos\theta \tag{91}$$

(more detail may be found in Chapter 5). From this follows the inequality of Section 5-4,

$$R \leqslant \left(\frac{q_1 - q_2}{q_1 + q_2}\right)^2; \tag{92}$$

any non-absorbing *monotonic* profile cannot reflect more than a step profile between the same limiting values of potential, at the same energy and angle of incidence.

An alternative approach is to deal directly with the reflection amplitude, by setting

$$\psi = f\,\mathrm{e}^{i\phi} + g\,\mathrm{e}^{-i\phi}, \qquad \psi' = iq(f\,\mathrm{e}^{i\phi} - g\,\mathrm{e}^{-i\phi}), \tag{93}$$

where ϕ is the phase integral defined in (73). On using $\phi' = q$ we find that f and g satisfy the equations

$$f' + \frac{q'}{2q}(f - g\,\mathrm{e}^{-2i\phi}) = 0, \tag{94}$$

$$g' + \frac{q'}{2q}(g - f\,\mathrm{e}^{2i\phi}) = 0. \tag{95}$$

If the lower limit in ϕ is chosen so as to make $\phi \to q_1 z$ as $z \to -\infty$, the ratio g/f tends to the reflection amplitude r as $z \to -\infty$, and in fact may be interpreted as the reflection amplitude $r(z)$ at any z, for a profile truncated at z as explained in Chapter 5. The equation satisfied by $r(z)$ is

$$r'(z) = \frac{q'}{2q}\left(e^{2i\phi} - r^2(z)\, e^{-2i\phi}\right). \tag{96}$$

The reflection amplitude r of the entire profile is thus

$$r = -\int_{-\infty}^{\infty} dz\, \frac{q'}{2q}\left(e^{2i\phi} - r^2(z)\, e^{-2i\phi}\right). \tag{97}$$

The Rayleigh or weak reflection approximation is obtained by neglecting the term proportional to $r^2(z)$ in the integrand:

$$r^R = -\int_{-\infty}^{\infty} dz\, \frac{q'}{2q}\, e^{2i\phi}; \tag{98}$$

its long-wave limit is

$$r^R \to \tfrac{1}{2} \log \frac{q_1}{q_2}. \tag{99}$$

The Rayleigh approximation works well at all wavelengths, provided the reflection is weak. It fails whenever the reflection is strong, for example at grazing incidence. Note that the factor q'/q in the integrand of both the exact and the approximate formulae (97) and (98) for the reflection amplitude can be written as

$$\frac{q'}{q} = \frac{(q^2)'}{2q^2} = \frac{-m\, dV/dz}{\hbar^2 q^2}. \tag{100}$$

The contributions to the reflection amplitude are thus weighted by the local value of the ratio of the force $-dV/dz$ to the kinetic energy of motion normal to the interface, $\hbar^2 q^2/2m$.

A-6 Reflection of short waves

In Section 6-2 we saw that the Liouville-Green functions

$$\psi^+ = \left(\frac{q_1}{q}\right)^{1/2} e^{i\phi}, \quad \psi^- = \left(\frac{q_2}{q}\right)^{1/2} e^{-i\phi}, \quad \phi = \int^z d\zeta\, q(\zeta) \tag{101}$$

are approximate solutions of the wave equation (5). In fact $\psi^\pm$ satisfy

$$\frac{d^2\psi^\pm}{dz^2} + q^2\left\{1 + \frac{1}{2}\frac{d\gamma}{d\phi} + \frac{\gamma^2}{4}\right\}\psi^\pm = 0, \tag{102}$$

where the dimensionless function $\gamma(z)$ is given by

$$\gamma = \frac{dq}{q^2\, dz} = \frac{dq}{q\, d\phi}. \tag{103}$$

If $\mathrm{d}\gamma/\mathrm{d}\phi$ $(=q^{-1}\,\mathrm{d}\gamma/\mathrm{d}z)$ and γ^2 are small compared to unity, the functions $\psi^{\pm}$, and approximations to r and t resulting from their use, are expected to be accurate. Since

$$\gamma \;=\; \frac{1}{2q^3}\frac{\mathrm{d}q^2}{\mathrm{d}z} \;=\; \frac{-m\,\mathrm{d}V/\mathrm{d}z}{\hbar^2 q^3}, \tag{104}$$

where $-\mathrm{d}V/\mathrm{d}z$ is the force, $\hbar^2 q^2/2m$ the kinetic energy and $2\pi/q$ the effective wavelength (all relating to change or motion in the z direction),

$$4\pi\gamma \;=\; \frac{\text{force} \times \text{wavelength}}{\text{kinetic energy}}. \tag{105}$$

Thus $4\pi\gamma$ is the ratio of the potential energy change in one wavelength to the local kinetic energy of motion in the z direction.

Short wavelength approximations all depend on γ^2 and $\mathrm{d}\gamma/\mathrm{d}\phi$ being small, at least for most of the range of z. They generally fail at grazing incidence $(q_1 \to 0)$, and special techniques are needed at discontinuities in the slope of the potential, and at turning points (where q^2 passes through zero). We will summarize, without proof, some of the results of Chapter 6.

The Rayleigh approximation (98), which can be written as

$$r^R \;=\; -\frac{1}{2}\int_{-\infty}^{\infty} \mathrm{d}\phi\; \gamma\, \mathrm{e}^{2i\phi} \;=\; \frac{1}{4}\int_{-\infty}^{\infty} \mathrm{d}z\, \frac{\mathrm{d}V/\mathrm{d}z}{\mathscr{E} - V - \hbar^2K^2/2m}\,\mathrm{e}^{2i\phi}, \tag{106}$$

turns out to be closely related to the first order perturbation theory result

$$r^{(1)} \;=\; -\frac{1}{2}\int_{-\infty}^{\infty} \mathrm{d}\phi\; (\gamma - \gamma^2/4i)\, \mathrm{e}^{2i\phi}, \tag{107}$$

based on the approximate solutions $\psi^{\pm}$ and a Green's function constructed from them.

When the potential $V(z)$ has discontinuities in its gradient at the profile boundaries, but is otherwise smooth (as is the case for the linear, exponential and Rayleigh profiles), (106), (107), or (20) with $F = \psi^+$ and $G = \psi^-$, all give the reflection amplitude

$$r \;\simeq\; \tfrac{1}{4}\,\mathrm{e}^{i(\phi_1+\phi_2)}\,\{\gamma_1\,\mathrm{e}^{-i\Delta\phi} - \gamma_2\,\mathrm{e}^{i\Delta\phi}\}, \tag{108}$$

where ϕ_1 and ϕ_2 are the values of $\phi(z)$ at z_1 and z_2, $\Delta\phi = \phi_2 - \phi_1$ is the change in the accumulated phase across the profile, and the function γ changes from zero to γ_1 at z_1, and from γ_2 to zero at z_2. Exponentially small terms, originating from the smooth variation in $V(z)$ other than at the end points, are omitted from (108). The resulting reflectivity is

$$R \simeq \tfrac{1}{16}\{\gamma_1^2 + \gamma_2^2 - 2\gamma_1\gamma_2\cos 2\Delta\phi\}. \tag{109}$$

The dominant part of the reflectivity thus depends quadratically on the discontinuities in the potential gradient, and shows oscillatory decay with increasing energy.

Regions where $q^2(z) < 0$, which are classically inaccessible since the kinetic energy of motion in the z direction is then negative, occur where

$$\mathscr{E} - V < (\mathscr{E} - V_1)\sin^2\theta_1. \tag{110}$$

The locations where $q^2(z) = 0$ are called *turning points*. When $V_2 > V_1$ and $\theta_1 \geqslant \theta_c = \arcsin[(\mathscr{E} - V_2)/(\mathscr{E} - V_1)]^{1/2}$, total reflection occurs. Interest then centres on the phase of the reflection amplitude, which determines the time of arrival and the shape of reflected pulses. We write $r = e^{i\delta}$; the phase δ is then given by (in the short wave limit, on reflection from a profile with a single turning point)

$$\delta \simeq 2(\phi_0 - \phi_-) - \pi/2, \tag{111}$$

where ϕ_0 is the value of the phase integral at the turning point, and ϕ_- is defined by

$$\phi(z) \to q_1 z + \phi_- \quad \text{as } z \to -\infty. \tag{112}$$

The result (111) is derived in Section 6-7. A simpler version follows if one takes the (so far unspecified) lower limit of integration in the definition of ϕ to be the turning point z_0, in which case $\phi_0 = 0$. If also the origin $z = 0$ is chosen such that $q = q_1$ for $z < 0$, (111) becomes

$$\delta \simeq 2\int_0^{z_0} dz\, q(z) - \pi/2. \tag{113}$$

In the case of a potential barrier, (110) may hold in an interval between two turning points, z_1 and z_2. The wave penetrates the classically forbidden region where $q^2 < 0$, and a part of it *tunnels* through to beyond z_2 where $q^2 > 0$. For well separated turning points, and smooth potential energy barriers with $V_1 = V_2$, the reflection and transmission probabilities are given in Section 6-8:

$$R \simeq \tanh^2(\Delta\Phi + \log 2), \qquad T \simeq \operatorname{sech}^2(\Delta\Phi + \log 2). \tag{114}$$

Here $\Delta\Phi$ is the increment in the imaginary part of the phase between the turning points (and thus gives the exponent of the change in amplitude across the barrier),

$$\Delta\Phi = \int_{z_1}^{z_2} dz\, |q(z)|. \tag{115}$$

For large $\Delta\Phi$ the results (114) have the limiting forms

$$R \to 1 - e^{-2\Delta\Phi}, \qquad T \to e^{-2\Delta\Phi}. \tag{116}$$

A-7 Absorption, the optical potential

As we saw in Section 1-5, a medium containing scatterers can be approximated by an effective potential $V(z)$. When there is absorption, as for example in the case of neutrons by means of a nuclear reaction, or in the case of electrons by trapping or by "annihilation" with hole quasiparticles, the interaction with the medium can be approximated by a complex potential. This is in close analogy with the electromagnetic case, where absorption is represented by an imaginary part in the dielectric function or refractive index. Because of the analogy, the complex potential is referred to as the *optical potential* in nuclear and atomic physics.

For a medium with z-stratification, the propagation normal to the interface is characterized by the wave equation (5), with

$$q^2(z) = \frac{2m}{\hbar^2}(\mathscr{E} - V(z)) - K^2. \tag{117}$$

In the presence of absorption the potential has a negative imaginary part, as we shall see. Accordingly we set

$$V = V_r - iV_i, \tag{118}$$

with $V_i \geqslant 0$. The normal component of the wavevector is also complex, $q = q_r + iq_i$. From the real and imaginary parts of (117) we get

$$q_r^2 - q_i^2 = \frac{2m}{\hbar^2}(\mathscr{E} - V_r) - K^2, \tag{119}$$

$$2q_r q_i = \frac{2m}{\hbar^2} V_i. \tag{120}$$

In a uniform medium the transmitted wave is proportional to $\exp i(Kx + qz) = \exp i(Kx + q_r z) \exp(-q_i z)$. Both q_r and q_i are non-negative, and hence so is V_i. The real and imaginary components of q, for incidence at angle θ_1 from a medium with real potential V_1, are found from (119) and (120) to be given by

$$q_r^2 = \frac{m}{\hbar^2}\{\mathscr{E} - V_r - (\mathscr{E} - V_1)\sin^2\theta_1 + [(\mathscr{E} - V_r - (\mathscr{E} - V_1)\sin^2\theta_1)^2 + V_i^2]^{1/2}\}, \tag{121}$$

$$q_i = \frac{mV_i}{\hbar^2 q_r}. \tag{122}$$

($V_i = 0$ is a degenerate case; then either $q_i = 0$ or $q_r = 0$, depending on whether θ_1 is less than or greater than the critical angle θ_c, given by (22).) In a uniform absorbing medium the surfaces of constant amplitude are planes parallel to the interface, while surfaces of constant real phase are the planes $Kx + q_r z =$ constant. The normal to these planes is inclined at an angle θ_2' to the normal to the interface, where $\theta_2' = \arctan(K/q_r)$, with $\hbar^2 K^2/2m = (\mathscr{E} - V_1)\sin^2\theta_1$. In general the angle of refraction (θ_2), defined by Snell's Law (4), is complex. It is equal to the real angle θ_2' only at normal incidence, or when V is real.

At a sharp boundary between a medium with real potential V_1 and an absorbing medium with $V_2 = V_r - iV_i$, the reflection amplitude and reflectivity are

$$r = e^{2iq_1 z_1}\frac{q_1 - q_r - iq_i}{q_1 + q_r + iq_i}, \quad R = \frac{(q_1 - q_r)^2 + q_i^2}{(q_1 + q_r)^2 + q_i^2} \tag{123}$$

(z_1 being the location of the interface). The formulae for the uniform absorbing potential barrier are obtained from those of Section A-2 by making q complex. The same is true for the hyperbolic tangent profile, for which formulae in the absorbing case are given in Section 8-7.

The conservation law (11) no longer holds in the presence of absorption, since particles are removed and the probability density current thus decreases into the absorbing medium. However, the reciprocity law (15) remains valid, and so the transmittance through an arbitrary inhomogeneous absorbing interface (between two non-absorbing media) is the same in either direction:

$$T_{12} = \frac{q_2}{q_1}|t_{12}|^2 = \frac{q_1}{q_2}|t_{21}|^2 = T_{21}. \tag{124}$$

The result that $r \to -1$ at grazing incidence also holds in the presence of absorption.

A non-absorbing film on an absorbing substrate can give zero reflection (and thus total absorption) at an angle given by

$$(q^2 - q_1 q_r)(q_r - q_1) = q_1 q_i^2, \tag{125}$$

where q_1, q, and $q_r + iq_i$ are the normal components of the wavevector in the first medium, the film, and the substrate. This is discussed in Section 8-3. Formulae for the reflectance and transmittance of an absorbing layer on a non-absorbing substrate are given in Section 8-4.

Thin absorbing films between two unlike media ($V_1 \neq V_2$) can either decrease or increase the reflectivity, depending on whether the particles go up or down in potential. This follows from the general expression for the reflection amplitude to first order in the film thickness, which (from (78)) is

$$r = r_0 - \frac{2iq_1}{(q_1 + q_2)^2}\frac{2m\mu_1}{\hbar^2} + \dots . \tag{126}$$

The corresponding reflectivity is

$$R = \left(\frac{q_1 - q_2}{q_1 + q_2}\right)^2 - \frac{4q_1(q_1 - q_2)}{(q_1 + q_2)^3}\frac{2m}{\hbar^2}\int_{-\infty}^{\infty} \mathrm{d}z\, V_i(z) + \dots . \tag{127}$$

Since V_i is non-negative, we see that an absorbing film will (to first order in the film thickness) decrease the reflectance if $V_1 < V_2$, and increase the reflectance if $V_1 > V_2$. The transmittance is always decreased by absorption, on the other hand. From (16) with $\tilde{V} = V_0$ and $\tilde{\psi} = \psi_0$ we find

$$t = t_0 - \frac{2iq_1}{(q_1 + q_2)^2}\frac{2m\mu_1}{\hbar^2} + \dots , \tag{128}$$

where $t_0 = 2q_1/(q_1 + q_2)$. Thus the transmittance to first order in the film thickness is

$$T = \frac{q_2}{q_2}|t|^2 = \frac{4q_1q_2}{(q_1 + q_2)^2}\left\{1 - \frac{2}{q_1 + q_2}\frac{2m}{\hbar^2}\int_{-\infty}^{\infty} \mathrm{d}z\, V_i(z) + \dots\right\}. \tag{129}$$

In the presence of absorption the flux conservation law $R + T = 1$ does not hold. One can define an absorptance A such that $R + T + A = 1$; A is the probability of a particle being absorbed within the film. From (127) and (129) we find that, to first order in the film thickness,

$$A = \frac{4q_1}{(q_1 + q_2)^2} \frac{2m}{\hbar^2} \int_{-\infty}^{\infty} \mathrm{d}z \; V_i(z) + \ldots . \tag{130}$$

Reflection at a gradual transition between a non-absorbing medium (potential V_1) and an absorbing medium (potential $V_2 = V_r - iV_i$), which would be total when $V_i = 0$ for $\theta_1 > \theta_c$ in the $V_1 < V_r$ case, is less than total in the presence of absorption. The decrease from unity is greater the thicker the transition region, as a result of the greater probability of particle penetration into the absorbing region. The formulae derived for the tanh profile in Section 8-7 apply directly to the particle case.

A-8 Inversion of a model reflection amplitude

An exact inversion, due to Gelfand, Levitan and Marchenko, is possible if the reflection amplitude is known for all wavenumbers. (References are given in Chapter 9.) The exact inversion involves the solution of an integral equation. An approximate inversion, based on the Rayleigh approximation and Fourier analysis, was given in Section 9-3. This is adapted here to the particle case. The Rayleigh approximation to the scattering amplitude is, from (98),

$$r \simeq - \int_{-\infty}^{\infty} \mathrm{d}z \, \frac{\mathrm{d}q/\mathrm{d}z}{2q} \, \mathrm{e}^{2i\phi}, \qquad \phi = \int^{z} \mathrm{d}\zeta \; q(\zeta). \tag{131}$$

This may be written as the Fourier transform $r(q_1)$ of the function $-(\mathrm{d}q/\mathrm{d}x)/2q$, where $x = \phi/q_1$:

$$r(q_1) \simeq - \int_{-\infty}^{\infty} \mathrm{d}x \, \frac{\mathrm{d}q/\mathrm{d}x}{2q} \, \mathrm{e}^{2iq_1 x}. \tag{132}$$

The Fourier inverse of (132) is

$$-\frac{\mathrm{d}q/\mathrm{d}x}{2q} \simeq \frac{1}{2\pi} \int_{-\infty}^{\infty} \mathrm{d}q_1 \, \mathrm{e}^{-2iq_1 x} \, r(q_1) \equiv F(2x), \tag{133}$$

where the reflection amplitude is analytically continued to negative q_1 via $r(-q_1) = r^*(q_1)$. Thus, on integrating (133) from $-\infty$ to x,

$$q(x) \simeq q_1 \exp\left[-2 \int_{-\infty}^{2x} \mathrm{d}y \; F(y)\right]. \tag{134}$$

The square of (134) gives, on using $q^2 = (2m/\hbar^2)(\mathscr{E} - V) - K^2$,

$$\frac{\mathscr{E} - V(x)}{\mathscr{E} - V_1} \simeq \sin^2\theta_1 + \cos^2\theta_1 \exp\left[-4 \int_{-\infty}^{2x} \mathrm{d}y \; F(y)\right]. \tag{135}$$

From the definition of F,

$$\int_{-\infty}^{\infty} \mathrm{d}y \; F(y) = r(q_1 \to 0), \tag{136}$$

the long wave limit of the reflection amplitude. A check on the accuracy of the solution (135) is thus provided in the limit as $x \to \infty$. The left side of (135) then tends to $(\mathscr{E} - V_2)/(\mathscr{E} - V_1)$; the right side tends to the same value if the long wave limit (99) of the Rayleigh approximation, $r \to 1/2 \log (q_1/q_2)$, is substituted. But if the correct limit $(q_1 - q_2)/(q_1 + q_2)$ is substituted in the $x \to \infty$ limit of (135), the two sides differ by a term of order $(V_1 - V_2)^3$.

The approximate inversion formula (135), which can be expected to work well if the reflection is weak at all wavenumbers, must be supplemented by a relationship between the variable x and the physical depth z. This is obtained by integrating $d\phi = q\, dz = q_1\, dx$, using (134):

$$z(x) = \int_0^x dx_1 \frac{q_1}{q(x_1)} \simeq \int_0^x dx_1 \exp\left(2\int_{-\infty}^{2x_1} dx_2\, F(x_2)\right) \tag{137}$$

The inverse relation is

$$x(z) = \phi(z)/q_1 = q_1^{-1}\int_0^z d\zeta\, q(\zeta).$$

A-9 Reflection of wavepackets

We consider the reflection of a wavepacket made up by superposition of plane wave energy eigenstates, each with time dependence $e^{-i\mathscr{E}t/\hbar}$, which we write as $e^{-i\omega t}$. If

$$\psi_i(t) = \int_{-\infty}^{\infty} d\omega\, f(\omega)\, e^{-i\omega t} \tag{138}$$

represents the incident wavepacket at some reference plane (say $z = 0$) then, because of the linearity of Schrödinger's equation, the reflected wave at the same plane is made up of a similar superposition, each energy component having reflected with its own reflection amplitude:

$$\psi_r(t) = \int_{-\infty}^{\infty} d\omega\, r(\omega) f(\omega)\, e^{-i\omega t}. \tag{139}$$

When the energy distribution of the incident wavepacket is strongly peaked about some value $\mathscr{E}_0$, the wavepacket is nearly sinusoidal, with $\psi_i(t) = A(t)\, e^{-i\omega_0 t}$ and an amplitude function $A(t)$ which varies slowly over most of its range. (An example of such a pulse was given in Section 10-1.) From the Fourier inverse of (138), the energy (or frequency) distribution function $f(\omega)$ is given by

$$f(\omega) = \frac{1}{2\pi}\int_{-\infty}^{\infty} dt\, A(t)\, e^{i(\omega-\omega_0)t}, \tag{140}$$

and thus the reflected wavepacket is

$$\psi_r(t) = \frac{1}{2\pi}\int_{-\infty}^{\infty} d\omega\, r(\omega)\, e^{-i\omega t}\int_{-\infty}^{\infty} d\tau\, A(\tau)\, e^{i(\omega-\omega_0)\tau} \tag{141}$$

An explicit form for the reflected wavepacket can be obtained on the assumption that $r(\omega) = |r(\omega)|\,\mathrm{e}^{i\delta(\omega)}$ is well represented (within the dominant range of energies which make up the incident wavepacket) by

$$|r(\omega)| \simeq |r(\omega_0)|, \qquad \delta(\omega) \simeq \delta_0 + (\omega - \omega_0)\delta_0', \tag{142}$$

where $\delta_0 = \delta(\omega_0)$ and δ_0' is the derivative $\mathrm{d}\delta/\mathrm{d}\omega = \hbar\,\mathrm{d}\delta/\mathrm{d}\mathscr{E}$ evaluated at $\mathscr{E}_0$. Then (141) gives

$$\psi_r(t) \simeq |r(\omega_0)|\,\mathrm{e}^{i\delta_0}\,A(t - \delta_0')\,\mathrm{e}^{-i\omega_0 t}. \tag{143}$$

The amplitude function of the reflected wavepacket is thus unchanged in shape, but the wavepacket is delayed by

$$\Delta t = \delta_0' = \hbar\left[\frac{\mathrm{d}\delta}{\mathrm{d}\mathscr{E}}\right]_{\mathscr{E}_0}. \tag{144}$$

We will give some applications of this time-delay formula. The simplest case is that of partial reflection at a potential step, located at $z = z_1$. At normal incidence the reflection amplitude is

$$r = \mathrm{e}^{2ik_1z_1}\,\frac{k_1 - k_2}{k_1 + k_2}, \tag{145}$$

and the time-delay formula gives

$$\Delta t = 2z_1\hbar\,\frac{\mathrm{d}k_1}{\mathrm{d}\mathscr{E}} = \frac{2z_1}{u_1}, \tag{146}$$

where $u_1 = \hbar k_1/m$ is the group speed $\mathrm{d}\omega/\mathrm{d}k = \mathrm{d}\mathscr{E}/\hbar\,\mathrm{d}k$ in the first medium. The reflected pulse is delayed by just the time it takes to propagate to the barrier and back, at the group or wavepacket speed.

When $\mathscr{E} < V_2$ there is total reflection. Again considering reflection at a step located at z_1, at normal incidence, we have

$$r = \mathrm{e}^{2ik_1z_1}\,\frac{k_1 - i|k_2|}{k_1 + i|k_2|} = \exp 2i\left(k_1z_1 - \arctan\frac{|k_2|}{k_1}\right), \tag{147}$$

where

$$k_1^2 = \frac{2m}{\hbar^2}(\mathscr{E} - V_1), \qquad |k_2|^2 = \frac{2m}{\hbar^2}(V_2 - \mathscr{E}). \tag{148}$$

The time-delay formula (144) gives

$$\Delta t = \frac{2z_1}{u_1} + \hbar[(\mathscr{E} - V_1)(V_2 - \mathscr{E})]^{-1/2} = \frac{2}{u_1}(z_1 + |k_2|^{-1}). \tag{149}$$

The reflected wavepacket is thus delayed by more than the travel time $2z_1/u_1$ to the potential step; the increase can be interpreted in terms of penetration to the depth $|k_2|^{-1}$ into the barrier. (This penetration depth diverges as $\mathscr{E}$ tends to V_2 from below, but (144) is not valid as $|k_2| \to 0$ because the square root singularity in δ cannot be approximated by (142).)

For total reflection at a gradually rising potential barrier, the phase shift can be approximated by the short wave formula

$$\delta \simeq 2 \int_0^{z_0} \mathrm{d}z\, k(z, \mathscr{E}) - \pi/2 \tag{150}$$

(this is the normal incidence form of (113)). The time-delay can then be written in terms of the local value of the wavepacket speed, $u = \mathrm{d}\mathscr{E}/\hbar\, \mathrm{d}k = \hbar k/m$:

$$\Delta t \simeq 2 \int_0^{z_0} \frac{\mathrm{d}z}{u(z, \mathscr{E})}. \tag{151}$$

(The turning point z_0 is also a function of $\mathscr{E}$, but in the differentiation of δ the term $\mathrm{d}z_0/\mathrm{d}\mathscr{E}$ is multiplied by $k(z_0, \mathscr{E})$, which is zero.) The interpretation of (151) is that the wavepacket travels up to the classical turning point z_0 (where $\mathscr{E} = V(z)$) and back, at the group velocity $u(z, \mathscr{E})$. In contrast to (149), there is negligible penetration into the classically forbidden region where $\mathscr{E} < V(z)$. For a linear variation of V with z in $z_1 \leqslant z \leqslant z_1 + \Delta z$, with V rising between V_1 and $V_2 = V_1 + \Delta V$, (151) gives

$$\Delta t \simeq \frac{2z_1}{u_1} + 2[2m(\mathscr{E} - V_1)]^{1/2} \frac{\Delta z}{\Delta V}. \tag{152}$$

Note that the correction to $2z_1/u_1$ is the same as the classical time delay

$$2 \int_{z_1}^{z_0} \mathrm{d}z/v(z), \quad \text{where} \quad \tfrac{1}{2}mv^2(z) = \mathscr{E} - V(z).$$

The time-delay discussed above is based on the approximation $\delta \simeq \delta_0 + (\omega - \omega_0)\delta_0'$. Higher order terms in the Taylor expansion of δ about ω_0 lead to pulse spreading and distortion; references to these are given in Section 10-1. (The intrinsic spreading of the wavepacket with time has been neglected here. This spreading takes place even in a uniform medium, but can be made small for highly monoenergetic wavepackets.)

References

Many references to quantum mechanical theory have been made in the main body of the book. Here we give a few additional references to specific particle reflection and transmission problems

D. O. Edwards, P. Fatouros, G. G. Ihas, P. Mrozinski, S. Y. Shen, F. M. Gasparini and C. P. Tam (1975) "Specular reflection of ^{4}He atoms from the surface of liquid ^{4}He", Phys. Rev. Lett. **34**, 1153–1156.

D. O. Edwards and P. P. Fatouros (1978) "Theory of atomic scattering at the free surface of liquid ^{4}He", Phys. Rev. **B17**, 2147–2159.

J. B. Hayter, R. R. Highfield, B. J. Pullman, R. K. Thomas, A. I. McMullen and J. Penfold (1981) "Critical reflection of neutrons", J. Chem. Soc. Faraday Trans. (1) **77**, 1437–1448.

R. R. Highfield, R. P. Humes, R. K. Thomas, P. G. Cummins, D. P. Gregory, J. Mingins, J. B. Hayter and O. Schaerpf (1984) "Critical reflection of neutrons from a soap film", J. Colloid and Interface Science **97**, 367–373.

E. Burstein and S. Lundquist (eds) (1969) "Tunneling phenomena in solids", Plenum.

References

L. Solymar (1972) "Superconductive tunnelling and applications", Chapman and Hall.

A. Barone and G. Paternò (1982) "Physics and applications of the Josephson effect", Wiley.

J. Tersoff and D. R. Hamann (1985) "Theory of the scanning tunneling microscope", Phys. Rev. **B31**, 805–813.

G. Binnig and H. Rohrer (1986) "Scanning tunneling microscopy", IBM. Journal of Res. and Dev., July issue.

Author index

Subject index